CPSS Power Electronics Series

This series comprises advanced textbooks, research monographs, professional books, and reference works covering different aspects of power electronics, such as Variable Frequency Power Supply, DC Power Supply, Magnetic Technology, New Energy Power Conversion, Electromagnetic Compatibility as well as Wireless Power Transfer Technology and Equipment. The series features leading Chinese scholars and researchers and publishes authored books as well as edited compilations. It aims to provide critical reviews of important subjects in the field, publish new discoveries and significant progress that has been made in development of applications and the advancement of principles, theories and designs, and report cutting-edge research and relevant technologies. The CPSS Power Electronics series has an editorial board with members from the China Power Supply Society and a consulting editor from Springer.

Readership: Research scientists in universities, research institutions and the industry, graduate students, and senior undergraduates.

More information about this series at http://www.springer.com/series/15422

Deshang Sha · Guo Xu

High-Frequency Isolated Bidirectional Dual Active Bridge DC–DC Converters with Wide Voltage Gain

Springer

Deshang Sha
Advanced Power Conversion Center,
School of Automation
Beijing Institute of Technology
Beijing
China

Guo Xu
Advanced Power Conversion Center,
School of Automation
Beijing Institute of Technology
Beijing
China

ISSN 2520-8853 ISSN 2520-8861 (electronic)
CPSS Power Electronics Series
ISBN 978-981-13-0258-9 ISBN 978-981-13-0259-6 (eBook)
https://doi.org/10.1007/978-981-13-0259-6

Library of Congress Control Number: 2018939940

Printed on acid-free paper

This Springer imprint is published by the registered company Springer Nature Singapore Pte Ltd.
part of Springer Nature
The registered company address is: 152 Beach Road, #21-01/04 Gateway East, Singapore 189721, Singapore

Preface

Bidirectional DC–DC converters are widely used as power interfaces for many applications that need both changing and discharging of the energy storage components, such as batteries, super-capacitors. Among various types of bidirectional DC–DC converters, dual active bridge (DAB) converter has become popular and preferred topology during these years for its attractive advantages including symmetrical structure, zero-voltage switching, bidirectional power transfer capability, and high-power density.

This book presents different DAB converter topologies and power control strategies to achieve better power conversion performances for applications requiring both bidirectional power flow and galvanic isolation. The studied topologies can be used for applications needing a relatively wide voltage ranges such as energy storage system, automotive applications, solid-state transformer (SST)-based DC fast charger. The converter working principles, power transfer characteristics, converter losses are analyzed in this book. Based on that, the power control strategies are applied to achieve better performance aimed at different control goals. This book will benefit the researchers and engineers in the field of topology and control for dual active bridge isolated bidirectional DC–DC converters. The outcomes will enable effective development and high-quality mass production of novel high-performance DC–DC converters for the aforementioned bidirectional power applications. The main objectives are as follows,

1) To review and summarize the existed mainstream topologies and control methods for dual active bridge converters which are viewed as challenging research and application topic in the field of electrical engineering.
2) To develop new topologies for the applications that require wide voltage gain range due to the interfacing with the batteries. These applications include energy storage system, automotive applications, SST-based DC fast charger.
3) To develop effective and simple control methods to achieve reduced current stress and wide ZVS ranges for the bidirectional dual active bridge DC–DC converters.

4) To present useful methodologies and philosophies to develop new topologies and controls for isolated bidirectional DC–DC converter to achieve wide voltage range.

This book consists of 12 Chapters based on our several research projects, and covering the aspects of converter topologies and power control strategies for DAB converters. Like most books, this book starts with an introduction in Chap. 1 to present a brief introduction focusing on the applications and classifications of bidirectional DC–DC converters, review of the control methods for dual active bridge control, and key issues of DAB converters. For the rest of the contents, from the topology point of view, this book can be divided into two parts. The first part is focused on topology and control for voltage fed dual active bridge converters which are from Chaps. 2 to 6, and the second part presents the topologies and controls for current-fed dual active bridge converters which are included from Chaps. 7 to 12. The detail organization of the book content is summarized in Chap. 1.

The authors wish to express their sincere thanks to Prof. Zhiqiang Guo, Beijing Institute of Technology for his contribution on the three-level DAB converter research and other contributions to this book. The author would also like to acknowledge the contributions of Mr. Yaxiong Xu and Dr. Jiankun Zhang on the voltage-type DAB converters, Dr. Xiao Wang, Mr. Deliang Chen, Mr. Lingyu Xu, Mr. Wenqi Yuan on the current-fed DAB converters.

The author would like to thank for the support from the National Natural Science Foundation of China under Grant 51577012, from State Key Laboratory of Alternate Electrical Power System with Renewable Energy Sources under Grant LATS17019, from Key Laboratory of Solar Thermal Energy and Photovoltaic System of Chinese Academy of Sciences.

The author would also like to thank their families who have given tremendous support all the time. Finally, the authors are extremely grateful to Springer and the editorial staff for the opportunity to publish this book and help in all possible manners.

Beijing, China

Deshang Sha
Guo Xu

Contents

List of Figures

Chapter 1
Introduction

Abstract This chapter presents a brief introduction focusing on the applications and classifications of bidirectional DC–DC converters, review of the control methods for dual active bridge control, and key issues of DAB converters. Bidirectional power flow, galvanic isolation, and wide voltage gain range are necessary for some applications such as energy storage system, automotive applications, SST-based DC fast charger. To achieve wide ZVS ranges and low current stresses for the conversion stages of these applications, the topology candidates and the corresponding power control methods have been a challenge in both industry and research communities. This chapter provides a basic foundation for the whole work, and it gives the goal of this book which is to provide valuable knowledge, advanced control methods, and practical design guides for the high-frequency isolated bidirectional dual active bridge DC–DC converters with wide voltage gain.

Keywords Bidrectional DC-DC converter · Energy storage system
Solid State transformer · Currrent fed dual active bridge converter
Voltage fed dual active bridge converter

1.1 Application of Bidirectional DC–DC Converter

The fact that batteries can store electricity has been well known for most people all over the world, because they are widely used in our daily appliances such as cameras, shavers, laptops. On a larger scale, batteries are also used to store energy for the electrical grid, electric vehicles, and DC microgrid. The energy storage of these batteries plays an important role for the system stability. It has the following values:

(a) Energy storage can be used as a flexible resource for the power grid.
(b) Achieve the peak shaving function for the grid.
(c) Frequency regulation, transmission capacity relief, and VAR supporting.
(d) Reduce the energy-related emissions.
(e) Increase the efficiency and reliability for microgrid.

D. Sha and G. Xu, *High-Frequency Isolated Bidirectional Dual Active Bridge DC–DC Converters with Wide Voltage Gain*, CPSS Power Electronics Series,
https://doi.org/10.1007/978-981-13-0259-6_1

The battery stores the energy through charging and transfers the energy out through discharging. The charging or discharging power transfer should be controlled by a bidirectional converter with a DC interface. This bidirectional converter could be a bidirectional DC–AC converter or a bidirectional DC–DC converter. In this book, it is focused on the bidirectional DC–DC converter.

1.1.1 Energy Storage System for Microgrid or Smart Grid

Battery energy storage systems (BESS) [1–5] can be installed in electricity distribution networks, homes, remote area power supplies, and commercial/industrial installations. Figure 1.1 shows the structure of a DC home grid system [6] for a net-zero energy building. The whole system mainly includes: (1) renewables which could be top solar panels as photovoltaic (PV) system or wind turbine; (2) utility grid connection through a power control center (PCC); (3) energy storage devices, such as battery energy storage devices and the plug-in electric vehicle; (4) load, all kinds of home appliances. The DC bus voltages for this home grid system are suggested to be constant 400 V, and a bidirectional DC–DC converter provides a power link between these energy storage devices and the DC bus. When the power generated by the renewables exceeds the demanded power, the bidirectional DC–DC converter will charge the battery and thus stores this energy for backup. On the other case, when the power is not enough to support the facilities, the bidirectional DC–DC converter would transfer the energy out by discharging the battery. All these power flow regulations are controlled by PCC unit. The performance of the bidirectional DC–DC converter will have impact on the system efficiency, cost, dynamic response, and reliability.

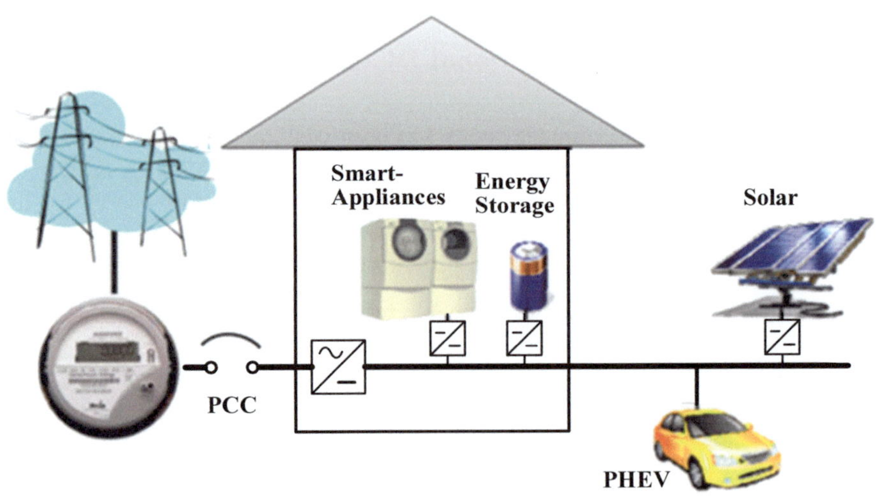

Fig. 1.1 Energy storage in a DC home grid system

1.1.2 Automotive Applications

The bidirectional DC–DC converter is also widely needed for the automotive applications, such as electric vehicle charger and motor drive system that need to absorb the reverse power and stabilize the DC voltage bus. Figure 1.2 shows the typical power system architecture for HEV [7]. There are two DC voltage buses. The high-voltage (HV) DC bus is mainly employed to manage the power transfer between the HV batteries and the propulsion system. Besides, it also supplies the high power load such as power steering and air conditioner. For the DC–DC converter connected between HV battery and HV DC bus, the power transfer is designed to be bidirectional. The low-voltage (LV) DC bus is added here to supply the low-power load, because the conventional low-power automotive loads are mainly designed based on 14 V standards. The LV voltage battery is connected to the LV bus to make the DC bus more robust. A DC–DC converter is needed for transferring the power from HV DC bus to LV DC bus in normal operation. Even though unidirectional power of the DC–DC converter can meet the requirements for most of the operation conditions, some practical reasons still support the needing of

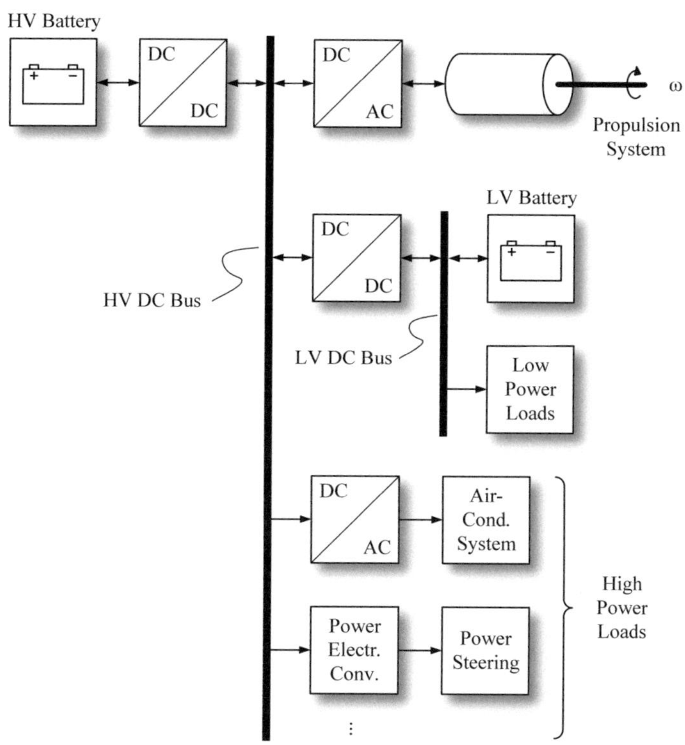

Fig. 1.2 Typical HEV power system architecture

bidirectional power follow, because the LV battery may need to provide the power to the HV bus under some particular occasions such as jump starting the car, charging the HV battery, and failure of HV battery.

1.1.3 SST Application

The structure of Future Renewable Electric Energy Delivery and Management (FREEDM) System [8] was proposed by Dr. Huang from the NSF FREEDM System Center, which is regarded as a roadmap for a future automated and flexible electric power distribution system. The conceptualized grid interface of a future home in the FREEDM system is shown in Fig. 1.3. The distributed medium voltage is directly connected to 20 kVA solid-state transformer (SST), and the transferred power is controlled by the Distributed Grid Intelligence (DGI) unit. SST generates two different voltage buses including 400 V direct current (DC) bus and the 120 V alternating current (AC) bus. These two bus interfaces provide a plug-and-play capability for the loads, storages, and the generating units. As seen, SST is the key element for controlling the power in this system. Figure 1.4 shows an example of electrical topology for the SST unit [9]. As shown, the medium voltage is directly connected to a modular cascaded H-bridge. Instead of using the bulky conventional line-frequency transformer to achieve the galvanic isolation, the high-frequency-based bidirectional DC–DC converter is adopted. The outputs of all the isolated DC–DC converters are connected in parallel to provide the 400 V DC bus voltage. Then, a three-phase bidirectional inverter is connected to the DC bus voltage to supply for the AC loads. For this system, the voltage of each cascaded H-bridge

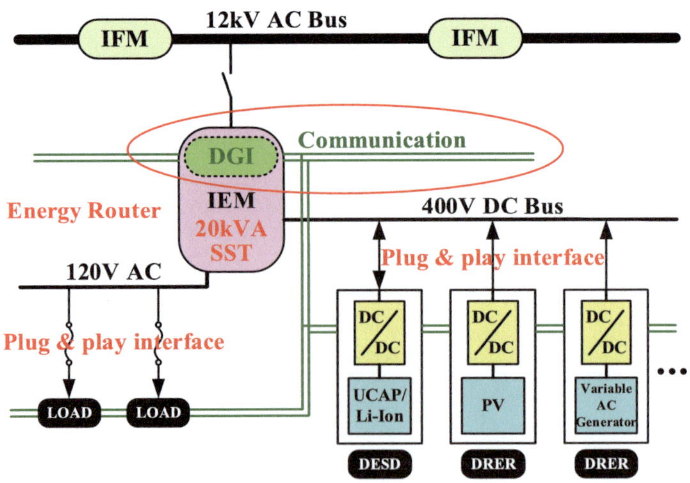

Fig. 1.3 Interface of a future home in the FREEDM system

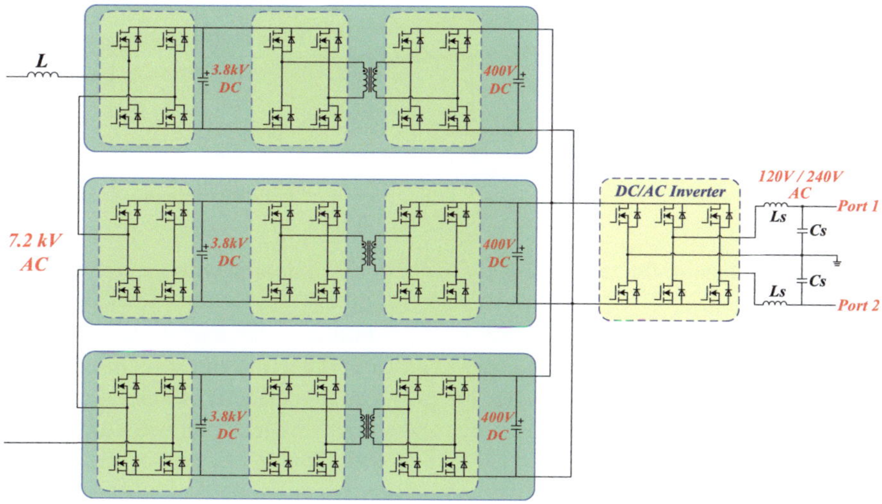

Fig. 1.4 Topology for the SST

(CHB) unit should be balanced, and the power should also be equally shared among each DC–DC converter. The power control methods have been studied in detail in [10–12]. In addition, this system can be extended to a three-phase solid-state system. The power and voltage balance control for a three-phase solid-state transformer is studied in [13]. Actually, the voltage balance control method could also be referred from static synchronous compensator (STATCOM) using CHB topology. For instance, in [14], Dr. Xu proposed a voltage balance control based on ADRC controller, which has been successfully applied to a 10 kV/2 Mvar CHB STATCOM converter.

In these two SST structures, the bidirectional DC–DC converter with isolation is necessary to provide the bidirectional power control and isolation from the medium-voltage input.

1.2 Classification of Bidirectional DC–DC Converter

1.2.1 Non-isolated and Isolated DC–DC Converter

From the galvanic isolation point of view, the bidirectional DC–DC converter can be divided into two categories, which are non-isolated bidirectional DC–DC converters and isolated bidirectional DC–DC converters. Compared with non-isolated converters, the isolated bidirectional DC–DC converter can not only achieve isolation, but also can make the conversion gain more flexible. On the other hand, because the isolated converter needs to invert the DC voltage to high-frequency AC voltage to interface the transformer and then rectify the high-frequency AC voltage

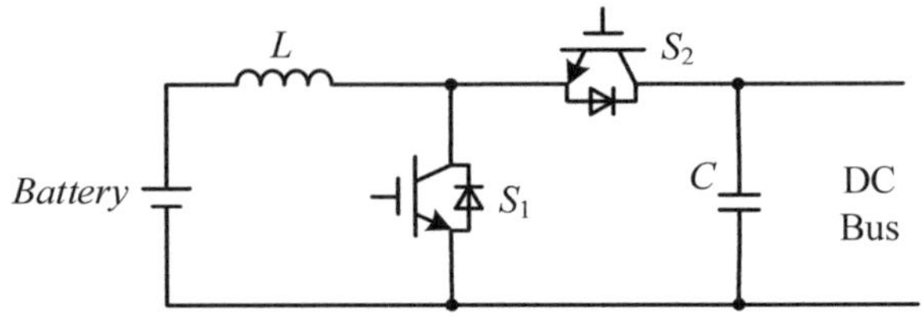

Fig. 1.5 Bidirectional buck/boost converter topology

to DC voltage again, the topology is more complex compared with the non-isolated converters.

Figure 1.5 shows the typical non-isolated bidirectional buck/boost converter topology. The converter works as a boost converter when the battery is discharging, and works as a buck converter when the power is reversed. This topology is very simple. It only needs two power switches and a high-frequency inductor to build up the topology. For this converter, zero-voltage switching (ZVS) cannot be achieved for both of the two switches during continuous conduction mode (CCM) operation, leading to high switching loss when operating at high frequency. To increase the efficiency and also the loss density, some ZVS techniques can be used, for example quasi-resonant technic [15] and critical conduction mode operation [16]. Besides this buck/boost converter, there are other types of non-isolated bidirectional DC–DC converters, like the universal bidirectional DC–DC converter [17], interleaved high-conversion ratio bidirectional DC–DC converter [18].

The typical bidirectional DC–DC converter with isolation is shown in Fig. 1.6. It is an isolated full-bridge converter with synchronous rectification [19, 20]. When power transfers from the V_1 side to V_2 side, the bridge at V_2 side works as a synchronous rectifier, while when the power transfers from V_2 side to V_1 side, the bridge at V_1 side works as a synchronous rectifier (SR). As a result, the pulse width modulation (PWM) logic needs to be shifted under different power transfer conditions. Also, the PWM for the SR can be given based on the zero-crossing instant of the transformer current to reduce the conduction loss. Because the output has an inductor in series, high current turn-off voltage spikes may occur for the switches at V_2 side. In order to solve this issue, the full-bridge converter with RCD snubber [21], CD2 snubber [22], and active clamping method [23] can be adopted.

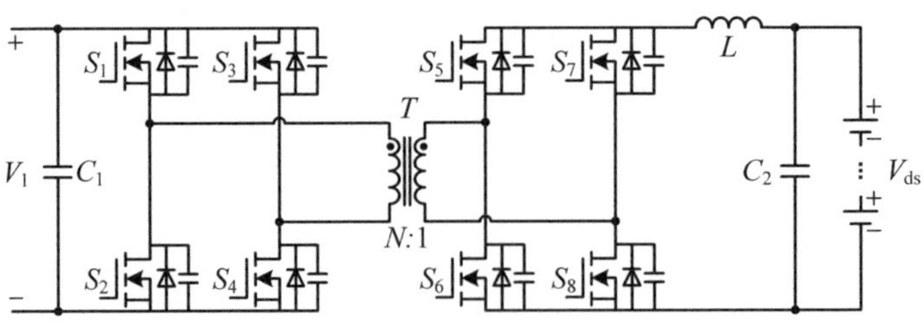

Fig. 1.6 Bidirectional isolated full-bridge DC–DC converter

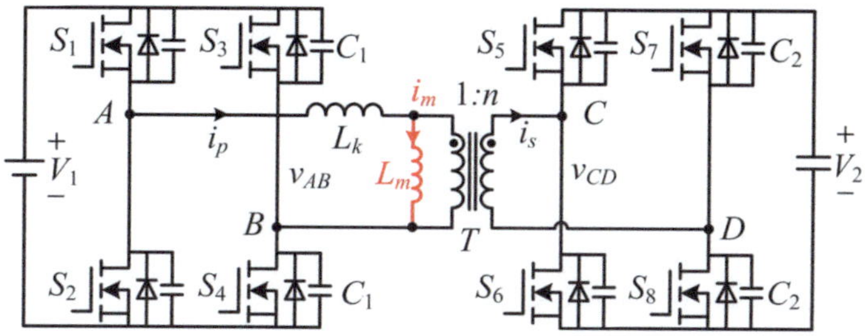

Fig. 1.7 Bidirectional isolated dual active bridge converter

Another type of DC–DC converter that is widely used for bidirectional power industrial applications is the dual active bridge converter [24, 25] as shown in Fig. 1.7. The dual active bridge consists of two active full bridges, which is symmetrical from both sides. The leakage inductance of the transformer L_k works as a power link between the input side and output side. Through controlling of the voltage induced crossing two ports of the leakage inductance, the transferred power and the power direction can be regulated. DAB converters have many advantages: wide ZVS range for all the power switches over wide load variation, naturally bidirectional power flow control and symmetrical architecture in topology.

Galvanic isolation of the converter is actually required for many applications due to some safety reasons. In addition, in some applications when the converter conversion ratio is very high, the transformer can be used to adjust the gain of the converter to extend to optimal operating range. This book is focused on the bidirectional DC–DC converter with isolation.

1.3 Isolated Bidirectional DC–DC Converter

1.3.1 PWM Controlled, Frequency Controlled and Phase Shift Controlled Bidirectional DC–DC Converter

In view of the power control for the isolated bidirectional DC–DC converter, they can be classified into three categories, which are PWM-controlled converter, frequency-controlled converter, and phase shift-controlled converter. The bidirectional full-bridge DC–DC converter shown in Fig. 1.6 is an example of this type converter. The PWM-controlled converter uses pulse width modulation technic to adjust the converter output voltage, thus regulating the power. The output voltage of the converter is controlled by the duty cycle of the PWM signal of the power switches.

Unlike the PWM-controlled converter, the frequency-controlled converter uses the operating frequency to control the power transfer. This type of converter usually contains a resonant tank which can suppress the undesired frequency components to regulate the output gain. An example of this kind of converter is shown in Fig. 1.8. It is a resonant CLLC-type converter [26]. The duty ratios of the PWM for all the switches are fixed as 50%, and the operating frequency is changing according to conversion gain of the converter. Resonant converter is a very good candidate for the fixed voltage gain applications since it can achieve very high efficiency when it operates at the resonant frequency. However, as the operating frequency goes far from the resonant frequency, the converter will suffer from high reactive current leading to high current-related loss.

The third type is phase shift-controlled bidirectional DC–DC converter. The phase shift control is mainly used for different kinds of dual active bridge converter, such as the one previously shown in Fig. 1.7. The main feature of this type of converter is the existing of relatively larger leakage inductance compared with the other converters. It controls the phase angle between the two voltages induced at the two port of the leakage inductance, like v_{AB} and v_{CD} in Fig. 1.7. If the duty cycle for the two active bridges is fixed as 50%, the power will be transferred from V_1 side to V_2 side when the phase of v_{AB} is leading v_{CD} and the power will be reversed when the phase of v_{AB} is lagging v_{CD}. This method is named as conventional phase shift (CPS) control [27]. Furthermore, duty cycles for the two active bridges can also be used to improve the performance of DAB converters.

Compared with the other two types of converter, DAB converter can have the following two advantages:

(a) It can achieve nature bidirectional power transfer. For DAB converter, bidirectional power is controlled by the phase shift angle. When phase shift angle is positive the power is positive, otherwise the power would be reversed. As a result, the transfer between the positive power flow and negative power flow would be naturally and smoothly achieved. While, for the other two types of converter, under different power transfer direction, the PWM logic needs to be changed since the bridge that works as a rectifier needs to be shifted to the other side.

(b) It does not need the SR driving. Generally, the SR driving is done by sensing the drain–source voltage of the switches in the rectifier. It uses the voltage drop of the body diodes to determine the current zero-crossing instant. Because the voltage drop of body diode is low (for example 1.2 V), it is sensitive to the switching noise. As a result, the SR driving will need additional voltage sensing circuit and need to take care of the noise which is complex in practice.

Additionally, two of these control methods can be combined together to achieve better performance of DAB converter in sacrifice of the control complexity. These controls can be called as hybrid control.

1.3.2 Current-Fed DAB Converter

In view of the type of sources at the two sides of the DAB converter, it can also be divided into two parts: voltage-fed dual active bridge and current-fed dual active bridge. The converter belongs to the category of "voltage fed", if the input and output are voltage sources and is directly connected to the switch nodes. Otherwise, it can be regarded as a current-fed DAB converter. For instance, the DAB converter previously shown in Fig. 1.6 is a typical voltage-fed converter, because both the input and output are directed connected to a voltage source (or capacitor). There are also some other voltage-fed DAB converters that are similar to this converter, such as the DAB converter based on two half bridges [28], the DAB converter based on a full bridge and a half bridge [29], and the DAB converter based on voltage-doubler rectifier [30].

Figure 1.9 shows an example of a current-fed dual active bridge converter [31]. In this topology, the DC bus is connected in series with an inductor and then connected to the full-bridge DAB converter. For this connection, due to the fact that inductor current cannot be interrupted as open loop, the PWM signals for the two switches in the same bridge should be overlapped to provide the circulating current

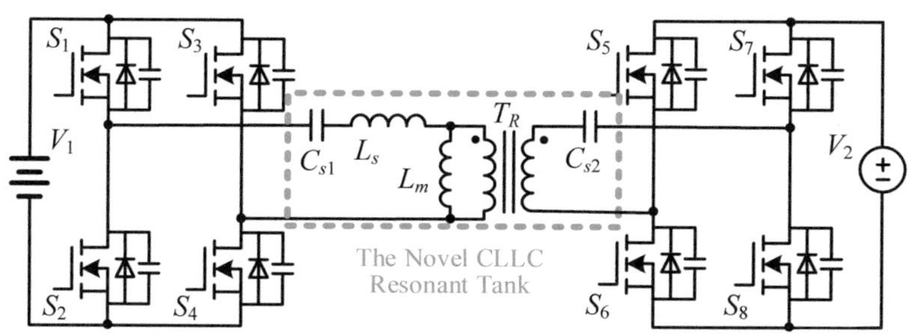

Fig. 1.8 Bidirectional CLLC resonant converter

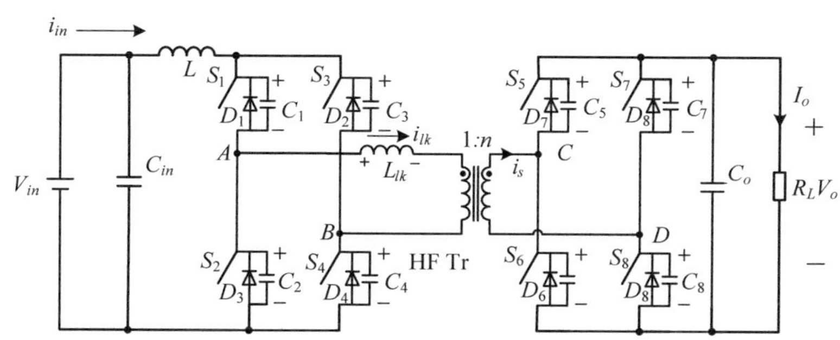

Fig. 1.9 ZCS current-fed dual active bridge

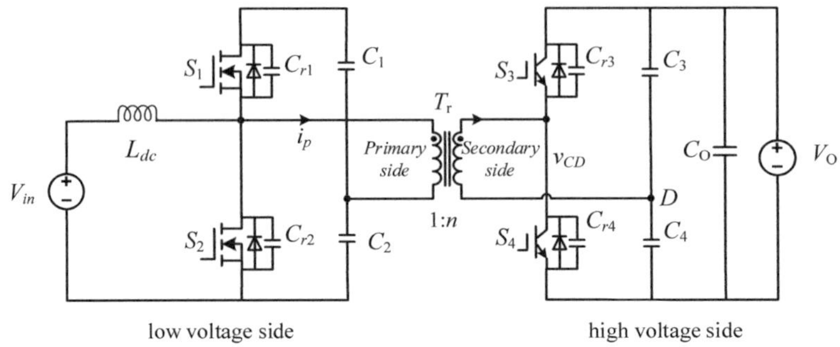

Fig. 1.10 Current-fed dual active bridge converter

path. As pointed out in [31], zero-current switching (ZCS) for the primary side switches can be achieved.

In high-voltage and low-current applications, actually ZVS matters more in the aspect of switching loss. To achieve wide ZVS range for the power switches, Fig. 1.10 shows another example of a current-fed dual active bridge converter [32]. As seen, the voltage source at the low-voltage side is connected to the switch node through an inductor. In the middle part of the converter, it is actually a DAB converter with two half bridges. By combining the boost converter and the half-bridge converter together, the LVS side has double functions which serve as:

(1) A boost converter to step up the voltage to adjust the clamping voltage of the low-voltage side.
(2) An inverter to generate the high-frequency AC voltage to pass through the transformer for isolation.

To reduce the current ripple of the low-voltage side, the interleaved current-fed DAB topology [33] can be employed as shown in Fig. 1.11. The topology at

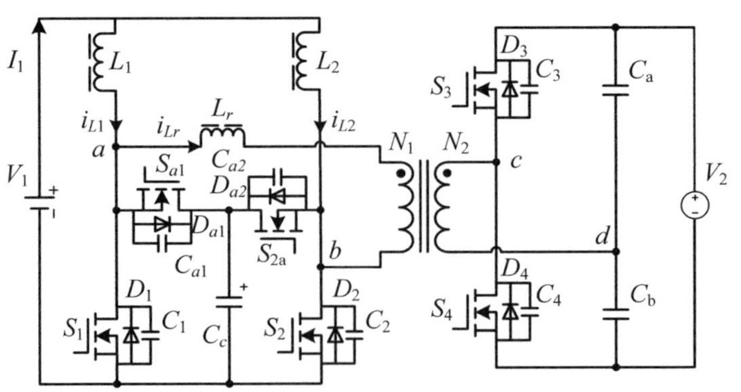

Fig. 1.11 Interleaved current-fed dual active bridge converter

low-voltage side consisted of two boost converters with input parallel and output parallel connection. The PWM signals for the two boost converters are 180° phase shift with each other. This topology can achieve low current ripple at the low-voltage side and also wide ZVS ranges for all the power switches, which makes it popular in low-voltage to high-voltage conversion applications.

1.3.3 Multi-level DAB DC–DC Converter

From the voltage levels at the transformer input/output side (or the devices voltage rating) point of view, the aforementioned DAB topologies are two-level DAB converters. To reduce the voltage rating of the device which is adorable in high-voltage applications, the multi-level DAB converter can be adopted. Also, the multi-level technique offers another control degree of freedom for the high-frequency voltage modulation, which gives a possibility to further improve the performance of DAB converters.

The voltage-fed three-level DAB converter with flying capacitor [34] is shown in Fig. 1.12. In the input side, it is a three-level full-bridge inverter with flying capacitor. The voltage stresses for all the eight switches are half of the input voltage. The phase shift control can still be applied to this converter through controlling the phase between v_1 and v_2 as denoted in the figure.

In Fig. 1.13, the current-fed three-level DAB converter is shown. Two interleaved inductors, in series with the DC voltage sources at the low-voltage side, can reduce the current ripple. At the high-voltage side, the three-level topology was employed to reduce the device voltage rating to be half of the high side voltage. In this converter topology, both the capacitor and diodes are used for the voltage clamping. The phase shift angle between v_{AB} and v_{CD} is used to control the power flow and achieve smooth transfer between different power directions. This topology is suited for the low-voltage battery input and high-voltage output applications.

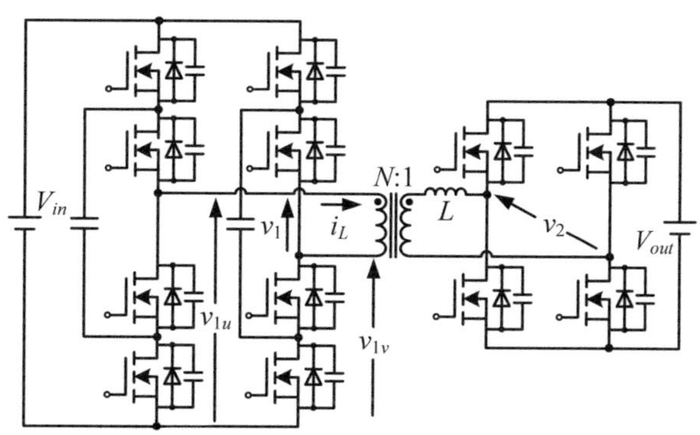

Fig. 1.12 Voltage-fed three-level DAB converter with flying capacitor

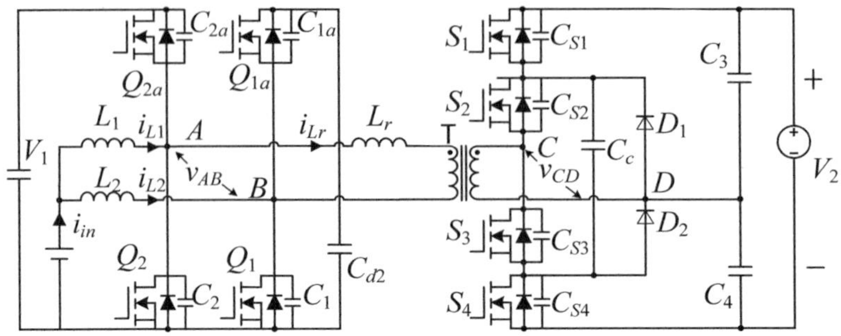

Fig. 1.13 Current-fed three-level DAB converter with active diodes and flying capacitors

1.4 Research Literature of DAB Converters

1.4.1 Basic Principle of DAB Converters

The dual active bridge was firstly proposed in the 1990s [24]. Nowadays, it has become a preferred topology for power conversion applications that require galvanic isolation and bidirectional power flow due to its many advantages, such as high power density, bidirectional power transfer capability, zero-voltage switching (ZVS), and symmetric structure. The basic working model of DAB converter can be shown in Fig. 1.14. v_{AB} is the voltage generated by the bridge at V_1 side, and v_{CD} is the voltage generated by the bridge at V_2 side. n is the transformer turns ratio from V_1 side to V_2 side, and L_k is the leakage inductance. The power transfer is related to the shapes of v_{AB} and v_{CD}, and the phase shift angle between the two voltages. v_{AB} and v_{CD} can be the waveforms with two levels or multi-levels according to different topologies and different PWM modulation methods.

Figure 1.15 shows illustrative basic operation waveforms under four different PWM modulation control strategies for a two-level voltage-fed DAB converter. In this figure, T_{s} is the switching period, and D_1 and D_2 are named as duty ratios for the two full active bridges. D_1T_{s} is the duration of high-voltage level in a switching period for v_{AB}, and D_2T_{s} is the duration of high-voltage level in a switching period for v_{CD}. In Fig. 1.15a, no zero-voltage level exists in the two voltage waveforms,

Fig. 1.14 Basic model for DAB converter

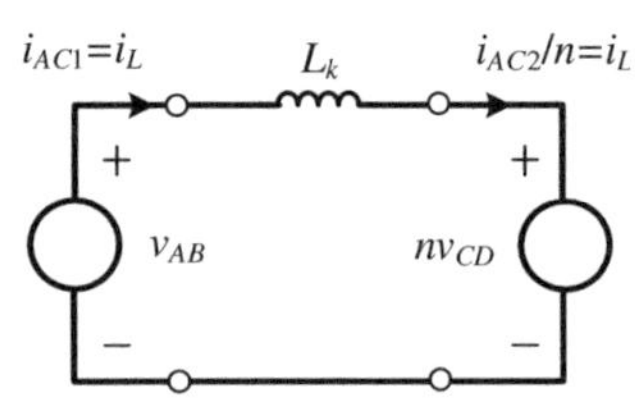

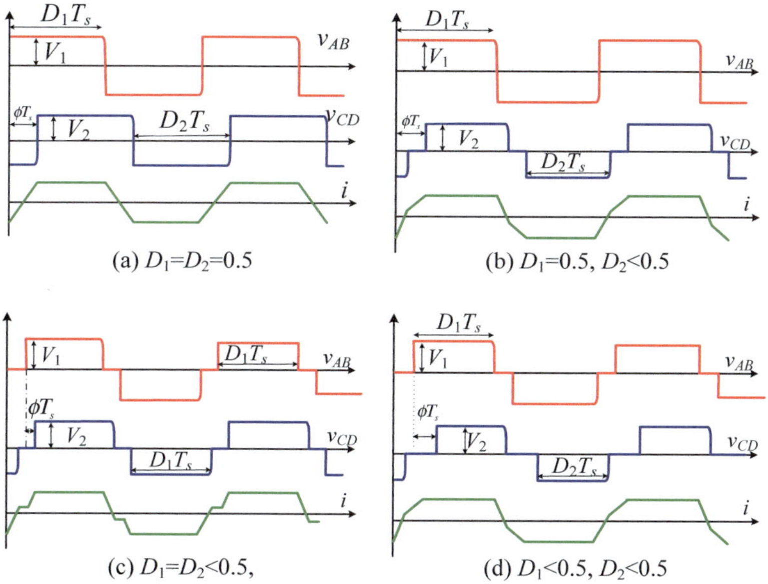

Fig. 1.15 Four different PWM modulation strategies for dual active bridge converter

and $D_1 = D_2 = 0.5$. Since both D_1 and D_2 are fixed, the power transfer is only related to the phase shift angel Φ. Consequently, this control method only has one control degree of freedom, which is called as conventional phase shift control (CPS). The control degree of freedom can be increased to obtain a better converter performance through adjusting one duty ratio for the two bridges such as the cases in Fig. 1.15b, c. In Fig. 1.15b, D_1 is fixed to be 0.5 and D_2 is a changeable variable. While D_1 and D_2 are both changeable variables with a constraint that D_1 is equal to D_2 in Fig. 1.15d. For both of the two cases, the transferred power is controlled by phase shift angel Φ and one duty ratio. As a result, these two controls have two control degrees of freedom. The one with three control degrees of freedom is shown in Fig. 1.15b. D_1 and D_2 are both not fixed, and they are smaller than 0.5. In the control loop, D_1 D_2 and Φ can be all utilized to achieve different control goals for DAB converters according to different instant input voltage, output voltage, and the load.

1.4.2 Control of Voltage-Fed DAB Converters

SPS control for voltage-fed DAB converter has been intensively studied in [35, 36], whose typical waveforms have been shown in Fig. 1.15a. It only has one control degree of freedom and is very easy to be implemented. This control is suited for the application that the input and output voltages are relatively fixed. The converter can achieve wide ZVS range, low current stress, and high dynamics under that condition. As a result, SPS control is popular for the DAB converters used in applications such as: SST application [37, 38] and high-voltage direct current transmission (HVDC) application [39]. When the voltage conversion ratio varies, the conversion efficiency using SPS could be limited because of the high conduction loss due to the high non-active power and the large switching loss caused by limited ZVS ranges [40].

To improve the converter performance under changeable voltage conversion gain, the extended-phase shift (EPS) control can be adopted. Its typical waveforms have been shown in Fig. 1.15b, which has two control degrees of freedom. In [41–44], the working principle, the power transfer characteristics, ZVS ranges, and current stress for EPS control are discussed. Compared with SPS control in facing unmatched voltage gain, EPS can obtain higher global efficiency because of the reduced current stress, wider ZVS ranges, and improved regulating flexibility. EPS control utilizes the duty cycle from one bridge to reduce the mismatching between the two-side voltages. Because the duty cycle cannot be larger than 1/2, when V_1 is larger than V_2/n, the duty cycle in V_1 side should be adjusted. However, when V_1 is smaller than V_2/n, the duty cycle in V_2 side should be adjusted. Consequently, EPS control needs to exchange the PWM generations logic under different input voltages and the transfer between the two cases cannot be smoothly achieved.

Another type of control strategy for DAB converter is DPS control, [45–50], which also has two control degrees of freedom similar to EPS control. Like EPS control, DPS can also help to reduce the circulating current and conduction loss, and to extend ZVS ranges compared with SPS control.

The optimized performance of DAB converter can be obtained with fully utilizing the three control degrees of freedom. These controls are called as triple-phase shift (TPS) control [51–57]. For these control methods, the power transfer is controlled by three independent control variables, which are two bridge inner phase shift ratios and the phase shift ratio between the two bridges. To optimize the selection of these three independent control variables, the optimization schemes are based on mathematical analysis with different optimization goals, such as optimized efficiency [52], minimizing conduction loss [53]. These optimizations need complex power calculation, working mode classification and may also highly depend on the system parameters.

For TPS control, there are three independent variables including two duty cycles and one phase shift angle. In practice, a linear interpolation can be used to determine the selection of the three variables based on data stored in the digital signal processor (DSP) memory to achieve the closed-loop control, so the system dynamic response may be affected. Besides, if the transformer two-side voltages are not matched well, the current stress during the power transfer stage is high. This could be a drawback of the voltage-fed DAB converter.

1.4.3 Control of Current-Fed DAB Converters

Actually, for applications that are connected to the batteries, during the charging or discharging process, the voltage range of the battery is very wide. The current ripple of the voltage-fed DAB port is also comparatively high and the battery lifetime may be affected [58]. The aforementioned attributes of the voltage-fed DAB make it unsuitable to be used in charging/discharging a battery whose voltage variation range is wide. Therefore, an inductor can be used between the battery and voltage-fed DAB to reduce the battery charging/discharging current ripple [59].

In contrast, current-fed DAB is a good option for battery charging/discharging situations. A novel current-fed DAB characterized by reducing one-side current ripple dramatically was introduced for battery application [32]. The current ripple reduction makes it very attractive for battery charging/discharging. When the conversion ratio varies, the conduction loss during the power transfer stage is not minimized. To adapt to the wide voltage conversion ratio situations, a PWM plus phase shift control was proposed, in which the battery side is PWM modulated while duty cycles for the high-voltage side (HVS) switches are fixed as 50%. Since the active clamp voltage can be regulated based on the HVS voltage value and the turns ratio, the current stress during the power transfer stage can be suppressed as the voltage conversion ratio varies widely by employing PWM control [60–62]. Although minimum conduction loss during the power transfer stage can be obtained, for the non-power transfer stage, especially when the battery voltage is rather low, the current spike and circulation loss may be high, causing conversion efficiency reduction and reliability issues. Of course, the current stress can be suppressed by adding an additional inductor in series connected with the transformer to increase the equivalent leakage inductance value, but the maximum transferred power capability is limited due to the large inductance value. Besides, the circulation stage length increases and results in low conversion efficiency. PWM control can be also employed for HVS switches to reduce the circulation loss [63–65]. To ensure ZVS for HVS switches, unequal duty cycles for both side switches are a very good option with unmatched voltage control [63]. But the current stress is relatively high due to the unmatched voltage control. For voltage-matched control, although identical duty cycle control can achieve ZVS for all the power switches in current-fed semi-DAB converter employing diodes [64], ZVS cannot be ensured for all the switches in current-fed DAB. The duty cycle relationship between both sides has to vary according to the specific working conditions. This complicates the control system design [65].

1.5 Key Issues of DAB Converter

1.5.1 ZVS Range

As discussed above, ZVS range of DAB converter is dependent on the load, input voltage, and output voltages. Under light load [66, 67] and unmatched voltages, it is very complex for DAB converter to achieve ZVS. To reduce the complexity in control without losing wide ZVS range, the fundamental-optimal strategy based on a universal model [68] is used. This control strategy is simple and can be easily implemented from a power control point of view. In addition, modifications can also be made for the DAB converter topologies. In [69–71], a high-frequency capacitor is added in series with the transformer to improve the soft switching ranges of DAB converter. A CLLC-type resonant topology was developed based on DAB converter in [72, 73], which can operate over large input voltage variation while maintaining soft switching capability. Based on the resonant tank concept, there are many other derived resonant DAB topologies such as LLL type [74], switched-inductor-augmented resonated type [75], and tunable resonant tank [76]. Even though the resonant tank could improve the ZVS performance of DAB converters, it brings additional undesired features, such as high inrush current during start-up and difficult to build up a model for control. Besides using resonant tank, the series injection method is used to achieve ZVS range enhancement [77].

1.5.2 Non-active Power and Current Stress

The non-active current [78, 79] causes additional conduction loss for DAB converters. Non-active power loss is defined in [79] and used as the optimization object to reduce the conduction loss with TPS control. The unified triple-phase shift control [80, 81] is used to minimize current stress. For these controls, the selection of the control variables is also not unified and hard to be implemented in real-time calculation. The triangular modulation (TRM) [82] has no non-active power issue and can achieve zero-current switching (ZCS) for four switches. But the ZVS is lost for these ZCS switches. Even though the modified TRM control is proposed in [83], it still suffers from high peak current and large RMS current. These issues can be avoided with the hybrid modulation [84] using CPS at heavy load and the modified TRM control at light load, while smooth transfer between the two modes is a challenge. Compared with TRM, the trapezoidal modulation (TZM) [85] can make the active bridge work with large duty cycles and result in lower peak current and RMS value.

1.5.3 *Wide Voltage Gain*

For voltage-fed DAB converter, the fundamental duty modulation [86], based on the fundamental optimization method in [68], is aimed at reducing circulating current and simplifying of the controller design. Just as the other DAB converters, it has been pointed out that ZVS for all the switches can be obtained when input voltage V_1 equals to nV_2, while ZVS range of the converter is limited in other cases. Instead of optimizing the modulation scheme, in [87], the authors creatively made the modification in DAB converter topology and also proposed a hybrid phase shift control for wide input voltage range. However, although hybrid control can lead to wide ZVS range, smooth transfer between the two control modes is a challenge in practice. To obtain better performance, studying new DAB converter topology could be a solution.

The current-fed DAB converter studied in [33] is considered to be suited for wide voltage gain applications, because it can adjust the clamp voltage to make the voltages at the two sides of the transformer to be matched. In addition, since it adds two inductors to the switching nodes of the active bridge, it could change the ZVS conditions and may gain some benefits of ZVS ranges if the inductor value is properly designed. However, for this converter, it adds two DC inductors which would bring additional inductor losses and increase the size of converter.

1.6 Organization of the Book

From the topology point of view, this book can be divided into two parts. The first part is focused on topology and control for voltage-fed dual active bridge converters which is from Chaps. 2–6, and the second part presents the topologies and controls for current-fed dual active bridge converters which are included from Chaps. 7–12. The content for each chapter is organized as follows.

Chapter 2 is focused on the improvement of conventional voltage-fed dual active bridge converter. The fixed duty cycle compensation and magnetizing current design method are introduced to achieve ZVS over full load ranges for the voltage-fed dual active bridge converter using trapezoidal modulation (TZM) control. Also, the boundary TZM control method is presented to achieve small current stress of the converter.

In Chaps. 3 and 4, two DAB converter topologies are studied for wide voltage gain application. To simply the control of DAB converter aimed at achieving wide ZVS range and low current stress, the DAB converter topology can be modified. In Chap. 3, two active switches and two capacitors are added in the conventional DAB converter topology. The modified topology is called as hybrid-bridge-based DAB converter. In Chap. 4, two capacitors and a transformer are added. The new topology is called as dual-transformer-based DAB converter. The working principle, power transfer characteristics, and control methods for the two converters are analyzed, respectively, in Chaps. 3 and 4.

Chapter 5 presents a blocking-cap-based DAB converter to improve the converter performance over wide input range. The DC blocking capacitor is utilized for the blocking-cap-based converter. Through proper PWM modulation strategies, this converter can change the operating state from full bridge to half bridge according to the input voltages. The working modes of the blocking-cap-based DAB converters, ZVS conditions, and power transfer characters are studied in Chap. 5.

Chapter 6 aims to develop a three-level voltage-fed DAB converter. From the topology point of view, an auxiliary inductor is added at the transformer secondary side to achieve ZVS over full load range. The operating modes, ZVS region analysis, power control method, and parameter design of the converter are discussed in detail in this chapter.

From Chaps. 7–12, the current-fed dual active bridge converters are discussed.

Chapters 7 and 8 study the conventional two-level current-fed DAB converters. In Chap. 7, a double PWM plus double phase shift control is presented for the current-fed DAB converter. The benefits of this control include reduced circulating current and also wide ZVS range. The converter working principles under different power directions are discussed in details. The performance of the converter is analyzed and compared with the conventional DC–DC converters. In Chap. 8, all the operation modes of the current-fed DAB converter are discussed. Based on the ZVS conditions for all the possible working modes, the optimized switching patterns to achieve full load ZVS range are presented and analyzed in details.

Chapters 9 and 10 are focused on the bidirectional three-level DAB converter. The converter working principles are discussed in both chapters. In Chap. 9, a direct current slew rate control of leakage inductance is presented to minimize the conduction loss and current stress in facing the load variation, the mismatch of turns-ration, and circuit parasitic parameters. In Chap. 10, the double PWM plus double phase shifted control utilizing voltage matching concept is presented, which is aimed at reducing circulating current loss and full load ZVS range.

In Chap. 11, a current-fed hybrid dual active bridge converter is presented. With the topology, the high-frequency input current ripple can be reduced to minimum because the input-side switches are always switched at 50% duty cycles in spite of the fuel cell voltage and the load variation. In addition, all of the power devices can achieve zero-voltage switching by the proposed control strategy.

In Chap. 12, the performance of parallel connected bidirectional DC–DC converter is presented. The current sharing is ensured with the presented current-sharing strategy. In addition, to improve the dynamic performance, feed-forward control employing optimized feed-forward coefficient based on the small-signal analysis is implemented.

It should be noted that the effectiveness of all the converter topologies and control methods are verified with experimental results.

References

1. Khaligh A, Li Z (2010) Battery, ultracapacitor, fuel cell, and hybrid energy storage systems for electric, hybrid electric, fuel cell, and plug-in hybrid electric vehicles: state of the art. IEEE Trans Veh Technol 59(6):2806–2814
2. Ribeiro PF, Johnson BK, Crow ML, Arsoy A (2001) Energy storage systems for advanced power applications. Proc IEEE 89(12):1744–1756
3. Lukic SM, Cao J, Bansal RC, Rodriguez F, Emadi A (2008) Energy storage systems for automotive applications. IEEE Trans Ind Electron 55(6):2258–2267
4. Vazquez S, Lukic SM, Galvan E, Franquelo LG, Carrasco JM (2010) Energy storage systems for transport and grid applications. IEEE Trans Ind Electron 57(12):3881–3895
5. Cao J, Emadi A (2012) A new battery/ultracapacitor hybrid energy storage system for electric, hybrid, and plug-in hybrid electric vehicles. IEEE Trans Power Electron 27(1):122–132
6. Zhang W (2015) Energy management system in DC future home, Master's thesis. Virginia Polytechnic Institute and State University, Blacksburg
7. Florian KK (2010) Modeling and optimization of bidirectional dual active bridge DC–DC converter topologies, Doctor's thesis. Power Electronic Systems Laboratory (PES): Swiss Federal Institute of Technology Zurich (ETH Zurich)
8. Huang AQ, Crow ML, Heydt GT, Zheng JP, Dale SJ (2011) The future renewable electric energy delivery and management (FREEDM) system: the energy internet. Proc IEEE 99 (1):133–148
9. Shi J, Gou W, Yuan H, Zhao T, Huang AQ (2011) Research on voltage and power balance control for cascaded modular solid-state transformer. IEEE Trans Power Electron 26(4): 1154–1166
10. She X, Huang AQ, Wang G (2011) 3-D space modulation with voltage balancing capability for a cascaded seven-level converter in a solid-state transformer. IEEE Trans Power Electron 26(12):3778–3789
11. Zhao T, Wang G, Bhattacharya S, Huang AQ (2012) Voltage and power balance control for a cascaded H-bridge converter-based solid-state transformer. IEEE Trans Power Electron 28 (4):1523–1532
12. She X, Huang AQ, Ni X (2014) Current sensorless power balance strategy for dc/dc converters in a cascaded multilevel converter based solid state transformer. IEEE Trans Power Electron 29(1):17–22
13. Wang L, Zhang D, Wang Y, Wu B, Athab HS (2016) Power and voltage balance control of a novel three-phase solid-state transformer using multilevel cascaded h-bridge inverters for microgrid applications. IEEE Trans Power Electron 31(4):3289–3301
14. Xu R, Yu Y, Yang R, Wang G, Xu D, Li B, Sui S (2015) A novel control method for transformerless H-bridge cascaded STATCOM with star configuration. IEEE Trans Power Electron 30(3):1189–1202
15. Lee YS, Cheng GT (2006) Quasi-resonant zero-current-switching bidirectional converter for battery equalization applications. IEEE Trans Power Electron 21(5):1213–1224
16. Huang X, Lee FC, Li Q, Du W (2016) High-frequency high-efficiency GaN-based interleaved CRM bidirectional buck/boost converter with inverse coupled inductor. IEEE Trans Power Electron 31(6):4343–4352
17. Waffler S, Kolar JW (2009) A novel low-loss modulation strategy for high-power bidirectional buck + boost converters. IEEE Trans Power Electron 24(6):1589–1599
18. Wang YF, Xue LK, Wang CS, Wang P, Li W (2016) Interleaved high-conversion-ratio bidirectional dc–dc converter for distributed energy-storage systems—circuit generation, analysis, and design. IEEE Trans Power Electron 31(8):5547–5561
19. Garcia O, Flores LA, Oliver JA, Cobos JA, De IPJ (2005) Bi-directional DC/DC converter for hybrid vehicles. In: IEEE 36th conference on power electronics specialists (PESC) 2005, pp 1881–1886

20. Ramachandran R, Nymand M (2016) A 98.8% efficient bidirectional full-bridge isolated dc-dc GaN converter. In: IEEE applied power electronics conference and exposition, pp 609–614
21. Zhu L (2006) A novel soft-commutating isolated boost full-bridge ZVS-PWM dc–dc converter for bidirectional high power applications. IEEE Trans Power Electron 21(2):422–429
22. Kim ES, Joe KY, Choi HY, Kim YH (1998) An improved soft switching bi-directional PSPWM FB DC/DC converter. In: Proceedings of the 24th annual conference of the IEEE industrial electronics society, IECON, vol 2, pp 740–743
23. Wang K, Lee FC, Lai J (2000) Operation principles of bi-directional full-bridge dc/dc converter with unified soft-switching scheme and soft-starting capability. In: Fifteenth annual IEEE applied power electronics conference and exposition (APEC), pp 111–118
24. Dedoncker RW, Kheraluwala MH, Divan DM (1991) Power conversion apparatus for dc/dc conversion using dual active bridges. US Patent 5,027,264
25. Kheraluwala MH, Gascoigne RW, Divan DM, Baumann ED (1992) Performance characterization of a high-power dual active bridge dc-to-dc converter. IEEE Trans Ind Appl 28 (6):1294–1301
26. Chen W, Rong P, Lu Z (2010) Snubberless bidirectional dc–dc converter with new CLLC resonant tank featuring minimized switching loss. IEEE Trans Ind Electron 57(9):3075–3086
27. Inoue S, Akagi H (2007) A bidirectional dc–dc converter for an energy storage system with galvanic isolation. IEEE Trans Power Electron 22(6):2299–2306
28. Xu X, Khambadkone AM, Oruganti R (2007) A soft-switched back-to-back bi-directional dc/dc converter with a FPGA based digital control for automotive applications. In: IEEE conference of the industrial electronics society, IECON, pp 262–267
29. Morrison R, Egan M (1998) A new single transformer, power factor corrected UPS design. IEEE Trans Ind Appl 36(1):237–243
30. Zhan X, Wu H, Xing Y, Ge H (2016). A high step-up bidirectional isolated dual-active-bridge converter with three-level voltage-doubler rectifier for energy storage applications. In: IEEE applied power electronics conference and exposition, pp 1424–1429
31. Pan X, Rathore AK (2013) Novel bidirectional snubberless soft-switching naturally clamped zero current commutated current-fed dual active bridge (CFDAB) converter for fuel cell vehicles. IEEE energy conversion congress and exposition, pp 1894–1901
32. Peng FZ, Li H, Su GJ, Lawler JS (2004) A new ZVS bidirectional dc–dc converter for fuel cell and battery application. IEEE Trans Power Electron 19(1):54–65
33. Xiao H, Xie S (2008) A ZVS bidirectional dc–dc converter with phase-shift plus pwm control scheme. IEEE Trans Power Electron 23(2):813–823
34. Higa H, Itoh JI (2015) Derivation of operation mode for flying capacitor topology applied to three-level DAB converter. In: IEEE international future energy electronics conference, pp 1–6
35. Inoue S, Akagi H (2007) A bidirectional dc–dc converter for an energy storage system with galvanic isolation. IEEE Trans Power Electron 22(6):2299–2306
36. Costinett D, Maksimovic D, Zane R (2013) Design and control for high efficiency in high step-down dual active bridge converters operating at high switching frequency. IEEE Trans Power Electron 28(8):3931–3940
37. Zhang Z, Zhao H, Fu S, Shi J, He X (2016) Voltage and power balance control strategy for three-phase modular cascaded solid stated transformer. In: IEEE applied power electronics conference and exposition, pp 1475–1480
38. Liu C, Sun P, Lai JS, Ji Y, Wang M, Chen CL et al (2012) Cascade dual-boost/buck active-front-end converter for intelligent universal transformer. IEEE Trans Ind Electron 59 (12):4671–4680
39. Zhao B, Song Q, Li J, Liu W (2017) A modular multilevel dc-link front-to-front dc solid-state transformer based on high-frequency dual active phase shift for HVDC grid integration. IEEE Trans Ind Electron 64(11):8919–8927
40. Oggier GG, García GO, Oliva AR (2011) Modulation strategy to operate the dual active bridge dc–dc converter under soft switching in the whole operating range. IEEE Trans Power Electron 26(4):1228–1236

41. Zhao B, Yu Q, Sun W (2012) Extended-phase-shift control of isolated bidirectional dc–dc converter for power distribution in microgrid. IEEE Trans Power Electron 27(11):4667–4680
42. Oggier GG, Ledhold R, Garcia, GO, Oliva, AR, Balda JC, Barlow F (2006) Extending the ZVS operating range of dual active bridge high-power dc–dc converters. In: IEEE xplore power electronics specialists conference (PESC), pp 1–7
43. Demetriades GD, Nee HP (2008) Characterization of the dual-active bridge topology for high-power applications employing a duty-cycle modulation. In: IEEE power electronics specialists conference (PESC), pp 2791–2798
44. Oggier GG, García GO, Oliva AR (2009) Switching control strategy to minimize dual active bridge converter losses. IEEE Trans Power Electron 24(7):1826–1838
45. Zhao B, Song Q, Liu W (2012) Power characterization of isolated bidirectional dual-active-bridge dc–dc converter with dual-phase-shift control. IEEE Trans Power Electron 27(9):4172–4176
46. Bai H, Nie Z, Mi CC (2010) Experimental comparison of traditional phase-shift, dual-phase-shift, and model-based control of isolated bidirectional dc–dc converters. IEEE Trans Power Electron 25(6):1444–1449
47. Kim M, Rosekeit M, Sul SK, Doncker RWAAD (2011) A dual-phase-shift control strategy for dual-active-bridge dc–dc converter in wide voltage range. In: IEEE, international conference on power electronics and ECCE Asia, vol 10, pp 364–371
48. Zhao B, Song Q, Liu W, Sun W (2013) Current-stress-optimized switching strategy of isolated bidirectional dc–dc converter with dual-phase-shift control. IEEE Trans Ind Electron 60(10):4458–4467
49. Zhao B, Song Q, Liu W (2013) Efficiency characterization and optimization of isolated bidirectional dc–dc converter based on dual-phase-shift control for dc distribution application. IEEE Trans Power Electron 28(4):1711–1727
50. Bai H, Mi C (2008) Eliminate reactive power and increase system efficiency of isolated bidirectional dual-active-bridge dc–dc converters using novel dual-phase-shift control. IEEE Trans Power Electron 23(6):2905–2914
51. Krismer F, Kolar JW (2008) Accurate small-signal model for an automotive bidirectional dual active bridge converter. In: IEEE xplore the workshop on control and modeling for power electronics, pp 1–10
52. Krismer F, Kolar JW (2012) Closed form solution for minimum conduction loss modulation of dab converters. IEEE Trans Power Electron 27(1):174–188
53. Krismer F, Kolar JW (2012) Efficiency-optimized high-current dual active bridge converter for automotive applications. IEEE Trans Ind Electron 59(7):2745–2760
54. Zhou H, Khambadkone AM (2009) Hybrid modulation for dual-active bridge bidirectional converter with extended power range for ultracapacitor application. IEEE Trans Ind Appl 45(4):1434–1442
55. Du Y, Lukic SM, Jacobson BS, Huang AQ (2012) Modulation technique to reverse power flow for the isolated series resonant dc–dc converter with clamped capacitor voltage. IEEE Trans Ind Electron 59(12):4617–4628
56. Wu K, Silva CWD, Dunford WG (2012) Stability analysis of isolated bidirectional dual active full-bridge dc–dc converter with triple phase-shift control. IEEE Trans Power Electron 27 (4):2007–2017
57. Jain AK, Ayyanar R (2011) PWM control of dual active bridge: comprehensive analysis and experimental verification. IEEE Trans Power Electron 26(4):1215–1227
58. Yoo H, Sul SK, Park Y, Jeong J (2008) System integration and power-flow management for a series hybrid electric vehicle using supercapacitors and batteries. IEEE Trans Ind Appl 44(1): 108–114
59. Falcones S, Ayyanar R, Mao X (2013) A dc–dc multiport-converter-based solid-state transformer integrating distributed generation and storage. IEEE Trans Power Electron 28(5): 2192–2203
60. Xu D, Zhao C, Fan H (2004) A PWM plus phase-shift control bidirectional dc-dc converter. IEEE Trans Power Electron 19(3):666–675

61. Xiao H, Xie S (2008) A ZVS bidirectional dc–dc converter with phase-shift plus PWM control scheme. IEEE Trans Power Electron 23(2):813–823
62. Li W, Wu H, Yu H, He X (2011) Isolated winding-coupled bidirectional ZVS converter with PWM plus phase-shift (PPS) control strategy. IEEE Trans Power Electron 26(12):3560–3570
63. Shi Y, Li R, Xue Y, Li H (2015) Optimized operation of current-fed dual active bridge dc–dc converter for PV applications. IEEE Trans Ind Electron 62(11):6986–6995
64. Sha D, You F, Wang X (2016) A high-efficiency current-fed semi-dual-active bridge dc–dc converter for low input voltage applications. IEEE Trans Ind Electron 63(4):2155–2164
65. Ding Z, Yang C, Zhang Z, Wang C, Xie S (2014) A novel soft-switching multiport bidirectional dc–dc converter for hybrid energy storage system. IEEE Trans Power Electron 29(4):1595–1609
66. Karshenas HR, Daneshpajooh, H, Safaee A, Bakhshai A, Jain P (2011). Basic families of medium-power soft-switched isolated bidirectional dc-dc converters. In: IEEE conference on power electronics, drive systems and technologies, pp 92–97
67. Guidi G, Kawamura A, Sasaki Y, Imakubo T (2011) Dual active bridge modulation with complete zero voltage switching taking resonant transitions into account. In: Proceedings of the 2011 14th European conference on power electronics and applications, pp 1–10
68. Zhao B, Song Q, Liu W, Liu G, Zhao Y (2015) Universal high-frequency-link characterization and practical fundamental-optimal strategy for dual-active-bridge dc-dc converter under pwm plus phase-shift control. IEEE Trans Power Electron 30(12):6488–6494
69. Li X, Bhat AKS (2010) Analysis and design of high-frequency isolated dual-bridge series resonant dc/dc converter. IEEE Trans Power Electron 25(4):850–862
70. Ibanez F, Echeverria JM, Fontan L (2013) Novel technique for bidirectional series-resonant dc/dc converter in discontinuous mode. IET Power Electron 6(5):1019–1028
71. Corradini L, Seltzer D, Bloomquist D, Zane R, Maksimović D, Jacobson B (2014) Zero voltage switching technique for bidirectional dc/dc converters. IEEE Trans Power Electron 29 (4):1585–1594
72. Malan WL, Vilathgamuwa DM, Walker GR (2016) Modeling and control of a resonant dual active bridge with a tuned CLLC network. IEEE Trans Power Electron 31(10):7297–7310
73. Agamy MS et al (2017) A high power medium voltage resonant dual active bridge for MVDC ship power networks. IEEE J Emerg Select Topics Power Electron 5(1):88–99
74. Muthuraj SS, Kanakesh VK, Das P, Panda SK (2017) Triple phase shift control of an LLL tank based bidirectional dual active bridge converter. IEEE Trans Power Electron 32(10): 8035–8053
75. Yaqoob M, Loo KH, Lai YM (2016) A switched-inductor-augmented resonant DAB converter for achieving wide-range zero voltage switching. In: IEEE, international symposium on power electronics for distributed generation systems, pp 1–7
76. Yaqoob M, Loo KH, Lai YM (2017) Extension of soft-switching region of dual-active-bridge converter by a tunable resonant tank. IEEE Trans Power Electron 32(12):9093–9104
77. Tripathi AK, Mainali K, Madhusoodhanan S, Kadavelugu A, Vechalapu K, Patel DC et al (2017) A novel ZVS range enhancement technique of a high-voltage dual active bridge converter using series injection. IEEE Trans Power Electron 32(6):4231–4245
78. Harrye YA, Ahmed KH, Aboushady AA (2014) Reactive power minimization of dual active bridge dc/dc converter with triple phase shift control using neural network. In: IEEE international conference on renewable energy research and application, pp 566–571
79. Wen H, Xiao W, Su B (2014) Nonactive power loss minimization in a bidirectional isolated dc–dc converter for distributed power systems. IEEE Trans Ind Electron 61(12):6822–6831
80. Huang J, Wang Y, Li Z, Lei W (2016) Unified triple-phase-shift control to minimize current stress and achieve full soft-switching of isolated bidirectional dc–dc converter. IEEE Trans Industr Electron 63(7):4169–4179
81. Hou N, Song W, Wu M (2016) Minimum-current-stress scheme of dual active bridge dc–dc converter with unified phase-shift control. IEEE Trans Power Electron 31(12):8552–8561
82. Schibli N (1999) DC–DC converters for two-quadrant operation with controlled output voltage. In: EPE 1999: European conference on power electronics and applications

83. Hirose T, Takasaki M, Ishizuka Y (2013) A power efficiency improvement technique for a bidirectional dual active bridge dc–dc converter at light load. IEEE Trans Ind Appl 50(6): 4047–4055
84. Zhou H, Khambadkone AM (2009) Hybrid modulation for dual-active-bridge bidirectional converter with extended power range for ultracapacitor application. IEEE Trans Ind Appl 45(4):1434–1442
85. Krismer F, Round S, Kolar JW (2006) Performance optimization of a high current dual active bridge with a wide operating voltage range. In: IEEE power electronics specialists conference (PESC), pp 1–7
86. Choi W, Rho KM, Cho BH (2016) Fundamental duty modulation of dual-active-bridge converter for wide-range operation. IEEE Trans Power Electron 31(6):4048–4064
87. Shen Y, Sun X, Li W, Wu X, Wang B (2016) A modified dual active bridge converter with hybrid phase-shift control for wide input voltage range. IEEE Trans Power Electron 31 (10):6884–6900

Chapter 2
Unified Boundary Trapezoidal Modulation Control for Dual Active Bridge DC–DC Converter

Abstract The unified boundary trapezoidal modulation (TZM) control utilizing fixed duty cycle compensation and magnetizing current design for dual active bridge DC–DC converter is proposed in this chapter. The fixed duty cycle compensation and magnetizing current design are firstly introduced to achieve the zero-voltage switching (ZVS) of the power switches, which cannot be ensured with the conventional TZM control. As a result, all the power switches of dual active bridge DC–DC converter can achieve ZVS and four switches can be turned off with very low current. Besides, based on the revealed power transfer characteristic, the power control variables including the duty cycles and phase shift ratio can be unified without look-up tables or operation region division. With the proposed boundary TZM control, circulating current losses can be reduced and non-active power is significantly suppressed according to the mathematic analysis, resulting in decrease of the conduction loss. A 1.6 kW laboratory prototype is built to verify the theoretical analysis and effectiveness of the proposed control.

Keywords Dual active bridge · Fixed duty cycle compensation
Zero-voltage switching (ZVS) · Boundary trapezoidal modulation
(boundary TZM control)

2.1 Fixed Duty Cycle Compensation and Magnetizing Current Design for DAB DC–DC Converter with Trapezoidal Modulation

Figure 2.1 shows the topology of DAB DC–DC converter [1, 2]. It consists of two active bridges connected by a high-frequency transformer. The active bridge in the left side generates a high-frequency voltage v_{AB}, and the other one generates voltage v_{CD}. Bidirectional power transfer is controlled by the phase shift ratio between the two high-frequency voltages and their duty cycles.

In view of non-active power of this DAB converter, the triangular modulation (TRM) [3, 4] has no non-active power issue and can achieve zero-current switching

D. Sha and G. Xu, *High-Frequency Isolated Bidirectional Dual Active Bridge DC–DC Converters with Wide Voltage Gain*, CPSS Power Electronics Series,
https://doi.org/10.1007/978-981-13-0259-6_2

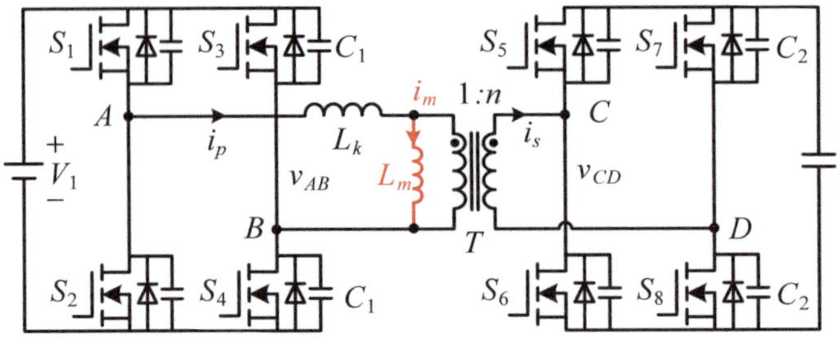

Fig. 2.1 Topology of DAB DC–DC converter

(ZCS) for four switches. But the ZVS is lost for these ZCS switches. Even though the modified TRM control is proposed in [5, 6], it still suffers from high peak current and large RMS current. These issues can be avoided with the hybrid modulation [5] using CPS at heavy load and the modified TRM control at light load. While, smooth transfer between the two modes is a challenge. Compared with TRM, the trapezoidal modulation (TZM) [4] can make the active bridge work with large duty cycles and result in lower peak current and RMS value.

2.1.1 Conventional Trapezoidal Modulation (TZM)

The operation principle of TZM is introduced in Fig. 2.2. Φ is the phase shift ratio between the voltages generated by two active bridges, and D_1 and D_2 represent the inner phase shift ratio of the two active bridges, respectively, which are also named as duty cycles. T_s is the switching period, and 1: n denotes the transformer turns ratio. i_p and i_s are the transformer currents in V_1 and V_2 sides, respectively, which are also denoted in Fig. 2.1. Unlike CPS and EPS control, neither D_1 nor D_2 are fixed as 0.5. They are both less than 0.5 and can be different when V_1 is not equal to V_2/n. The relationship between D_1 and D_2 for TZM is written as [6],

$$D_2 = (nV_1/V_2)D_1 \tag{2.1}$$

The power transfer under TZM is also given as,

$$P_o^{\mathrm{TZM}} = \frac{V_1}{L_k f_s}\left(D_1^2 V_1 - (D_1 - \Phi)^2 V_2/n\right) \tag{2.2}$$

where f_s is the switching frequency. Compared with CPS, EPS, and other control methods, TZM has its unique advantages such as no non-active power, zero-current switching, and changeable duty cycles when the input and output voltages are not matched.

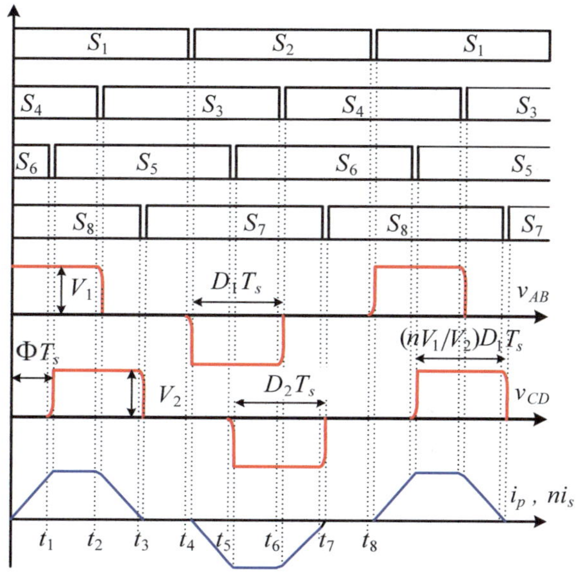

Fig. 2.2 Typical operation waveforms of TZM

2.1.2 *ZVS Conditions for DAB Converter with Conventional TZM Control*

To achieve ZVS of power switches for a conventional full-bridge converter with phase shift control, the energy in the inductive components must be high enough to charge/discharge their junction capacitors completely. The dead time of the two switches in the same leg should also be properly designed to ensure that the charge/discharge process is finished before switches are turned on. This principle is also applied to TZM DAB converter. Just like the conventional full-bridge converter, the switches in the lagging leg can achieve ZVS easily because the transformer leakage current is almost approaching its maximum value before these switches are turned on. As for the leading leg switches, the transformer leakage current is always zero when the switches are turned off if the magnetizing current is ignored. ZVS is impossible as there is no current to charge/discharge the junction capacitors. Consequently, for TZM DAB converter, S_3, S_4, S_5, and S_6 can achieve ZVS, while S_1, S_2, S_7, and S_8 cannot [4].

The leg composed of S_1 and S_2 is taken as an example to analyze ZVS conditions for these leading legs as shown in Fig. 2.3. Because the two switches are the same type, the junction capacitors are both denoted as C_1 in Fig. 2.3. When S_2 is turned off at time t_8 (t_8 is denoted in Fig. 2.2), none of the two switches conducts and the secondary side of the transformer is shorted. The corresponding equivalent circuit is shown in Fig. 2.3a, and related current and voltage waveforms after t_8 are shown in Fig. 2.3b.

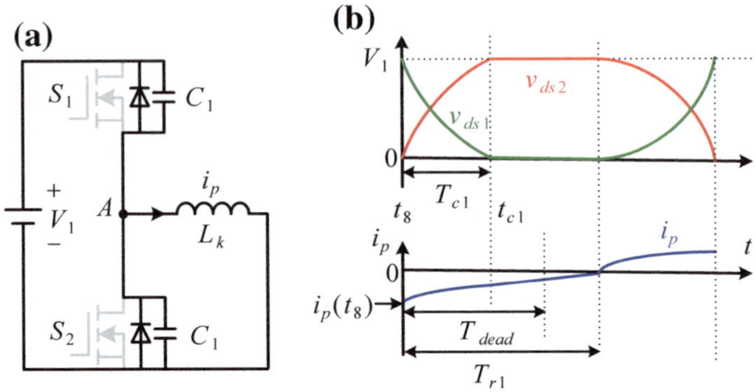

Fig. 2.3 ZVS conditions for S_1 and S_2. **a** Circuit. **b** Waveforms

To charge/discharge the junction capacitors, the energy stored in the inductor L_k should be larger than that in the two capacitors

$$0.5L_k i_p^2(t_8) \geq 0.5C_1 V_1^2 + 0.5C_1 V_1^2 \tag{2.3}$$

The current should charge the junction capacitor of S_2 before S_1 is turned on. Therefore, $i_p(t_8) < 0$. According to (2.3), it can be written by

$$i_p(t_8) \leq -V_1\sqrt{2C_1/L_k} \tag{2.4}$$

Meanwhile, according to the circuit in Fig. 2.3a, the relationship between i_p and v_{ds2} is shown as

$$\begin{cases} L_k \mathrm{d}i_p/\mathrm{d}t = v_{ds2} \\ i_p = C_1 \mathrm{d}(V_1 - v_{ds2})/\mathrm{d}t - C_1 \mathrm{d}v_{ds2}/\mathrm{d}t = -2C_1 \mathrm{d}v_{ds2}/\mathrm{d}t \end{cases} \tag{2.5}$$

where v_{ds1} and v_{ds2} are the junction voltage for S_1 and S_2, respectively.

The initial value of v_{ds2} is

$$v_{ds2}(t_8) = 0 \tag{2.6}$$

Based on (2.5) and (2.6), v_{ds2} can be expressed as

$$v_{ds2}(t - t_8) = \frac{-i_p(t_8)}{\sqrt{2C_1/L_k}} \sin\frac{t - t_8}{\sqrt{2C_1 L_k}} \tag{2.7}$$

The charging process is not finished until v_{ds2} reaches its maximum value V_1.

$$T_{c1} = t_{c1} - t_8 = \sqrt{2C_1L_k}\arcsin(-V_1\sqrt{2C_1/L_k}/i_p(t_8)) \tag{2.8}$$

Based on (2.7), the charging time T_{c1} can be written by (2.8), and S_1 should be turned on after the charging process is over. Meanwhile, when i_p reaches zero the junction capacitor of S_2 starts to discharge, causing loss of ZVS. As a result, S_1 should be turned on before i_p reaches zero. Therefore, as also illustrated in Fig. 2.3b

$$T_{c1} < T_{\text{dead1}} < T_{r1} \tag{2.9}$$

Based on (2.8),

$$i_p(t_{c1}) = -2C_1\frac{dv_{ds2}}{dt}(t_{c1}) = i_p(t_8)\cos\left((t_{c1} - t_8)/\sqrt{2C_1L_k}\right) \tag{2.10}$$

Since $L_k i_p(t_{c1})/T_{r1} = V_1$, T_{r1} can be expressed as

$$T_{r1} = \frac{V_1}{L_k i_p(t_8)\cos(T_{c1}/\sqrt{2C_1L_k})} + T_{c1} \tag{2.11}$$

To sum up, the minimum current for achieving ZVS of S_1 has been illustrated in (2.4). The limit for the dead time is also given in (2.8), (2.9), and (2.11).

Likewise, ZVS conditions for the other three switches (S_2, S_7, S_8) can also be summed up in Table 2.1. t_3, t_4, t_7, and t_8 are the switch turn-off instants denoted in Fig. 2.2; C_2 is the junction capacitor of switch S_7 or S_8; $L_{ks} = n^2L_k$, which is the equivalent transformer leakage inductance of V_2 side reflected from V_1 side.

2.1.3 *Proposed Fixed Duty Cycle Compensation*

The proposed fixed duty cycle compensation is shown in Fig. 2.4 to achieve ZVS for S_1 and S_2. As illustrated, instead of using the relationship of duty cycles in (2.1), a duty cycle compensation value D_c is added and expressed as

$$D_2 = (nV_1/V_2)D_1 - D_c \tag{2.12}$$

Table 2.1 ZVS conditions for the four switches

	S_1	S_2	S_7	S_8
Current range	$i_p(t_8) < -I_{1\min}$	$i_p(t_4) > I_{1\min}$	$i_s(t_3) < -I_{2\min}$	$i_s(t_7) > I_{2\min}$
Dead time	$T_{c1} < T_{\text{dead1}} < T_{r1}$		$T_{c2} < T_{\text{dead2}} < T_{r2}$	

$T_{c2} = \sqrt{2C_2L_{ks}}\arcsin\frac{V_2\sqrt{\frac{2C_2}{L_{ks}}}}{-i_s(t_3)}$, $t_{r2} = \frac{V_2}{L_{ks}i_s(t_3)\cos\frac{T_{c2}}{\sqrt{2C_2L_{ks}}}} + T_{c2}$, $I_{2\min} = V_2\sqrt{\frac{2C_2}{L_{ks}}}$, $I_{1\min} = V_1\sqrt{\frac{2C_1}{L_k}}$

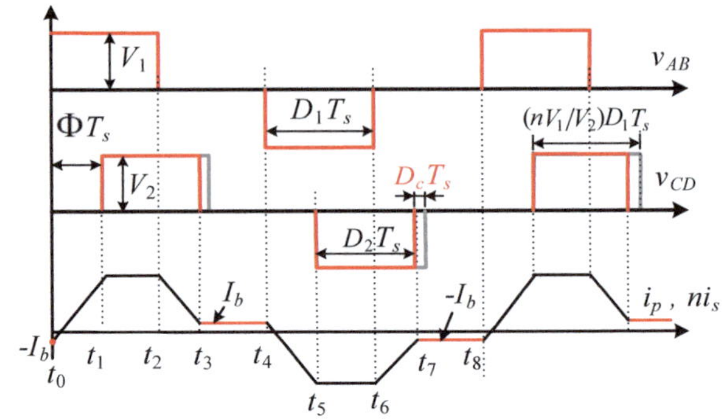

Fig. 2.4 Fixed duty cycle compensation

As shown in Fig. 2.4, unlike the conventional TZM method, a bias current (I_b) occurs in i_p at the moment (t_4 or t_8) when S_1 or S_2 turns off. If this bias current is properly designed, it can be used to charge/discharge the junction capacitor to achieve ZVS.

Since phase shift control is used, the polarity of the bias current during the interval t_3–t_4 should be reversed with the one during interval t_7–t_8. The relationship of $i_p(t_8)$ and $i_p(t_3)$ can be expressed as

$$i_p(t_3) = -i_p(t_8) = I_b \tag{2.13}$$

Meanwhile, $i_p(t_3)$ can also be calculated as

$$i_p(t_3) = T_s(D_1V_1 - D_2V_2/n)/L_k - I_b \tag{2.14}$$

Substitution of (2.12) and (2.13) into (2.14) yields

$$I_b = D_cT_sV_2/(2nL_k) \tag{2.15}$$

According to the ZVS conditions in the previous section, to achieve ZVS for S_1 and S_2, the bias current must be larger than the minimum charge current

$$I_b > I_{1\min} = V_1\sqrt{2C_1/L_k} \tag{2.16}$$

Then based on (2.15) and (2.16), D_c should be designed as

$$D_c > 2nV_1\sqrt{2C_1L_k}/(T_sV_2) \tag{2.17}$$

Equation (2.17) shows that a minimum duty cycle compensation value D_c can be designed to achieve ZVS for S_1 and S_2 regardless of the load. This implies the ZVS achievement is not affected by the output power. It is worth noting that the dead time can be designed according to its limit in Table 2.1.

2.1.4 Magnetizing Current Design to Achieve ZVS for S_7 and S_8

In the literature, the magnetizing current of the transformer is seldom considered for non-resonant DAB converters. Based on Table 2.1, to achieve ZVS of S_2, $i_p(t_4) > I_{1\min}$; while to achieve ZVS for S_7, $i_s(t_3) < -I_{2\min}$. However, when the magnetizing current is ignored, $i_p(t_4) = ni_s(t_3)$, which is illustrated in both Fig. 2.2 and Fig. 2.3. This leads to a conflict between ZVS conditions of S_2 and S_7, resulting in that only one of the two switches can achieve ZVS. Similarly, the conflict also exists between ZVS for S_1 and S_6. As a result, although ZVS of S_1 and S_2 can be achieved with the proposed duty cycle compensation, ZVS of S_7 and S_8 still cannot be obtained. To obtain the ZVS conditions: $i_s|_{t=t3} < -I_{2\min}$ and $i_s|_{t=t7} > I_{2\min}$ for S_7 and S_8, the magnetizing current i_m can be utilized as shown in Fig. 2.5.

Figure 2.5 illustrates the waveforms of i_p, the magnetizing current i_m, and i_s utilizing the magnetizing current. In this case, the waveform of i_p is the same as the one shown in Fig. 2.4, while the value of ni_s is written by

$$ni_s = i_p - i_m \tag{2.18}$$

where n is the turns ratio as mentioned before, and ni_s is the current of transformer in V_2 side reflected to V_1 side.

As shown in Fig. 2.5, at t_3, i_m is equal to $I_{\max}$, and the following can be obtained during the interval t_1–t_3

$$L_m \frac{2I_{\max}}{D_2T_s} = \frac{V_2}{n} \tag{2.19}$$

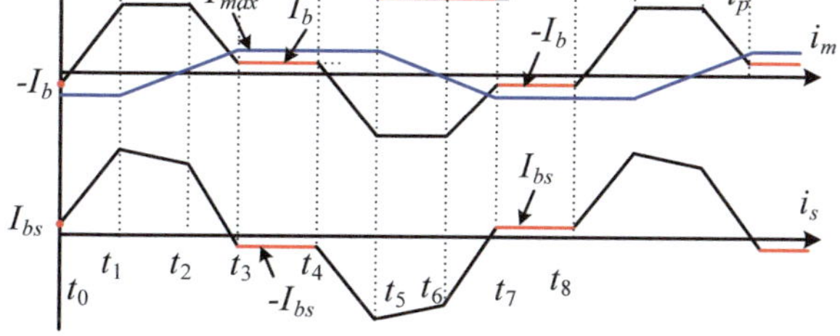

Fig. 2.5 Magnetizing current design to achieve ZVS for S_7 and S_8

Rearranging (2.19) yields

$$I_{\max} = V_2 D_2 T_s / (2nL_m) \tag{2.20}$$

At t_3, the bias current of i_s can be expressed as

$$I_{\text{bs}} = -(I_b - I_{\max})/n \tag{2.21}$$

Substituting (2.20) into (2.21), the expression of the current bias I_{bs} for V_2 side can be obtained as

$$I_{\text{bs}} = V_2 D_2 T_s / (2n^2 L_m) - I_b/n \tag{2.22}$$

where the expression of I_b is shown in (2.15).

To design a suitable bias current I_{bs} in order to obtain ZVS for S_7 and S_8 in Table 2.1, its value is written by

$$I_{\text{bs}} = V_2 D_2 T_s / (2n^2 L_m) - I_b/n > I_{2\min} \tag{2.23}$$

Once I_{bs} is certain, the dead time can also be designed according to the dead time limit in Table 2.1.

2.2 Power Transfer Characteristic and Selections of Duty Cycles and Phase Shift Ratio

In this section, the power transfer characteristic is studied for the unified control of duty cycles (D_1, D_2) and phase shift ratio (Φ). The case when $\Phi < 0$ is not discussed since DAB DC–DC converter is symmetrical in view of the topology and control. Since the bias current is low, it is neglected in this section. Even though D_2 can be expressed by D_1 using (2.1), the power transfer is still controlled by two control variables including both D_1 and Φ. Optimized selection of D_1 and Φ should be made since there are numerous combinations of them for the same output power.

Figure 2.6 presents the qualitative waveforms of three possible selections for the same output power transfer with TZM control. During the interval t_0–t_1, v_{CD} is zero. The transformer current does not contribute to the power transfer and only causes circulating current loss. Similarly, the circulating current loss also exists in the interval: t_4–t_5. Compared with (a) and (b), the duration lengths of these intervals are the shortest in (c) and the maximum current is lowest, indicating lower RMS current and lower circulating current loss.

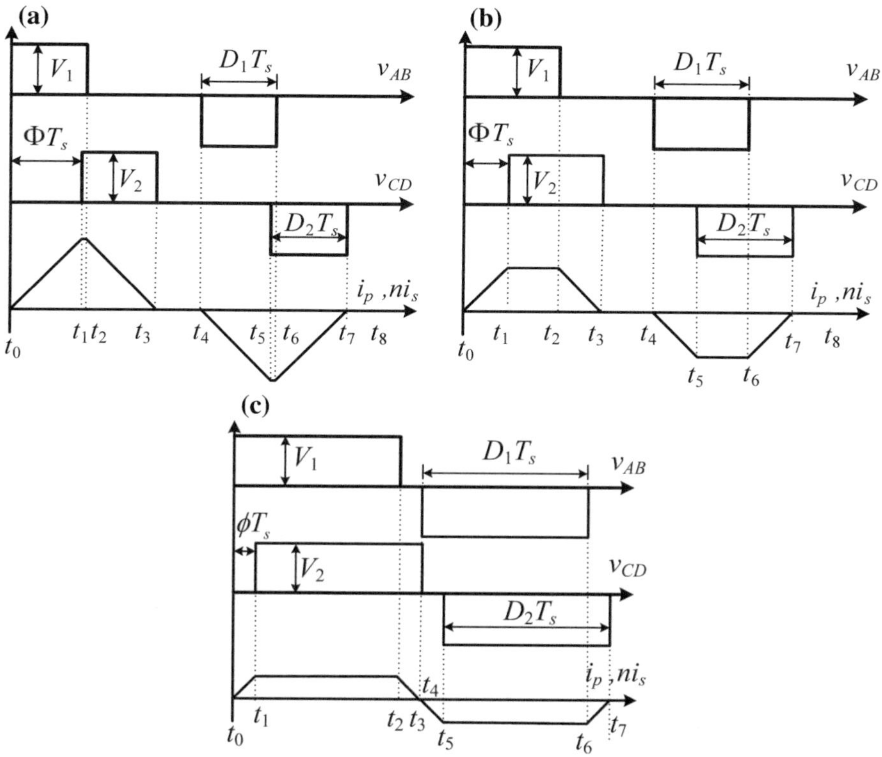

Fig. 2.6 Three conditions of D_1 and Φ for the same power transfer with TZM control. **a** Small duty cycles and large Φ. **b** Middle duty cycles and middle Φ. **c** Small Φ and large duty cycles under critical continuous conduction mode

2.2.1 Selections of Duty Cycles and Phase Shift Ratio for Minimum RMS Circulating Current

As mentioned before, different control points of D_1 and Φ generate different current waveforms, leading to different peak currents and circulating losses. Therefore, it is necessary to optimize duty cycles and the phase ratio.

Mathematically, based on the waveform in Fig. 2.6, $i(t_1)$ and $i(t_2)$ can be written by, respectively,

$$
\begin{aligned}
i(t_1) &= \int_{t_0}^{t_1} (V_1/L_k)\mathrm{d}t = V_1 \Phi T_s/L_k, \\
i(t_2) &= \int_{t_0}^{t_1} \frac{V_1}{L_k}\mathrm{d}t + \int_{t_1}^{t_2} \frac{V_1 - V_2}{L_k}\mathrm{d}t = \frac{V_2 T_s}{nL_k}\left(\Phi + \left(\frac{nV_1}{V_2} - 1\right)D_1\right)
\end{aligned}
\tag{2.24}
$$

The RMS circulating current in a switching period for one active bridge can be calculated as

$$
I_{c\_\mathrm{RMS}} = \sqrt{\left(\int_{t_0}^{t_1}\left(\frac{V_1}{L_k}t\right)^2 dt + \int_{t_5}^{t_4}\left(\frac{V_1}{L_k}t\right)^2 \mathrm{d}t\right)\Big/T_s} = \frac{V_1 T_s}{L_k}\sqrt{\frac{2\Phi^3}{3}}
\tag{2.25}
$$

Equation (2.24) and (2.25) indicate that small Φ results in lower circulating current and $i(t_1)$, causing low conduction loss and peak current. However, if Φ is too small, the power transfer may be limited since decreasing of Φ may reduce the output power. Thus, it is necessary to find the minimum Φ with the precondition of the ability to transfer the required power.

Let p, d_1, d_2, and φ denote the variables for the equilibrium values P, D_1, D_2, and Φ, respectively. Then, the power transfer function in (2.2) is

$$
p(d_1, \varphi) = V_1\left(d_1^2 V_1 - (d_1 - \varphi)^2 V_2/n\right)/(L_k f_s)
\tag{2.26}
$$

According to the working principle of TZM control, the ranges for two duty cycles can be shown as

$$
\varphi T_s < d_1 T_s < 0.5T_s \cap \varphi T_s + d_2 T_s < 0.5T_s
\tag{2.27}
$$

Substitution of $d_1 V_1 = d_2 V_2/n$ into (2.27) leads to

$$
\varphi \le d_1 \le 0.5 \cap \varphi + nV_1 d_1/V_2 \le 0.5
\tag{2.28}
$$

For a given fixed power P_o, φ can be regarded as a function of d_1. Thus,

$$
\frac{\partial P_o}{\partial d_1} = \frac{\partial\left(V_1(d_1^2 V_1 - (d_1 - \varphi)^2 V_2/n)/(L_k f_s)\right)}{\partial d_1} = 0
\tag{2.29}
$$

Simplification of (2.29) leads to

$$\frac{\partial \varphi}{\partial d_1} = \frac{d_1(V_2 - nV_1) - \varphi V_2}{(d_1 - \varphi)V_2} \tag{2.30}$$

If $V_1 \geq V_2/n$, $\partial\varphi/\partial d_1 < 0$, φ is monotone decreasing with d_1. Therefore, minimum φ can be obtained at the boundary condition in (2.28). When $\varphi = d_1$, minimum φ cannot be got because $I_{p\max}$ and I_{c_RMS} are both the largest as also illustrated in Fig. 2.6a. Therefore, the minimum φ for a given output power P_o exists at the boundary condition, which is described by

$$\varphi + nV_1 d_1/V_2 = 0.5 \quad (d_1 \leq 0.5) \tag{2.31}$$

Figure 2.7 shows an example for plot of φ versus d_1 for different output power when $V_1 = V_2$. As the output power changes from 0 to 2200 W, the point of minimum φ changes from point F to A and they all exist at the boundary, which agrees with the aforementioned analysis.

If $V_1 < V_2/n$, (2.30) is not a monotone decreasing function of d_1. The minimum point can be calculated by setting $\partial\varphi/\partial d_1 = 0$. Then,

$$\varphi_{\min} = d_{1\min}(V_2 - nV_1)/V_2 \tag{2.32}$$

Therefore, for a given power P_o, if $\varphi_{\min} + nV_1 d_{1\min}/V_2 > 0.5$, the minimum φ exists at the boundary condition: $\varphi + nV_1 d_1/V_2 = 0.5$; if $\varphi_{\min} + nV_1 d_{1\min}/V_2 < 0.5$, the minimum φ is $\varphi_{\min} = d_{1\min}(V_2 - nV_1)/V_2$.

Figure 2.8 plots φ versus d_1 with different output power when $V_1 < V_2$ for one case. As seen, when the output power changes from 500 to 2000 W, the point of minimum φ also occurs at the boundary. When the power is relatively lower such as 0 or 200 W, the points of minimum φ are point G and point H, respectively. But the boundary points for the two output power plots occur at point F and point E. In this case, the boundary condition is not suitable for achieving minimum φ. However, the calculation for $\varphi_{\min}$ and $d_{1\min}$ is complex, and the difference of φ between the

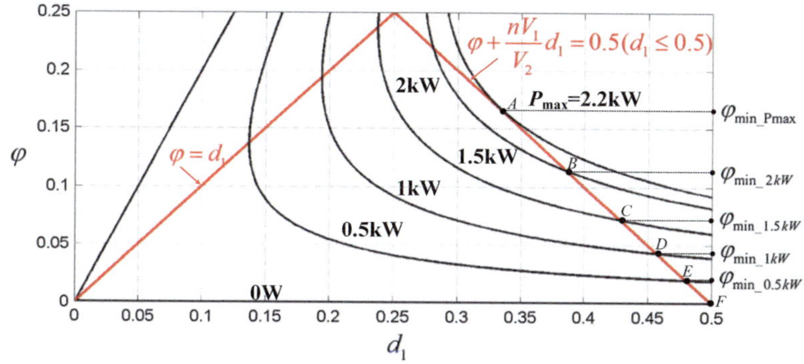

Fig. 2.7 Plot of φ versus d_1 for different output power (V_1 = 200 V, V_2 = 200 V, n = 1, L_k = 30 μH, f_s = 50 kHz)

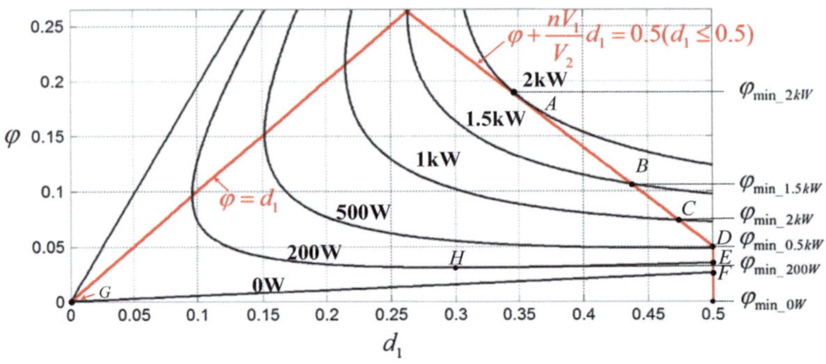

Fig. 2.8 Plot of φ versus d_1 for different output power (V_1 = 180 V, V_2 = 200 V, n = 1, L_k = 30 μH, f_s = 50 kHz)

point H and E (or G and F) is very small. Therefore, when $V_1 < V_2/n$ at light load, the boundary condition can still be used to select D_1 and Φ to simplify the calculation at the cost of slightly increased circulating current and RMS value.

2.2.2 *Maximum Power Transfer Point*

To investigate the combination of D_1 and Φ for maximum power transfer, the following function is defined as

$$\varphi + nV_1 d_1/V_2 = a, (0 \leq a \leq 0.5) \tag{2.33}$$

When a changes from 0 to 0.5, this function can traverse all the existed point (D_1, φ) for the power transfer function in (2.26). Substitution of (2.33) into (2.26) and elimination of φ leads to

$$p(a, d_1) = \frac{V_1}{L_k f_s}\left(d_1^2 V_1 - \left(d_1 - a + \frac{nV_1}{V_2} d_1\right)^2 \frac{V_2}{n}\right) \tag{2.34}$$

Then,

$$\frac{\partial p(a, d_1)}{\partial d_1} = \frac{2V_1}{nV_2 L_k f_s}\left(-(V_2^2 + n^2 V_1^2 + nV_1 V_2) d_1 + (V_2^2 + nV_1 V_2) a\right) \tag{2.35}$$

Let $\partial p(a, d_1)/\partial d_1 = 0$, since $-(V_2^2 + n^2V_1^2 + 2nV_1V_2) < 0$, the maximum power transfer occurs at the point when

$$d_{1\_p\max}(a) = \frac{(V_2^2 + nV_1V_2)}{(V_2^2 + n^2V_1^2 + nV_1V_2)}a, \quad (0 < a \leq 0.5) \tag{2.36}$$

Substitution of (2.36) into (2.33) yields

$$\varphi_{p\max}(a) = \frac{V_2}{V_2 + nV_1}d_{1\_p\max} = \frac{V_2^2}{(V_2^2 + n^2V_1^2 + nV_1V_2)}a \tag{2.37}$$

Then, substitution of (2.36), (2.37) into (2.34) yields

$$\begin{aligned} p_{\max}(a) &= V_1(d_{1\_p\max}^2 V_1 - (d_{1\_p\max} - \varphi_{p\max})^2 V_2/n)/(L_k f_s) \\ &= \frac{V_1^2}{L_k f_s}\frac{V_2^4 + n^2V_1^2V_2^2 + nV_1V_2^3}{(V_2^2 + n^2V_1^2 + nV_1V_2)^2}a^2, \quad (0 < a \leq 0.5) \end{aligned} \tag{2.38}$$

$p_{\max}(a)$ is a quadratic function of a, and its maximum value exists when $a = 0.5$. Therefore, the global maximum value of $p(d_1, \varphi)$ is

$$P_{\max}(0.5) = \frac{V_1^2}{4L_k f_s}\frac{V_2^4 + n^2V_1^2V_2^2 + nV_1V_2^3}{(V_2^2 + n^2V_1^2 + nV_1V_2)^2} \tag{2.39}$$

And the corresponding $D_{1\_p\max}$ and $\phi_{p\max}$ are, respectively, expressed as

$$D_{1\_p\max} = \frac{(V_2^2 + nV_1V_2)}{2(V_2^2 + n^2V_1^2 + nV_1V_2)}, \quad \phi_{p\max} = \frac{V_2^2}{2(V_2^2 + n^2V_1^2 + nV_1V_2)} \tag{2.40}$$

Note that $D_{1\_p\max}$ and $\phi_{p\max}$ satisfy the boundary condition in (2.31), which indicates that the maximum power occurs at the boundary.

2.3 Boundary TZM Control and Its Implementation

2.3.1 Boundary TZM Control

As discussed before, the control point with lower RMS current and circulating loss for an arbitrary given power occurs mostly at the boundary condition, and the maximum power point also exists at the boundary as mentioned in Sect. 2.2.2. If

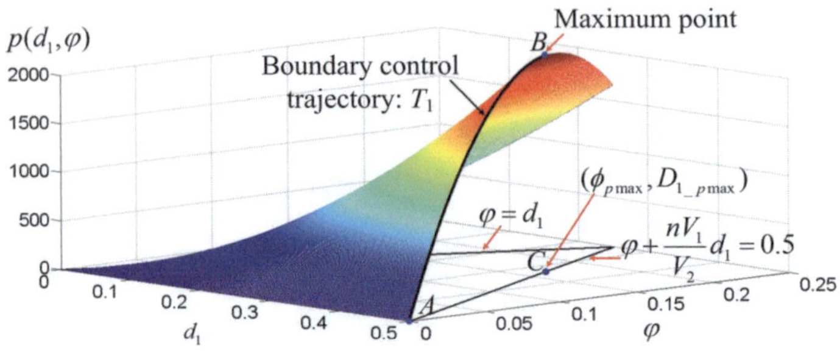

Fig. 2.9 Power transfer function curve (V_1 = 200 V, V_2 = 200 V, n = 1, L_k = 30 μH, f_s = 50 kHz)

this boundary condition is used for the control of d_1 and φ, two advantages can be obtained. One is that the conduction loss can be reduced; the other is that the maximum power transfer will not be limited as the global maximum power point exists at the boundary condition.

The curve of $p(d_1, \varphi)$ is plotted in Fig. 2.9. As illustrated, the maximum point exists at the boundary which agrees with the mathematical calculation in Sect. 2.2. Meanwhile, the boundary control trajectory T_1 is shown. This trajectory starts from the minimum point A and ends at the maximum point B. It cannot run across point B, because only one point should exist for the same output power, namely the control trajectory should be monotonous. As a result, the range for φ with the boundary condition can be obtained as

$$0 \leq \varphi \leq \phi_{pmax} \tag{2.41}$$

and the corresponding value range of d_1 is

$$D_{1\_pmax} \leq d_1 \leq 0.5 \tag{2.42}$$

where expressions of D_{1_pmax} and ϕ_{pmax} are shown in (2.40).

If φ is controlled through a *PI* regulator, based on the boundary condition and the value ranges in (2.41) and (2.42), d_1 can be obtained. Then, if d_1 is substituted into (2.1) d_2 can be calculated. Therefore, the optimized control for the power transfer can be realized by regulating the variable φ to achieve real-time control.

2.3.2 Implementation of Boundary TZM Control

Figure 2.10 shows the control diagram for the boundary TZM control. A *PI* regulator is used to control the voltage V_2 through control of φ. As shown, when $\varphi \geq 0$, the boundary condition is used to calculate d_1 and then d_2 is obtained based

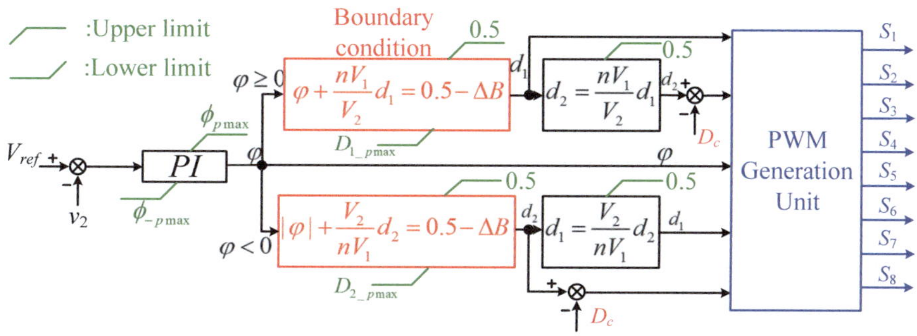

Fig. 2.10 Implementation of the boundary control using a PI regulator

on (2.1). A small constant ΔB is added in the boundary condition to ensure the duration length of the bias currents, which can charge/discharge the junction capacitor to achieve ZVS. Hence, ΔBT_s should be larger than the minimum charging/discharging time T_c. When $\varphi < 0$, since the control principle is symmetrical, the boundary condition can be obtained by interchanging V_1 with V_2/n and replacing d_1 with d_2 in (2.31). Using this boundary condition, d_2 can be calculated, and then d_1 is also obtained based on (2.1). The PWM generation unit generates all the switching signals based on both the output of *PI* regulator (φ) and the calculated control variables (d_1 and d_2).

2.4 Experimental Verification

A laboratory prototype was built to validate the feasibility of the proposed ZVS method and unified boundary control strategy.

The initial parameters of the laboratory prototype are: (a) V_1 = 180–240 V, V_2 = 200 V; (b) turns ratio: 1:1; (c) rated power P_o = 1.6 kW; (d) switching frequency f_s = 50 kHz; (e) IXFH69N30P is chosen as the power switches and its junction capacitor is 570 pF. Furthermore, the parameters including L_k, $I_{1\min}$, $I_{2\min}$, I_b, I_{bs}, D_c, and L_m are designed according to the following parameter design procedure.

(1) Calculation of inductor L_k

According to the maximum power transfer function in (2.39), L_k can be designed as

$$L_k \leq \frac{V_1^2}{4f_s} \frac{V_2^4 + n^2 V_1^2 V_2^2 + nV_1 V_2^3}{(V_2^2 + n^2 V_1^2 + nV_1 V_2)^2 P_{o\max}} \tag{2.43}$$

Here, P_{omax} is set as: $P_{omax} = (1 + 20\%)P_{\text{o}}$ to ensure a 20% margin of the output power. Therefore, based on (2.43)

$$L_k \leq 3.113 \times 10^{-5}(\text{H}) \tag{2.44}$$

The startup current of the converter decreases as L_k increases. L_k can be designed as large as it can be once (2.44) is satisfied. In this case, L_k is designed to be 30 μH.

(2) Calculation of $I_{1\min}$, $I_{2\min}$

The switches junction capacitor is 570 pF. The minimum charging/discharging current for achieving ZVS within full input voltage range can be got according to Table 2.1.

$$\begin{aligned} \max(I_{1\min}) &= V_{1\_\max}\sqrt{2C_1/L_k} = 1.479(\text{A}) \\ I_{2\min} &= V_2\sqrt{2C_2/L_{ks}} = 1.233(\text{A}) \end{aligned} \tag{2.45}$$

(3) The bias currents: I_b and I_{bs}

The current bias I_b of i_p in V_1 side does not change as the load varies. It is set as $(1+20\%)\cdot\max(I_{1\min})$ to obtain a 20% margin of the charging/discharging current for achieving ZVS of S_1 and S_2 within full input voltage range. However, the bias current I_{bs} of i_s in V_2 side is affected by the output power. Therefore, the minimum value of I_{bs}, which is denoted as $I_{\text{bs\_min}}$, should be larger than the minimum charging/discharging current $I_{2\min}$ with also a 20% margin. These two parameters can be written as

$$\begin{cases} I_b = (1+20\%)\cdot\max(I_{1\min}) = 1.775 \approx 1.8\,(\text{A}) \\ I_{bs\_\min} = (1+20\%)I_{2\min} = 1.479 \approx 1.5\,(\text{A}) \end{cases} \tag{2.46}$$

(4) Duty cycle compensation value: D_{c}.

Once L_k and I_b are certain, D_{c} can be obtained according to (2.15)

$$D_c = 2nL_kf_sI_b/V_2 = 0.027 \tag{2.47}$$

(5) Magnetization inductance: L_m.

When the boundary condition is applied, the transferred power p versus D_1 is written by

$$p(D_1) = \frac{V_1}{L_kf_s}\left(D_1^2V_1 - \left(D_1 + \frac{nV_1}{V_2}D_1 - 0.5\right)^2 V_2/n\right) \tag{2.48}$$

The range of D_1 is shown in (2.42), and the minimum value of D_1 is got at the highest transferred power. Assume $P(D_1) = P_o = 1600$ W, and solution of D_1 when $V_1 = 180$ V and $V_1 = 240$ V leads to the minimum value of D_1

$$D_{1\_min} = \begin{cases} 0.397 & \text{when} \quad V_1 = 240\,\text{V} \\ 0.436 & \text{when} \quad V_1 = 180\,\text{V} \end{cases} \tag{2.49}$$

Since $0.397 \times 240 > 0.442 \times 180 = 79.56$, the minimum value of D_1V_1 exists in the case when $V_1 = 180$ V and the minimum value is 79.56.

Meanwhile, based on (2.22), expression of the bias current I_{bs} can be written as

$$I_{bs} = V_2D_2T_s/(2n^2L_m) - I_b/n \tag{2.50}$$

L_m should be designed to make I_{bs} always larger than I_{bs_min} to achieve ZVS. Thus, the following must be met

$$\min(I_{bs}) = \min(V_1D_1)/2nL_mf_s - I_b/n \geq I_{bs\_min} \tag{2.51}$$

Therefore, L_m should be calculated as

$$L_m \leq \min(V_1D_1)/(2(I_b + nI_{bs\_min})f_s) = 2.411 \times 10^{-4}\ (\text{H}) \tag{2.52}$$

Since decreasing of L_m leads to higher peak current and RMS current in V_2 side, L_m should be designed as large as possible once (2.52) is satisfied. Therefore, in this case, L_m is 240 μH.

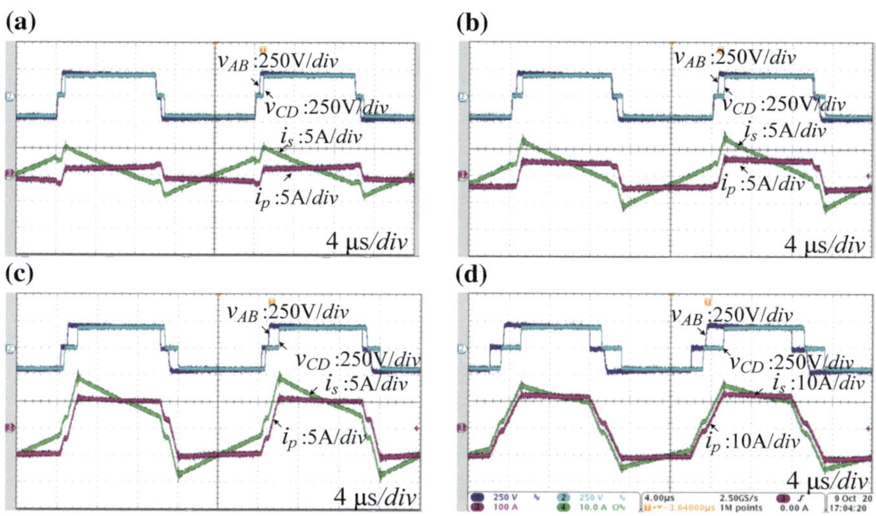

Fig. 2.11 Different load under $V_1 = V_2 = 200$ V condition. **a** 200 W. **b** 400 W. **c** 800 W. **d** 1600 W

The experimental results of different loads under $V_1 = V_2 = 200$ V condition are shown in Fig. 2.11. As seen, the two bias currents, which are properly designed by the fixed duty cycle compensation and magnetizing current design, occur at the boundary when both v_{AB} and v_{CD} are zero to create ZVS conditions for the aforementioned four switches. These two bias currents always exist despite the load. Under 200 W load condition, as shown in Fig. 2.11a, the duty cycles of v_{AB} and v_{CD} are the largest compared with other conditions. Besides, the duty cycles of v_{AB} and v_{CD} decrease while the phase shift ratio increases, which agrees with the analysis. Furthermore, as seen from the current waveforms, the transformer currents are always critical continuous for different loads if the two bias currents are neglected.

The ZVS when $V_1 = V_2 = 200$ V is shown in Fig. 2.12. Switches S_1 and S_7 are chosen to show the effectiveness of the proposed control. v_{gs} is the driver signals, and v_{ds} is the drain-to-source voltages. As shown in all the figures, ZVS can be obtained under both light load and full load.

Figure 2.13 shows different input voltages under rated load condition. As illustrated, the two bias currents also occur at the boundary and can be used to achieve ZVS as seen from the waveforms of v_{AB} and v_{CD}. Although the input and output voltage are not matched, the transformer current remains critical continuous if the two bias currents are ignored. This implies that the boundary condition can be ensured under different input voltages.

The experimental waveforms in reverse power flow under 400 and 1200 W conditions are shown in Fig. 2.14. As seen, v_{AB} lags v_{CD}, indicating that the phase shift ratio becomes negative to achieve reverse power flow. Meanwhile, the two

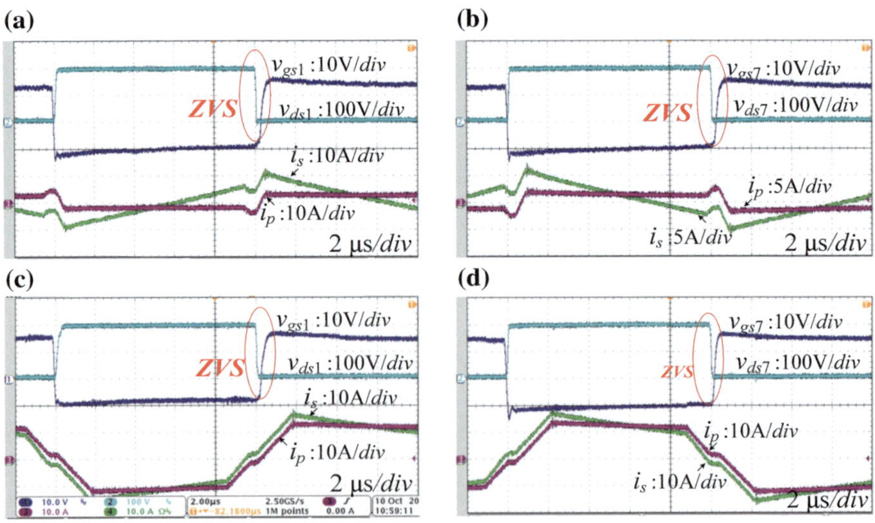

Fig. 2.12 ZVS under different loads ($V_1 = V_2 = 200$ V). **a** ZVS of S_1 under 200 W. **b** ZVS of S_7 under 200 W. **c** ZVS of S_1 under 1600 W. **d** ZVS of S_7 under 1600 W

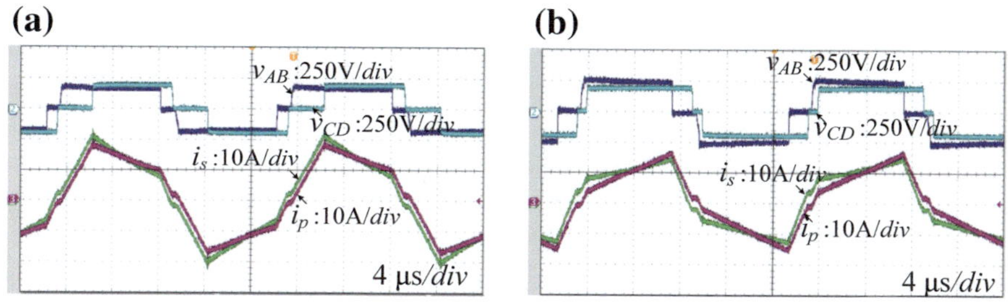

Fig. 2.13 Different input voltage under rated load condition. **a** 180 V input. **b** 240 V input

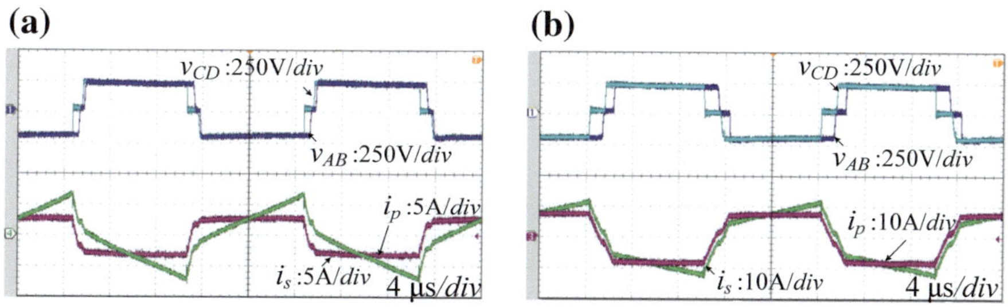

Fig. 2.14 Experimental waveforms when the power transfers backward from V_2 to V_1 under rated input voltage. **a** 400 W reverse power flow. **b** 1200 W reverse power flow

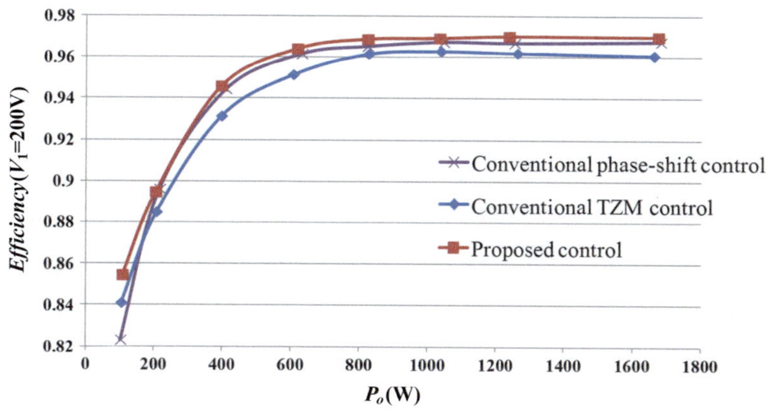

Fig. 2.15 Efficiency comparison when V_1 = 200 V

currents also remain critical continuous if the two bias currents are neglected, which shows the proposed control is also effective in reverse power flow condition. Meanwhile, no voltage spike or ringing occurs in v_{AB} and v_{CD}, indicating ZVS.

The efficiency of the converter with the proposed control method is measured and compared with the one using the conventional phase shift control and the conventional TZM control. Figure 2.15, $V_1 = 200$ V, which matches with the voltage in V_2 side. In this case, with the conventional TZM control, the efficiency is relatively lower because ZVS is lost for four power switches. Meanwhile, the proposed converter efficiency is slightly higher than the one with conventional phase shift control at heavy load due to very low turn-off current for four power switches.

While, when $V_2 = 240$ V as shown in Fig. 2.16, the efficiency of the converter with conventional phase shift control decreases significantly due to high peak current, large non-active power, and losing of ZVS. Besides, compared with

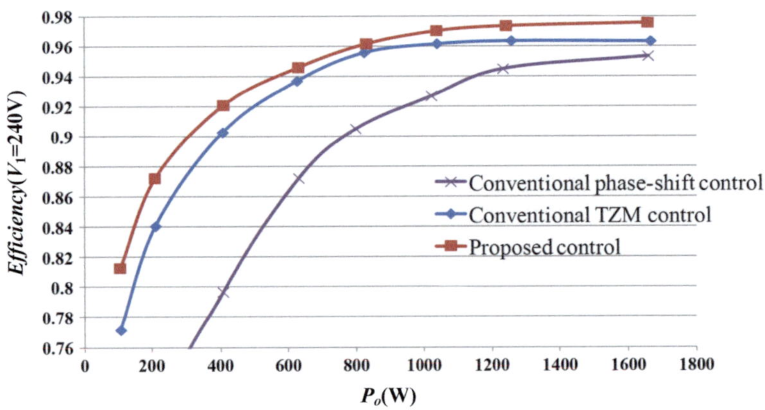

Fig. 2.16 Efficiency comparison when $V_1 = 240$ V

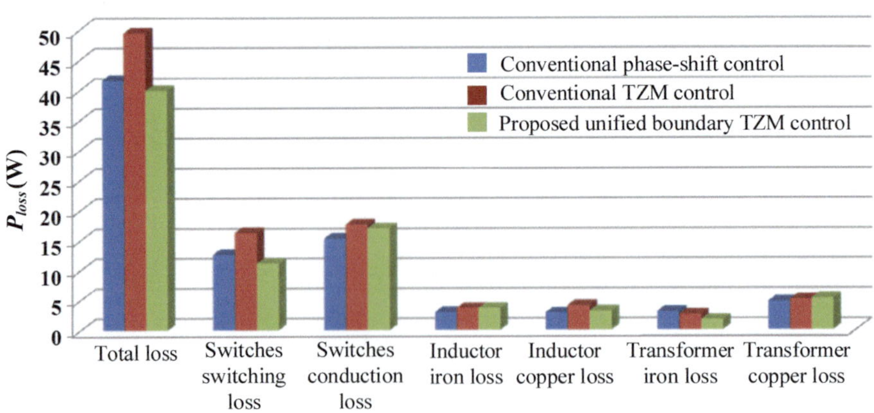

Fig. 2.17 Comparisons of the calculated power loss breakdown at the rated load when $V_1 = 200$ V

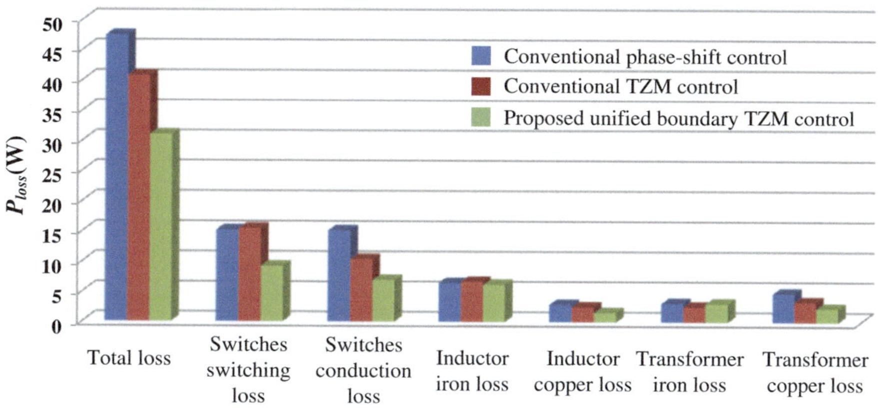

Fig. 2.18 Comparisons of the calculated power loss breakdown at the rated load when $V_1 = 240$ V

conventional TZM control, the efficiency with the proposed control has an improvement of 4% at light load (100 W) and the improvement at the heavy load is roughly 1.5%.

Figures 2.17 and 2.18 show the loss breakdown of the converter with the three control methods under rated power when $V_1 = 200$ V and $V_1 = 240$ V. As shown in Fig. 2.17, when the voltages of V_1 side and V_2 side are matched, the conventional TZM can lead to more loss due to higher switches switching loss and switches conduction loss. In this case, the total power loss of the system with the proposed control is slightly lower than the one with the conventional phase shift control because of lower switches switching loss. The total power loss difference between them is small since ZVS of all the eight power switches can also be achieved for conventional phase shift control when the voltages of V_1 side and V_2 side are matched. However, the total power loss difference between the conventional phase shift control and proposed control becomes larger when $V_1 = 240$ V as shown in Fig. 2.18. In this case, the switches peak current and circulating current are both increased for the conventional phase shift control, leading to higher switches switching loss and conduction loss compared with the proposed control. Meanwhile, compared with conventional TZM control, the proposed control has less switches loss and copper loss since the boundary condition is applied and ZVS of all the switches can be ensured with the fixed duty cycle compensation and magnetizing current design.

2.5 Conclusion

This paper proposes a unified boundary trapezoidal modulation (TZM) control for dual active bridge DC–DC converters. It utilizes the proposed fixed duty cycle compensation and magnetizing current design to achieve ZVS for all switches. With the proposed control, the selection of control variables for the power transfer is unified and can be implemented by one regulator. Besides, the converter has no non-active power issues if the two low bias currents are ignored. Using the introduced boundary control, the conduction loss can be further reduced because of no non-active power and lower circulating current. The fixed duty cycle compensation and the magnetizing current design can achieve ZVS for the left four power switches, which cannot be obtained with conventional TZM control. As a result, all the power switches can achieve ZVS and half of the switches can be turned off with ZCS if the small bias current is ignored. Experimental results from 1.6 kW laboratory prototypes have verified the theoretical analysis and the effectiveness of the proposed control.

References

1. Doncker RWAAD, Divan DM, Kheraluwala MH (1991) A three phase soft-switched high-power-density dc/dc converter for high-power applications. IEEE Trans Ind Appl 27 (1):63–73
2. Kheraluwala MH, Gascoigne RW, Divan DM, Baumann ED (1992) Performance characterization of a high-power dual active bridge dc-to-dc converter. IEEE Trans Ind Appl 28 (6):1294–1301
3. Schibli N (1999) DC-DC converters for two-quadrant operation with controlled output voltage. In: EPE 99: European conference on power electronics and applications, p 1–8
4. Krismer F, Round S, Kolar JW (2006) Performance optimization of a high current dual active bridge with a wide operating voltage range. In: Power electronics specialists conference. Pesc'06. IEEE, p 1–7
5. Hirose T, Takasaki M, Ishizuka Y (2014) A power efficiency improvement technique for a bidirectional dual active bridge dc–dc converter at light load. IEEE Trans Ind Appl 50 (6):4047–4055
6. Zhou H, Khambadkone AM (2009) Hybrid modulation for dual-active bridge bidirectional converter with extended power range for ultra capacitor application. IEEE Trans Ind Appl 45 (4):1434–1442

Chapter 3
Hybrid-Bridge-Based DAB Converter with Wide Voltage Conversion Gain

Abstract This chapter proposes a voltage match (VM) control for hybrid-bridge-based dual active bridge (DAB) converter in wide voltage conversion gain applications. With the addition of an auxiliary half-bridge circuit, the topology becomes an integration of a half-bridge and a full-bridge DAB converter. Unlike other PWM generation method for DAB converters, this converter utilizes four-level voltage at one port of the transformer to obtain matched voltage waveforms within the range of twice the minimum conversion gain. Wide conversion gain, decoupling of the two power control variables, and wide zero-voltage switching (ZVS) ranges can be achieved with the proposed voltage match control. Full load ranges of ZVS for the six main power switches can be achieved, and the two auxiliary switches can also operate in a wide ZVS range. In addition, the power control is done only using two control variables and its implementation is very simple, only needing a divider and a conventional voltage regulator. These characteristics and benefits of the proposed control are verified by experimental results from a 1 kW converter prototype.

Keywords Hybrid bridge · Dual active bridge · Voltage match control
Wide ZVS range · Wide conversion gain

3.1 Working Principle of Hybrid-Bridge-Based DAB Converter

The fundamental duty modulation [1], based on the fundamental optimization method in [2], is aimed at reducing circulating current and simplifying of the controller design. Just as other DAB converters, it has been pointed out that ZVS for all the switches can be obtained when input voltage V_1 equals to nV_2, while ZVS range of the converter is limited in other cases. Instead of optimizing the modulation scheme, [3] creatively made the modification in DAB converter topology and proposed a hybrid phase shift control for wide input voltage range. However,

D. Sha and G. Xu, *High-Frequency Isolated Bidirectional Dual Active Bridge DC–DC Converters with Wide Voltage Gain*, CPSS Power Electronics Series,
https://doi.org/10.1007/978-981-13-0259-6_3

although hybrid control can lead to wide ZVS range, smooth transfer between the two control modes is a challenge in practice.

In view of the converter topology, the dual-bridge DC–DC converter first proposed in [4] is a hybrid converter combining a full-bridge and a half-bridge converter. With this modification in full-bridge converter topology, wide operation range and small peak current can be achieved. The LLC resonate converter based on the dual bridge proposed in [5] is only suited for unidirectional power applications to obtain wide voltage gain and wide soft switching range. In [6], a T-type half bridge which is similar to the dual-bridge topology is utilized to achieve the wide voltage gain. The topology has more power switches in the primary side, and its control is also highly dependent on the system parameters.

3.1.1 Topology and Modulation Scheme for Hybrid-Bridge-Based DAB Converter

In this paper, to achieve wide voltage gain for bidirectional power flow applications, a voltage match (VM) control is proposed for hybrid-bridge-based DAB converter. The topology has been inspired by Song and Lehman [4] and is derived from adding two switches and two split capacitors to the DAB topology. The hybrid-bridge-based DAB converter is shown in Fig. 3.1. It can be divided into two parts: the conventional DAB circuit and the auxiliary half-bridge circuit. The DAB circuit is made up of an active full bridge consisting of switches S_1, S_2, S_3, and S_4 at the V_1 side and an active half bridge consisting of switches Q_1, Q_2 and capacitors C_{21}, C_{22} at the V_2 side. L_k is the transformer leakage inductance. Meanwhile, the auxiliary switches S_5, S_6 and capacitors C_{11}, C_{12} constitute the auxiliary half-bridge circuit. v_{AB} stands for the voltage across point A and B, v_{EF} denotes the voltage across point E and F, and the voltage across point C and D is denoted as v_{CD}.

The typical operation waveforms for this proposed converter are shown in Fig. 3.2. It should be noted that only the waveforms in the case of $0 < \Phi < D_1$ are taken as an example to illustrate the operation principle. Other cases will be analyzed in the next section. Besides, the item $m = nV_2/(2V_1)$ is assumed to be in the

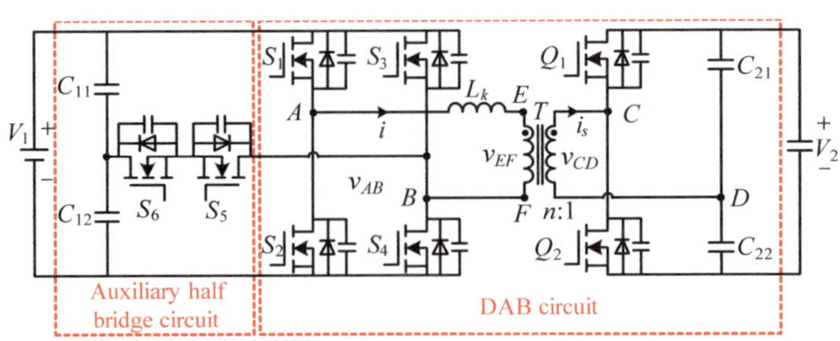

Fig. 3.1 Topology of hybrid-bridge-based DAB converter

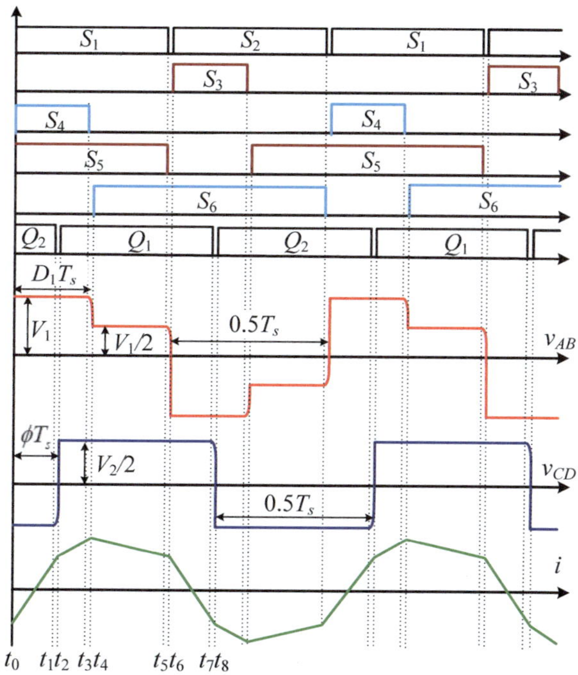

Fig. 3.2 Typical waveforms for the hybrid-bridge-based DAB converter when $0 < \Phi < D_1$ and $0.5 < m < 1$

range of [0.5, 1] to specify the rising or falling of the inductor current waveforms. The reason why this assumption can be made will be explained in the later section.

Switches S_1 and S_2 are operated complementarily with fixed 50% duty ratio. S_3 and S_5 are driven complementarily, and S_4 and S_6 are driven complementarily as well.

The duty ratio for both S_3 and S_4 is D_1 ($0 \leq D_1 \leq 0.5$). The turn-on instant for S_3 is the same as that of S_2. The turn-on instant for S_4 and S_1 is also the same. At V_2 side, Q_1 and Q_2 are also operated complementarily with fixed 50% duty ratio just as S_1 and S_2. However, PWM signal for Q_1 is phase shifted with a phase shift ratio Φ with respect to S_1. v_{AB} denotes the voltage generated by the hybrid active bridge at V_1 side, and it has four voltage levels: "V_1", "$V_1/2$", "$-V_1$", and "$-V_1/2$". The time duration for the voltage level "V_1" or "$-V_1$" is D_1T_s, and the time duration for "$V_1/2$" or "$-V_1/2$" is $0.5T_s-D_1T_s$. At extreme condition when $D_1 = 0$, the circuit at V_1 side works as a half-bridge converter to generate a high-frequency voltage v_{AB}, which has only two voltage levels: "$V_1/2$" and "$-V_1/2$". However, when $D_1 = 0.5$, the circuit works as a full-bridge converter to generate the high-frequency voltage which has two higher voltage levels: "V_1" and "$-V_1$". When D_1 varies from 0 to 0.5, the circuit at V_1 side works as a combination of both the half bridge and full bridge to generate v_{AB} with four-level voltages. This also explains why it is named as “hybrid bridge.” The high-frequency voltage v_{CD} generated by the half bridge at V_2 side is a two-level square wave with 50% duty ratio. Similar to conventional DAB converters, the bidirectional power flow is controlled by varying the phase shift ratio Φ between the two bridges.

3.1.2 Working Stages of the Converter

Since the PWM signals are symmetrical during two half switching cycles, the mode of operation is analyzed only within one half cycle from t_0 to t_6 as shown in Fig. 3.2. The corresponding six working modes are illustrated in Fig. 3.3.

Stage 1 (t_0, t_1) (Fig. 3.3a): Prior to t_0, S_5 and Q_2 are on, and S_1 and S_4 are off. At t_0, S_1 and S_4 are turned on. During this stage, the converter works in full-bridge mode, and the V_1 side high-frequency voltage v_{AB} is equal to V_1. The high-frequency voltage v_{CD} generated by the half bridge in V_2 side is equal to $-V_2/2$. The transformer current i starts to increase with a constant slew rate, and the following relationship is obtained

$$\begin{cases} v_{AB} = V_1, \quad v_{CD} = -V_2/2 \\ i(t) = i(t_0) + (V_1 + nV_2/2)(t - t_0)/L_k \end{cases} \tag{3.1}$$

where n:1 is the transformer turns ratio.

Stage 2 (t_1, t_2) (Fig. 3.3b): At time t_1, Q_2 is turned off. The transformer current at V_2 side (i_s) starts to discharge the junction capacitor of Q_1 and charge the junction capacitor of Q_2. When the drain–source voltage of Q_2 is charged to V_2, the drain–source voltage of Q_1 is zero and the body diode of Q_1 will be conducted, creating ZVS turn-on condition for Q_1. Since the leakage inductance is relatively large for DAB converters, the current required to charge/discharge the two drain–source capacitors is small and can be neglected to simplify the ZVS analysis. If the small charging current is ignored, ZVS condition for Q_1 is

$$i_s(t_1) = ni(t_1) > 0 \tag{3.2}$$

Stage 3 (t_2, t_3) (Fig. 3.3c): At t_2, Q_1 is turned on with ZVS. During this stage, v_{CD} is equal to $V_2/2$. In this case, $v_{AB} > nv_{CD}$ because m is assumed to be in the range of (0.5, 1). Consequently, the transformer leakage current i starts to increase with a relatively lower constant slew rate and its value can be expressed as

$$\begin{cases} v_{AB} = V_1, \quad v_{CD} = V_2/2 \\ i_L(t) = i(t_2) + (V_1 - nV_2/2)(t - t_2)/L_k \end{cases} \tag{3.3}$$

Stage 4 (t_3, t_4) (Fig. 3.3d): At time t_3, S_4 is turned off. The transformer current at V_1 side (i) starts to discharge the junction capacitors of both S_3 and S_6 and charge the junction capacitor of S_4. At the end of this charging process, the drain–source

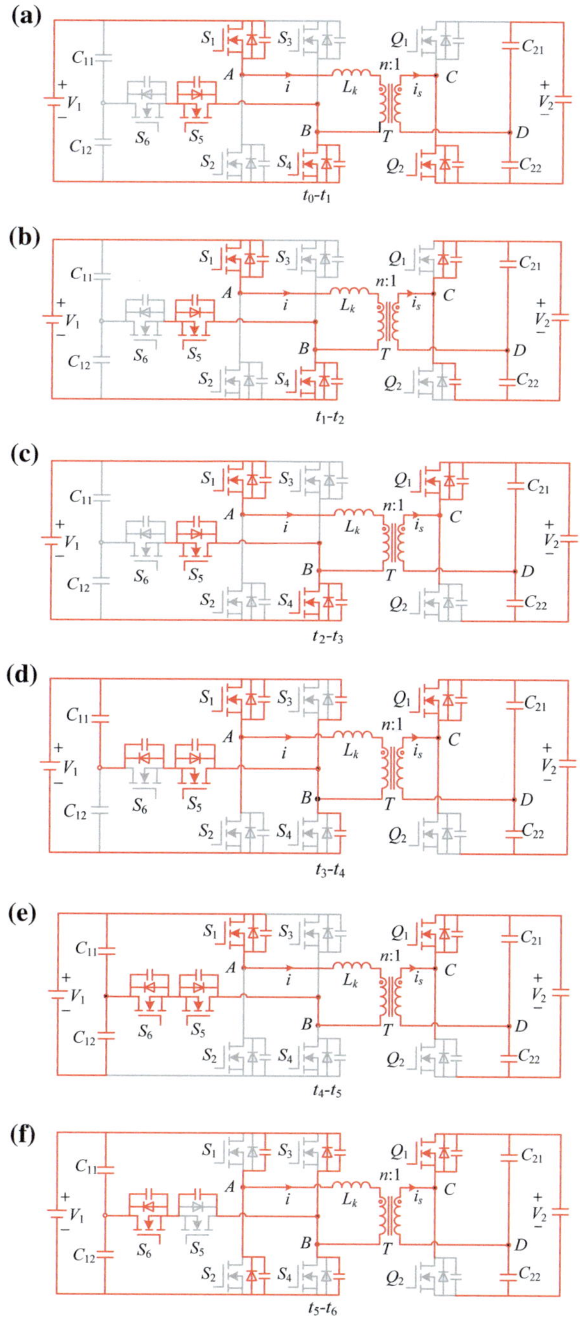

Fig. 3.3 Working modes of the proposed converter when $0 < \Phi < D_1$ and $0.5 < m < 1$. **a** Stage 1. **b** Stage 2. **c** Stage 3. **d** Stage 4. **e** Stage 5. **f** Stage 6

voltages of both S_3 and S_4 are $V_1/2$, and the drain–source voltage of S_6 reaches zero. The body diode conducts to create ZVS condition for S_6 in the next stage. If the charging energy stored in these capacitors is ignored, ZVS condition for S_6 is

$$i(t_4) > 0 \tag{3.4}$$

Stage 5 (t_4, t_5) (Fig. 3.3e): At t_4, S_6 is turned on under ZVS. During this stage, the converter works in half-bridge mode with v_{AB} equals to $V_1/2$ and v_{CD} equals to $V_2/2$. The transformer current i starts to increase with a constant slew rate, and the expression for this current is written as

$$\begin{cases} v_{AB} = V_1/2, \quad v_{CD} = V_2/2 \\ i(t) = i(t_4) + (V_1/2 - nV_2/2)(t - t_4)/L_k \end{cases} \tag{3.5}$$

Stage 6 (t_5, t_6) (Fig. 3.3f): S_1 and S_5 are both turned off at t_5. During this interval, the junction capacitors of S_1, S_4, and S_5 are charged and their drain–source voltages increase. At the same time, the junction capacitor of S_2 and S_3 is discharged. At the end of this interval, the drain–source voltage of S_5 is $V_1/2$, and the drain–source voltages of S_1 and S_4 are V_1. The drain–source voltages of S_2 and S_3 are both zero, leading to ZVS turned on for S_2 and S_3 at t_6. The ZVS conditions for S_2 and S_3 are described as

$$i(t_6) > 0 \tag{3.6}$$

3.2 ZVS Conditions and Power Control

3.2.1 *Current Range for ZVS*

Similar to many other DAB converters, the converter power transfer is controlled by not only one control variable. It is determined by two independent variables, which are the duty cycle D_1 and phase shift ratio Φ. Different combinations of these two variables can generate different voltage patterns and current waveforms, leading to different ZVS conditions. In the previous section, the case of $0 < \Phi < D_1$ is taken as an example to study the working modes and principles of the proposed converter. The realization of ZVS is only determined by the polarity of the transformer current when the switches are turned on as seen from the discussions of working stages 2, 4, and 6 in Fig. 3.3. In view of all the combinations of D_1 and Φ, they can be classified as four cases: $0 < \Phi < D_1$, $D_1 < \Phi < 0.5$, $D_1 - 0.5 < \Phi < 0$, and $-0.5 < \Phi < D_1 - 0.5$. The theoretical operation waveforms for these four operation patterns are illustrated, respectively, in Fig. 3.4.

Figure 3.4 shows the entire working patterns for all selections of D_1 and Φ. Figure 3.4a, b shows the operation when $\Phi > 0$, and Fig. 3.4c, d shows the cases when $\Phi < 0$. t_0, t_1, t_2, t_3, t_4, t_5, and t_6 are the switches turn-on instants for each

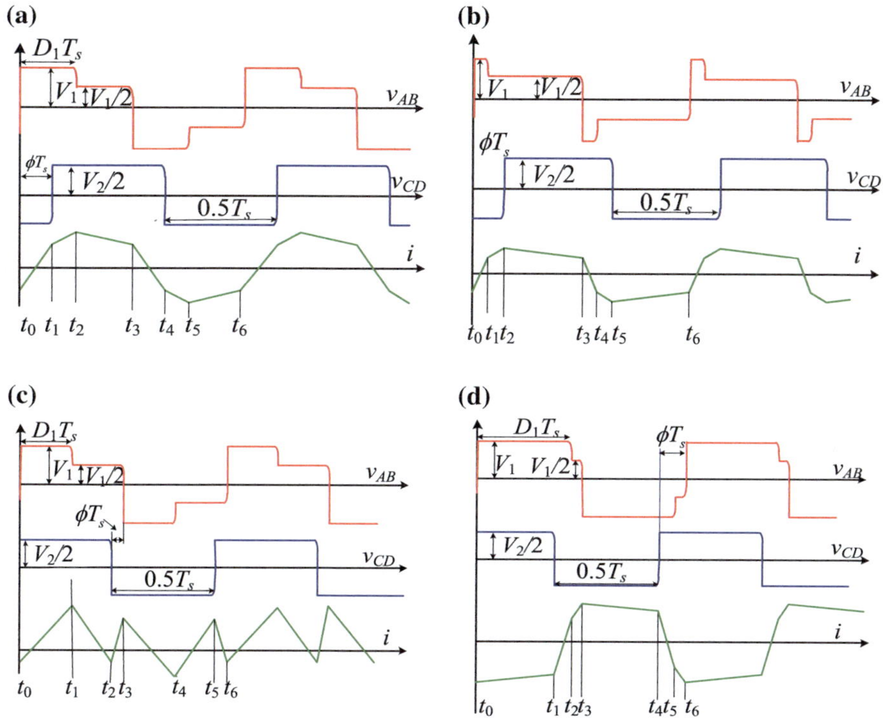

Fig. 3.4 Four operation patterns: **a** Pattern A: $0 < \Phi < D_1$. **b** Pattern B: $D_1 < \Phi < 0.5$. **c** Pattern C: $D_1 - 0.5 < \Phi < 0$. **d** Pattern D: $-0.5 < \Phi < D_1 - 0.5$

pattern. Since ZVS is related to the current value when power switches are turned on, calculation of the expression based on Fig. 3.4 for all these currents is necessary.

The pattern A in Fig. 3.4a is firstly taken as one case to calculate all these current values. Based on Fig. 3.4a, the current values at the instants when switches are turned on are shown as follows

$$\begin{cases} i(t_1) = -i(t_4) = i(t_0) + (V_1 + nV_2/2)(t_1 - t_0)/L_k \\ i(t_2) = -i(t_5) = i(t_1) + (V_1 - nV_2/2)(t_2 - t_1)/L_k \\ i(t_3) = -i(t_6) = i(t_2) + (V_1/2 - nV_2/2)(t_3 - t_2)/L_k \end{cases} \tag{3.7}$$

where t_1, t_2, t_3, t_4, t_5, and t_6 are the switches turn-on instants denoted in Fig. 3.4a, and $i(t)$ is the leakage inductance current value at time t.

Meanwhile, in a full switching cycle, the average value of the transformer current is zero

Table 3.1 Current limit for ZVS of all the switches for all the working patterns

	ZVS for S_1 & S_4	ZVS for S_2 & S_3	ZVS for Q_1 & Q_2	ZVS for S_5 & S_6
A: $0 < \Phi < D_1$	$i(t_0) < 0$	$i(t_0) < 0$	$i(t_1) > 0$	$i(t_2) > 0$
B: $D_1 < \Phi < 0.5$	$i(t_0) < 0$	$i(t_0) < 0$	$i(t_2) > 0$	$i(t_1) > 0$
C: $D_1 - 0.5 < \Phi < 0$	$i(t_0) < 0$	$i(t_0) < 0$	$i(t_2) < 0$	$i(t_1) > 0$
D: $-0.5 < \Phi < D_1 - 0.5$	$i(t_0) < 0$	$i(t_0) < 0$	$i(t_1) < 0$	$i(t_2) > 0$

$$\int_{t_0}^{t_6} i(t)\mathrm{d}t/T_s = 0 \tag{3.8}$$

Then, based on (3.7) and (3.8), current expressions for pattern A when switches are turned on are obtained

$$\begin{cases} i(t_0) = -i(t_3) = -(4\Phi nV_2 + (V_1 + 2V_1D_1 - nV_2))T_s/8L_k \\ i(t_1) = -i(t_4) = (8\Phi V_1 - (V_1 + 2V_1D_1 - nV_2))T_s/8L_k \\ i(t_2) = -i(t_5) = [8(\Phi nV_2/2 - nD_1V_2 + V_1D_1) \\ \qquad\qquad -(V_1 + 2V_1D_1 - nV_2)]T_s/8L_k \end{cases} \tag{3.9}$$

In addition, based on the working modes discussed in the last section, ZVS conditions for power switches when $0 < \Phi < D_1$ are summed up in the first row of Table 3.1. As illustrated, ZVS conditions for all the power switches are only determined by the polarity of three current values: $i(t_0)$, $i(t_1)$, and $i(t_2)$. According to the expressions for these current ranges shown in (3.9), ZVS range can be designed through proper selections of the phase shift ratio Φ and duty cycle D_1 to satisfy these current ranges. In addition, the current ranges for other three working patterns to achieve ZVS can also be shown in Table 3.1.

For pattern A, the expressions of $i(t_0)$, $i(t_1)$, and $i(t_2)$ are shown in (3.9). Based on the typical waveforms in Fig. 3.4, the current expressions for the other three working patterns along with the expression for pattern A are summarized in Table 3.2 through similar method.

Table 3.2 Instant leakage current expressions for all working patterns (Part A)

	$i(t_0)$	$i(t_1)$
A: $0<\phi<D_1$	$-0.5\ T_s\Phi nV_2/L_k - E$	$T_s\Phi V_1/L_k - E$
B: $\phi > D_1$	$-0.5T_s\Phi nV_2/L_k - E$	$0.5T_s(-\Phi nV_2 + nD_1V_2 + 2V_1D_1)/L_k - E$
C: $D_1 - 0.5<\phi<0$	$-0.5\ T_s\|\Phi\|nV_2/L_k - E$	$0.5T_s(-\|\Phi\|nV_2 - nD_1V_2 + 2V_1D_1)/L_k - E$
D: $\phi<D_1 - 0.5$	$-0.5T_s\|\Phi\|nV_2/L_k - E$	$0.25T_sV_1(-4\|\Phi\| - 2D_1 + 1)/L_k + E$

$E = (V_1 + 2V_1D_1 - nV_2)T_s/8L_k$

3.2.2 *Proposed VM Control to Ensure Wide ZVS Range*

Although Tables 3.2 and 3.3 have illustrated the ZVS constraints of the proposed converter, it is still difficult to find a suitable combination of Φ and D_1 for certain output power to ensure ZVS. The expressions in these tables are all affected by two independent control variables. However, in Table 3.2, all expressions of $i(t_0)$, $i(t_1)$, and $i(t_2)$ for different patterns share the same item: $E = (V_1 + 2V_1D_1 - nV_2)T_s/8L_k$. If this item is controlled to be zero, all these expressions can be simplified. For instance, when $E = 0$, expression $i(t_0) < 0$ for all the patterns will always be satisfied regardless of the two control variables, implying that ZVS of S_1, S_2, S_3, and S_4 can be achieved within the full range of voltage conversion gain and output power. Furthermore, when $E = 0$, ZVS conditions for Q_1 and Q_2 can also be achieved regardless of the two control variables.

Table 3.4 illustrates the ZVS results when $E = 0$ for all the power switches in the four working patterns. As seen, a full range of ZVS for S_1, S_2, S_3, S_4, Q_1, and Q_2 can be obtained regardless of the load and voltages. Besides, ZVS of S_5 and S_6 can be obtained within the full range when the converter works in pattern A. ZVS for S_5 and S_6 is depended on the voltages and load only in the case of pattern B, pattern C, and pattern D.

The realization and physical meaning of $E = 0$ for the converter are illustrated in Fig. 3.5a, where typical voltage waveforms for the proposed converter are shown, and v_{EF} is the transformer voltage referred from V_2 side to V_1 side as denoted in Fig. 3.1.

Table 3.3 Instant leakage current expressions for all working patterns (Part B)

	$i(t_2)$
A: $0<\phi<D_1$	$0.5T_s(\Phi nV_2 - nD_1V_2 + 2V_1D_1)/L_k - E$
B: $\phi > D_1$	$0.5T_s(\Phi + D_1)V_1/L_k - E$
C: $D_1 - 0.5<\phi<0$	$-0.5T_s\lvert\Phi\rvert V_1/L_k + E$
D: $\phi<D_1 - 0.5$	$0.5T_s((\lvert\Phi\rvert + D_1)nV_2 - V_1)/L_k + 3E$

$E = (V_1 + 2V_1D_1 - nV_2)T_s/8L_k$

Table 3.4 ZVS results when $E = 0$

	ZVS for S_1 & S_4	ZVS for S_2 & S_3	ZVS for Q_1 & Q_2	ZVS for S_5 & S_6
Pattern A: $0<\phi<D_1$	Always satisfied	Always satisfied	Always satisfied	Always satisfied
Pattern B: $\phi > D_1$	Always satisfied	Always satisfied	Always satisfied	Voltage- and load-dependent
Pattern C: $D_1 - 0.5<\phi<0$	Always satisfied	Always satisfied	Always satisfied	Voltage- and load-dependent
Pattern D: $\phi<D_1 - 0.5$	Always satisfied	Always satisfied	Always satisfied	Voltage- and load-dependent

$E - (V_1 + 2V_1D_1 - nV_2)T_s/8L_k$

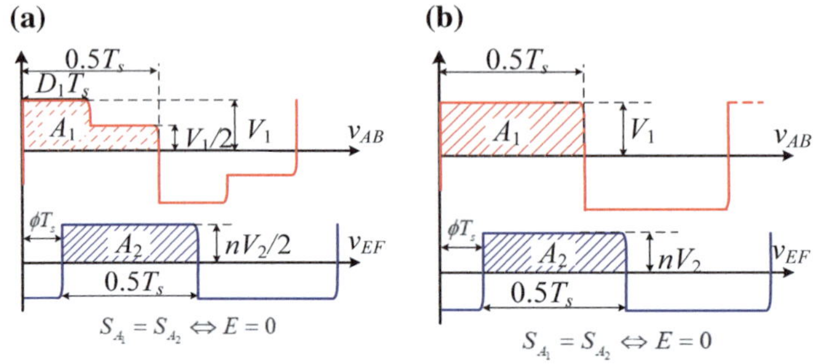

Fig. 3.5 Physical meaning of $E = 0$. **a** The proposed converter. **b** DAB converter with SPS control under optimal operating point

For the proposed converter, the shadow areas of A_1 and A_2 in Fig. 3.5a can be expressed as

$$\begin{cases} S_{A_1} = D_1 T_s V_1 + (0.5 - D_1) T_s V_1 / 2 \\ S_{A_2} = 0.5 T_s n V_2 / 2 \end{cases} \tag{3.10}$$

By setting the two areas to be equal, we can obtain

$$V_1 + 2V_1 D_1 - nV_2 = 0 \tag{3.11}$$

Substitution of (3.11) into expression of E leads to $E = 0$. Therefore, to make $E = 0$, the areas A_1 and A_2 need to be equal. This can be achieved by regulating the duty cycle D_1. Rearrangement of (3.11) leads to the expression of D_1

$$D_1 = 0.5nV_2 / V_1 - 0.5 \tag{3.12}$$

As seen in (3.12), D_1 is only determined by the voltages on both sides, and it is independent of another power control variable Φ. This decoupling can simplify the power control. Based on (3.12), D_1 can be used to ensure the wide ZVS ranges, while Φ can be used to control the bidirectional power transfer. This method is named as voltage match (VM) control since the control of D_1 is aimed at achieving the match for areas of A_1 and A_2 as shown in Fig. 3.5a.

For conventional DAB converters with SPS control, the optimal performance including full range of ZVS and low peak current is obtained at the optimal operating point where input voltage V_1 equals to nV_2. Actually, this operating point is equivalent to $E = 0$ in view of the voltage waveform. Figure 3.5b shows the waveforms for the optimal operating point. It can be easily obtained that the areas of A_1 and A_2 are equal if input voltage V_1 equals to nV_2. In other word, the equivalent optimal operating point can always be obtained for the proposed converter if the areas of A_1 and A_2 are controlled to be equal. This also explains why ZVS can be obtained within full input voltage range for switches S_1, S_2, S_3, S_4, Q_1, and Q_2.

3.3 Converter Performance with Proposed Voltage Match Control

3.3.1 Voltage Gain Under VM Control

According to (3.12), with VM control, D_1 changes with respect to m, where $m = nV_2/(2V_1)$ and n is the turns ratio as mentioned before. Then, substitution of m into (3.12) leads to

$$m = D_1 + 0.5 \tag{3.13}$$

As shown in (3.13), with the VM control, m should be limited in the range of [0.5, 1] since the value range for [0, 0.5]. This is why $0.5 \leq m \leq 1$ was assumed in Sect. 2. The typical waveforms for the proposed converter with the VM control for different values of m are shown in Fig. 3.6.

When $m = 0.5$, the corresponding value of D_1 is 0, and v_{AB} is a 50% square waveform with amplitude $V_1/2$. When $m = 1$, the corresponding value of D_1 is 0.5, and v_{AB} remains as a 50% square waveform. However, its amplitude becomes V_1 since the circuit now works as a full-bridge converter instead of half bridge. As seen in Fig. 3.6a, c, the waveforms are similar in both cases. At the two boundary m values, both the converters work at optimal operating point since the duty cycle of the two voltage waveforms is 50% and the shadow area of A_1 equals to A_2. Therefore, at the two boundary m values, the converter will have both a lower leakage inductor root-mean-square (RMS) current and peak current. This is quite different from the conventional DAB converters where leakage inductor RMS current and peak current are totally different in maximum voltage gain and minimum voltage gain situations.

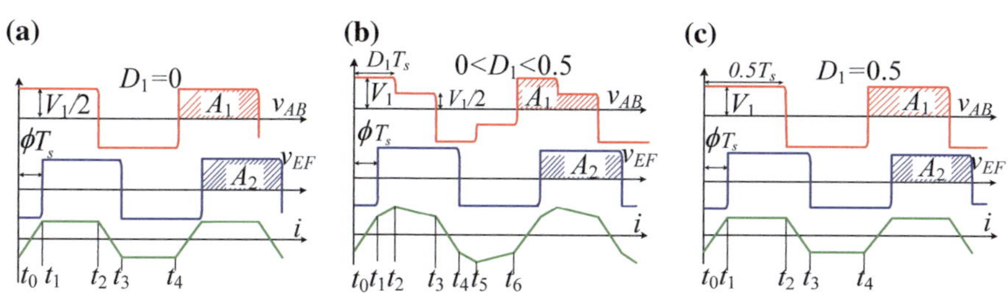

Fig. 3.6 Typical waveforms with different conversion gains. **a** $m = 0.5$, half-bridge mode. **b** $0.5 < m < 1$, hybrid bridge mode. **c** $m = 1$, full-bridge mode

3.3.2 Power Transfer Characteristics with VM Control

Based on the waveforms of each working pattern in Fig. 3.4, the transferred power can be expressed as

$$P_o(D_1, \Phi) = \frac{1}{T_s}\int_{t_0}^{t_6} v_{AB}(t)i(t)\mathrm{d}t \tag{3.14}$$

For pattern A, the transferred power can be written as

$$\begin{aligned} P_{o\_A}(D_1, \Phi) = \frac{1}{T_s}\Bigg[&\int_{t_0}^{t_2} V_1 i(t)\mathrm{d}t + \int_{t_2}^{t_3} 0.5V_1 i(t)\mathrm{d}t \\ &+ \int_{t_4}^{t_5} -V_1 i(t)\mathrm{d}t + \int_{t_5}^{t_6} -0.5V_1 i(t)\mathrm{d}t\Bigg] \end{aligned} \tag{3.15}$$

Then, according to the mathematical expression for the current in each time interval shown in (3.7), (3.8), and (3.9), the transferred power for pattern A can be calculated as

$$P_{o\_A}(D_1, \Phi) = \frac{-nV_2V_1T_s}{8L_k}(8\Phi^2 - 2\Phi - 4\Phi D_1 - D_1 + 2D_1^2) \tag{3.16}$$

Using the same method, the transferred power functions for the other three working patterns along with (3.16) are summarized as below

$$\begin{cases} P_{o\_A}(D_1, \Phi) = \frac{-nV_2V_1T_s}{8L_k}(8\Phi^2 - 2\Phi - 4\Phi D_1 - D_1 + 2D_1^2) \\ P_{o\_B}(D_1, \Phi) = \frac{-nV_2V_1T_s}{8L_k}(4\Phi^2 - 2\Phi + 4\Phi D_1 - D_1 - 2D_1^2) \\ P_{o\_C}(D_1, \Phi) = \frac{nV_2V_1T_s}{8L_k}(4\Phi^2 - 2\Phi - 4\Phi D_1 + D_1 - 2D_1^2) \\ P_{o\_D}(D_1, \Phi) = \frac{nV_2V_1T_s}{8L_k}(8\Phi^2 - 6\Phi_1 + 4\Phi D_1 + 2D_1^2 + 1 - 3D_1) \end{cases} \tag{3.17}$$

where P_{o_B} (D_1, Φ), P_{o_C} (D_1, Φ), and P_{o_C} (D_1, Φ) are power transfer functions for pattern B, pattern C, and pattern D, respectively.

When VM control is applied, with substitution of (3.12) and (3.13) into (3.17), the power transfer normalized by $P_{\text{base}} = n^2V_2^2T_s/16L_k$ between the phase shift ratio Φ for different voltage gain m is plotted in Fig. 3.7. The power is positive when it is transferred from V_1 side to V_2 side; otherwise, it is negative. As shown in Fig. 3.7a, when m changes from 0.5 to 1, the power transfer changes slightly, and the plots for the cases when m = 0.5 and m = 1 are overlapped. Just like the conventional DAB converters, the phase shift ratio Φ should be limited within a proper range to ensure that the power transfer P is monotonously increasing with the increase of Φ.

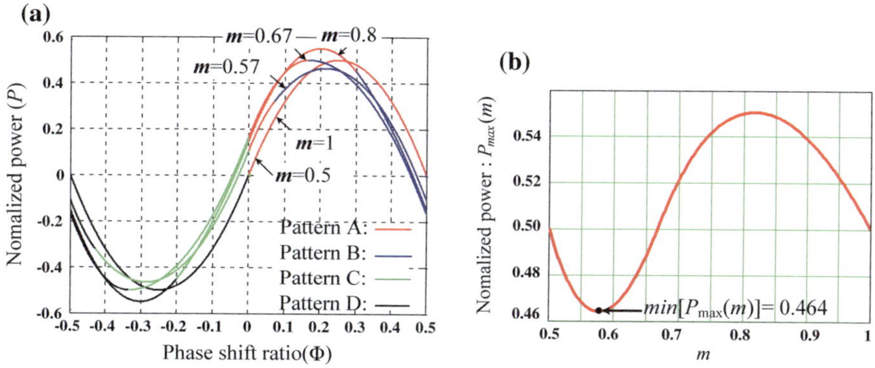

Fig. 3.7 Power transfer characteristics. **a** Normalized power transfer curve under different conversion gains. **b** Peak power transfer point for different conversion gain

The maximum power transfer for a particular m is symmetrical for the positive and negative power transfer. Therefore, the cases when power transfer is positive can be considered to obtain the maximum power transfer expression. The maximum power point for different m can be obtained based on (3.17) when $P > 0$, which is shown as

$$P_{\max}(m)\begin{cases} -n^2V_2^2T_s(1-3m+3m^2)/(16L_km) & \text{if } m<2/3 \\ -n^2V_2^2T_s(2-6m+3m^2)/(32L_km) & \text{else} \end{cases} \tag{3.18}$$

$P_{\max}(m)$ in (3.18) is plotted in Fig. 3.7b. When m changes from 0.5 to 1.0, the maximum power point varies. To ensure that the power can be achieved for all m, the maximum power transfer for the hybrid-bridge-based converter should be the minimum value of $P_{\max}(m)$, which is the valley point shown in Fig. 3.7b. The expression of the minimum value can be obtained based on (3.18).

$$\min\left[P_{\max}(m)\right] = T_sn^2V_o^2(2\sqrt{3}-3)/(16L_k) = 0.464P_{\text{base}} \tag{3.19}$$

Therefore, the global maximum transfer is $0.464P_{\text{base}}$ for the hybrid-bridge-based DAB converter for all m values. It should be pointed out that the V_1 side hybrid bridge works as a combination of full-bridge and half-bridge converters except for the two boundary conditions. In these cases, the effective voltage value across the transformer input is reduced due to a decreased duration of high-voltage level V_1 in v_{ab}. Consequently, the maximum power transfer is reduced when compared with full-bridge converter.

Meanwhile, based on (3.15) and (3.17), the operation regions for different patterns under different power transfer normalized by $P_{\_\text{base}} = n^2V_2^2T_s/16L_k$ and m are shown in Fig. 3.8. As shown, at different P and m, the converter mostly operates in the regions for pattern A and pattern C. Pattern B is observed at relatively lower

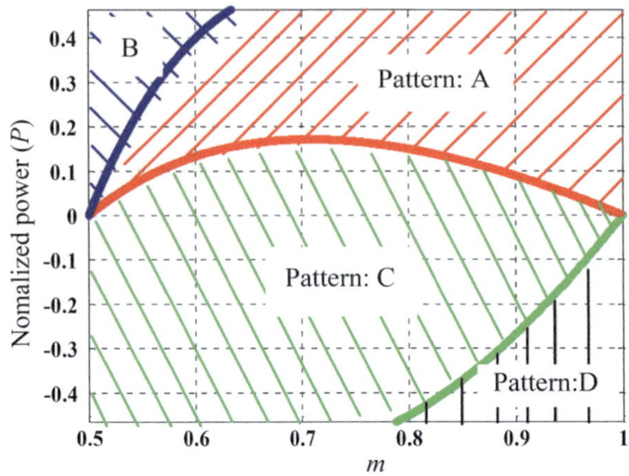

Fig. 3.8 Operation regions for different patterns under different power transfer and conversion gains

voltage gain when the power flow direction is positive. When the power transfer is negative, pattern D occurs at relatively higher voltage gain.

3.3.3 Switches ZVS Discussion

As discussed in the last section, ZVS for S_1, S_2, S_3, S_4, Q_1, and Q_2 can always be achieved for all the patterns regardless of the load and voltage gain with the VM control. Therefore, only the ZVS regions for the auxiliary switches S_5 and S_6 are discussed. Based on the current expressions and ZVS conditions for S_5 and S_6 in Tables 3.2 and 3.3, ZVS operation regions under different conversion gains and power transfer for S_5 and S_6 are plotted in Fig. 3.9. The dashed line as illustrated represents the soft switching boundaries, which is the locus of zero leakage current when S_5 or S_6 turns off. As shown, ZVS can be realized for most of the regions.

3.4 Implementation of the Proposed Control

With the VM control, the power control variables D_1 and Φ can be controlled separately. According to the power transfer characteristics discussed previously, voltage V_2 can be controlled by the phase shift ratio Φ with a traditional voltage feedback loop. The implementation of the proposed VM control is shown in Fig. 3.10. V_{ref} is the voltage reference of V_2 side voltage. v_1, v_2, d_1, and φ are the variables for the equilibrium values of V_1, V_2, D_1, and Φ, respectively. $\Phi_{-p\max}$ and $\Phi_{p\max}$ are the lower and upper limits of the voltage regulator, respectively.

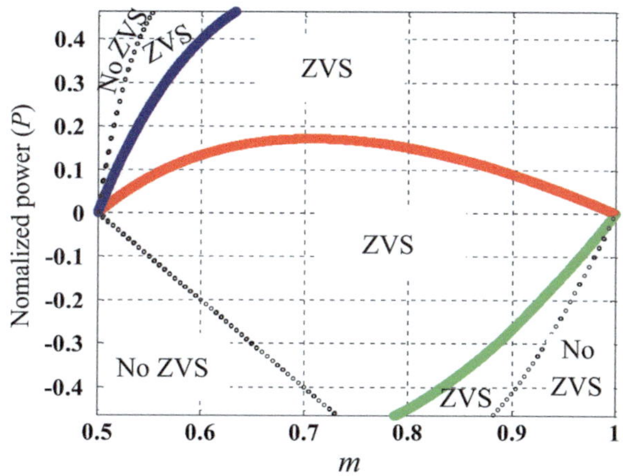

Fig. 3.9 ZVS regions for switches S_5 and S_6

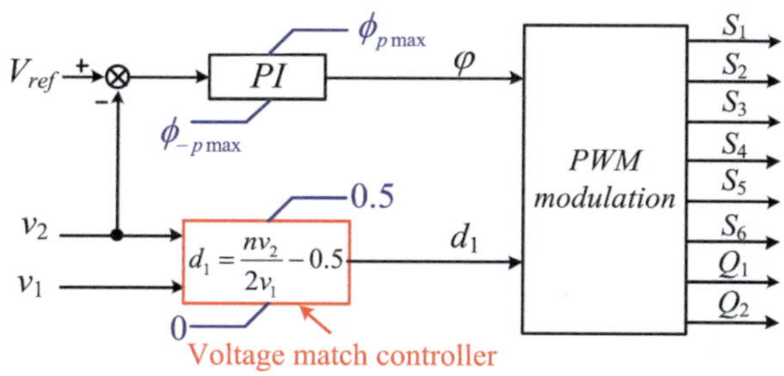

Fig. 3.10 Implementation of the proposed VM control

As shown in Fig. 3.10, the PI regulator is used to control voltage V_2 by adjusting the phase shift ratio φ. The VM control can be easily implemented by calculating the duty cycle value d_1 using voltages v_1 and v_2. This control loop is very simple and does not need any complicated calculations.

3.5 Comparison

3.5.1 General Comparisons

General comparisons of the topology and control are shown in Tables 3.5 and 3.6. As depicted, the hybrid-bridge-based DAB converter has the same number of power switches as other DAB converters. With VM control, wide ZVS range can be obtained using only two control degrees of freedom. Even though the hybrid-bridge-based DAB converter is not symmetrical, it is a unified control from the power control point of view. The PWM generation method does not change under different power levels and power directions. Device ratings for the

Table 3.5 General comparisons of the topology structure and control (Part A)

	Control degrees of freedom	Control complexity	Power switches number	Device voltage stress	
				V_1 side	V_2 side
SPS [7]	1	Lowest	8	V_1	V_2
EPS [8]	2	High	8	V_1	V_2
TPS [9]	3	Highest	8	V_1	V_2
VM	2	Low	8	V_1 or $V_1/2$	$V_2/2$

Table 3.6 General comparisons of the topology structure and control (Part B)

	Needing of split capacitor	Mode shifting in control	Symmetric in topology
SPS [7]	No	No	Yes
EPS [8]	No	Yes	Yes
TPS [9]	No	Yes	Yes
VM	Yes	No	No

hybrid-bridge-based DAB converter are also different from other DAB converters. At the V_1 side S_5 and S_6 only need to withstand half of its corresponding DC-link voltage and so is for the switches at V_2 side. Meanwhile, for other full-bridged DAB converters, all the eight switches have to withstand the full DC-link voltage.

3.5.2 Comparison of Inductor RMS Current and Total Conduction Loss

The RMS current and total conduction loss comparisons are evaluated based on parameters shown in Table 3.7. The conventional full-bridge converter employing

Table 3.7 Parameters related to the comparison

	Conventional DAB converter	Hybrid-bridge-based DAB converter
Power rating	1 kW	
Voltage rating	V_1 = 120–240 V, V_2 = 96 V	
Switching frequency	50 kHz	
Inductance value	20 μH	
Transformer turns ratio	15:8	10:4
Primary switches	FDA 59N30 * 4	S_1, S_2, S_3, S_4: FDA59N30 S_5, S_6: FDP 2614
Secondary switches	FDH 055N15A * 4	FDH 055N15A * 2

EPS method aimed at maximum ZVS range [10] is chosen as the comparison topology. The voltages and power ratings are the same for both of the two cases. Besides, the inductor and transformer parameters are also the same except for the transformer turns ratio of conventional DAB. It is optimized to be 15:8. With this optimized turns ratio, the conventional DAB works at its optimal operating point when $V_1 = 180$ V.

The inductor RMS currents for different voltages and power are shown in Fig. 3.11. As illustrated in Fig. 3.11a, when $V_1 = 120$ V the RMS current is smaller for the converter with the proposed VM control. When $V_1 = 140$ V, RMS current is smaller for VM control in most of its power range. However, when voltages are increased to 160, 180, and 200 V, the RMS current becomes relatively larger for

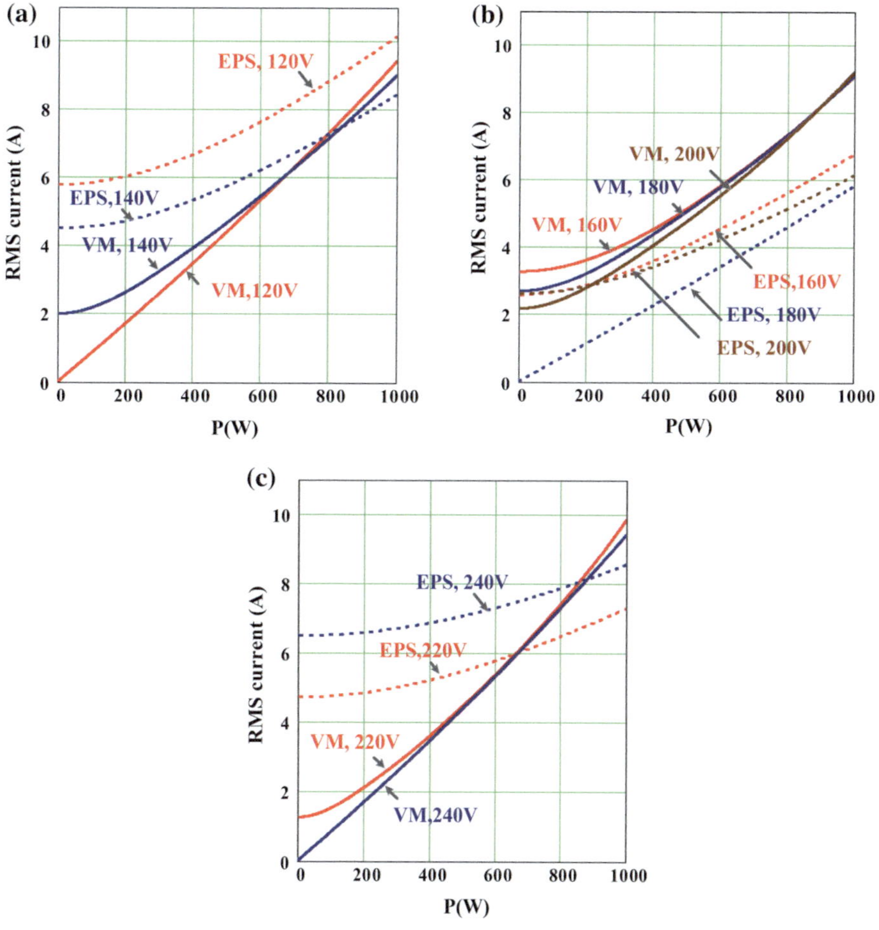

Fig. 3.11 Comparisons of inductor RMS currents for EPS control and proposed VM control under different voltages and power. **a** $V_1 = 120$ V and $V_1 = 140$ V. **b** $V_1 = 160$ V, $V_1 = 180$ V, and $V_1 = 200$ V. **c** $V_1 = 220$ V and $V_1 = 240$ V

converter with the proposed method. This is because the input voltage is near the optimal operation point (V_1 = 180 V) for EPS control. As the voltage increases to 220 and 240 V, the RMS value becomes larger again for most of the power range with EPS control. It is worth noting that RMS current values at light loads using EPS control are larger in most of cases when compared with VM control.

In Fig. 3.12, the total conduction loss of the hybrid-bridge-based converter with VM control and the conventional DAB converter with EPS control is compared. In Fig. 3.12a, the total conduction loss for two methods with V_1 = 120 and 140 V is shown. As seen, the total conduction loss for DAB converter with EPS is larger than that of the hybrid-bridge-based converter with VM control. In Fig. 3.12b, as the voltage V_1 increases to 160, 180, and 200 V, the total conduction loss for

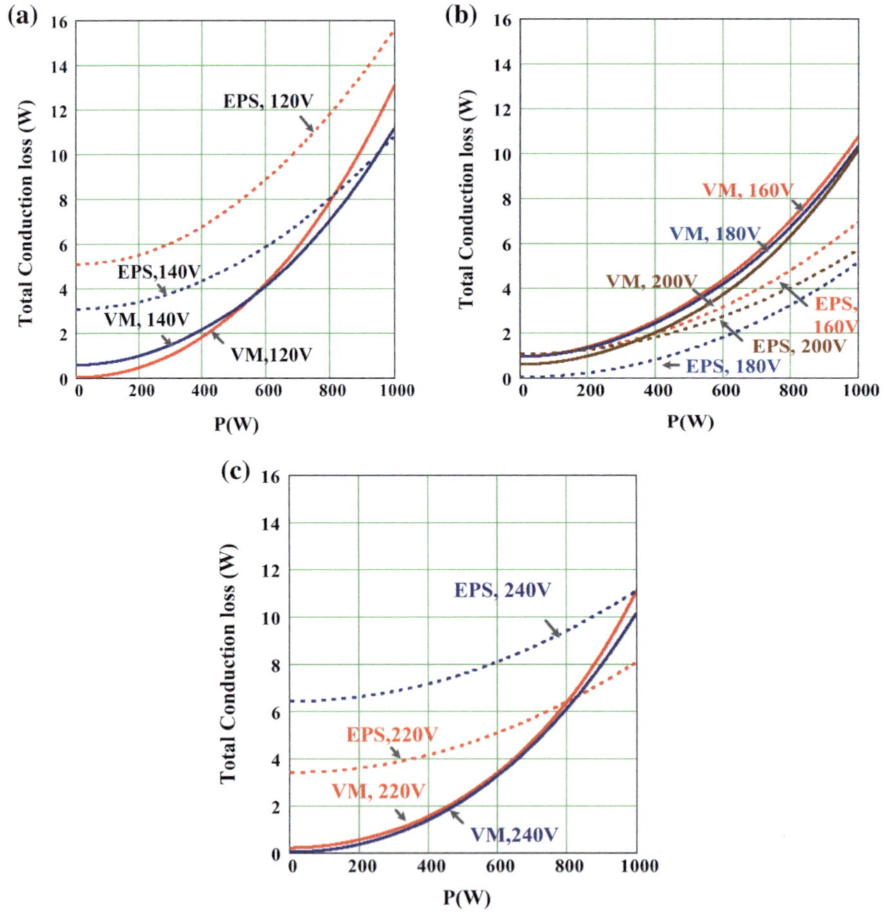

Fig. 3.12 Comparisons of total conduction loss for EPS control and proposed VM control under different voltages and power. **a** V_1 = 120 V and V_1 = 140 V. **b** V_1 = 160 V, V_1 = 180 V, and V_1 = 200 V. **c** V_1 = 220 V and V_1 = 240 V

conventional DAB converter is smaller than that of hybrid-bridge-based converter for most power levels. For the 240 and 220 V cases shown in Fig. 3.12c, the total conduction loss with EPS is always larger than the case with VM when V_1 = 240 V. When V_1 = 220 V, the total conduction loss using EPS is larger than that with VM control for power lower than 800 W. Therefore, from a conduction loss point of view, the converter with VM control has a better overall performance when the voltage range is large. Also, it is worth noting that the total conduction loss is much higher at light loads employing EPS compared to that with VM control when V_1 = 120 V, V_1 = 140 V, V_1 = 220 V, and V_1 = 240 V. This also shows that the hybrid-bridge-based DAB converters with VM control can reduce conduction loss for light load within a wide voltage gain range.

3.6 Experimental Verification

A laboratory prototype was built to validate the feasibility of such proposed VM control for hybrid-bridge-based DAB converters. The specifications are the same as listed in Table 3.7 for comparison section. In addition, the capacitance values for V_1 side half bridge are C_{11} = C_{12} =20 μF, and the capacitance values for V_2 side half bridge are C_{21} = C_{22} = 40 μF.

The experimental waveforms under rated power for different V_1 voltages are shown in Fig. 3.13. i_L is the measured leakage inductance current. As shown in Fig. 3.13a, d, the voltage and current waveforms are similar for the two cases when V_1 = 120 and V_1 = 240 with voltage gain m = 1 and m = 0.5, respectively. These operation waveforms are similar to those of DAB converters with SPS control running at optimal operating point where ZVS can be obtained easily. When V_1 = 160 V and 200 V with corresponding voltage gains m = 0.75 and m = 0.6, respectively, as seen in Fig. 3.13b, c, the duty cycle D_1 is larger than the phase shift ratio and they are both positive indicating that the converter is working in pattern A. This agrees well with the operation pattern waveforms shown in Fig. 3.4 and the pattern region division in Fig. 3.8. For these two cases, ZVS will be achieved for all the switches judging from the inductance current (i_L) polarity at each turn-on instant according to the current limit of ZVS for pattern A listed in Table 3.1.

Figure 3.14 shows the experimental waveforms of different voltage conversion gains at light-load condition (150 W). As seen in Fig. 3.14a and d, the current and voltage waveforms are similar when V_1 = 120 V and V_1 = 240 V, respectively. In these two cases, all the power switches can achieve ZVS similar to the DAB converter with SPS control at optimal operating condition. In Fig. 3.14b, c, when V_1 is 160 and 200 V, respectively, $D_1 - 0.5 < \Phi < 0$ is satisfied. This implies that the converter works in pattern C, which matches with operation pattern waveforms in Fig. 3.4 and the pattern region in Fig. 3.8. Judging from the current (i_L) polarity at each turn-on instant, ZVS for all the power switches can still be achieved for both cases based on ZVS condition in Tables 3.2 and 3.3.

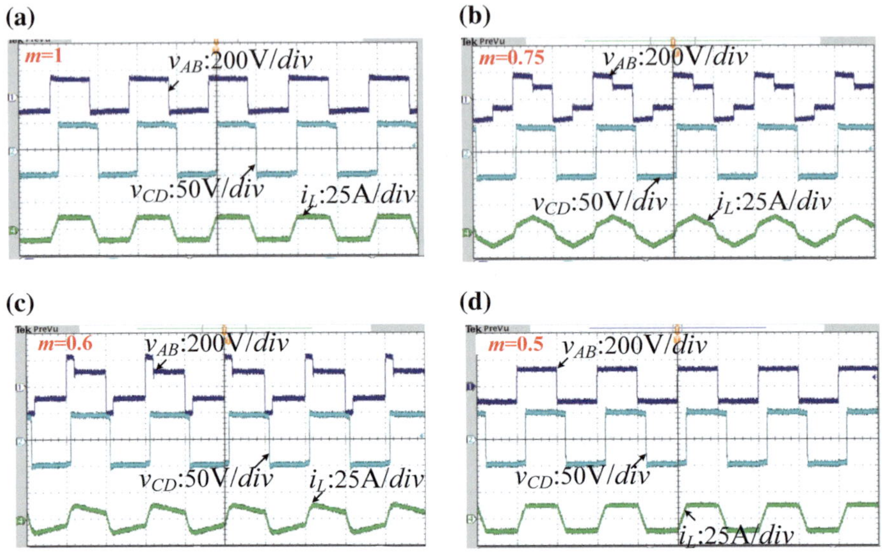

Fig. 3.13 Experimental waveforms of different input voltages when P_o = 1 kW. **a** V_1 = 120 V, m = 1. **b** V_1 = 160 V, m = 0.75. **c** V_1 = 200 V, m = 0.6. **d** V_1 = 240 V, m = 0.5

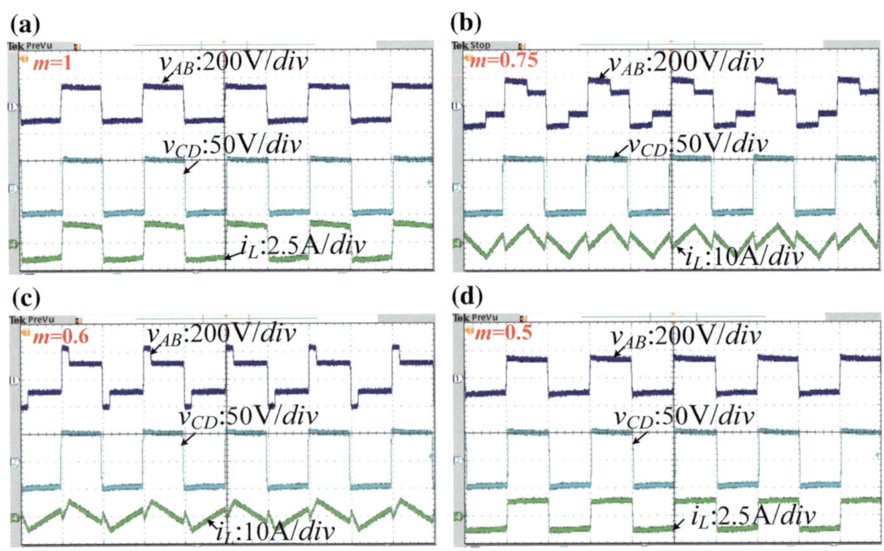

Fig. 3.14 Experimental waveforms of different input voltages when P_o = 150 W. **a** V_1 = 120 V, m = 1. **b** V_1 = 160 V, m = 0.75. **c** V_1 = 200 V, m = 0.6. **d** V_1 = 240 V, m = 0.5

When the power reverses from V_2 side to V_1 side, the experimental waveforms with different V_1 under −600 W are illustrated in Fig. 3.15. As seen in Fig. 3.15a, v_{AB} lags v_{CD} in phase implying that the power transfers backward. The waveforms in Fig. 3.15a are also similar to those in Fig. 3.15d. The converter works in pattern C when V_1 = 160 and 200 V, respectively, as shown in Fig. 3.15b, d, which corresponds to the operation pattern waveforms shown in Fig. 3.4 and the pattern region in Fig. 3.8. When V_1 = 200 V, ZVS of all the switches can also be achieved expect for S_5 and S_6, judging from the current polarity at each turn-on instant. In this case, ZVS for S_5 and S_6 is lost since the converter works in the no ZVS region of pattern C as shown in Fig. 3.9.

The efficiency curves of the hybrid-bridge-based converter with proposed VM control and conventional full-bridge DAB converter with EPS control are plotted in Fig. 3.16. The parameters for the two topologies are the same as those listed in Table 3.7. When V_1 is 120 and 240 V, the converter efficiency differs little with VM control since the voltage and current waveforms are similar. For these two cases, the efficiency using VM control is always higher than the efficiency employing EPS control, especially at light load. When V_1 is 160 and 200 V, the efficiency using these two controls is almost the same at heavy load condition. When the power is lower than 300 W, the efficiency is higher using proposed VM control compared to that with EPS control. For VM control, it can be seen that its efficiency does not change much for the four cases described above. With EPS control, however, the efficiency decreases greatly as V_1 voltage is varied from the

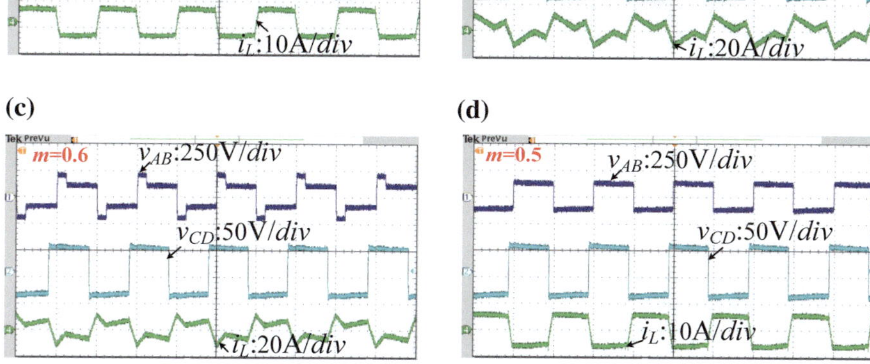

Fig. 3.15 Experimental waveforms of different input voltages when power transfer backward (P_o = −600 W). **a** V_1 = 120 V, m = 1. **b** V_1 = 160 V, m = 0.75. **c** V_1 = 160 V, m = 0.6. **d** V_1 = 240 V, m = 0.5

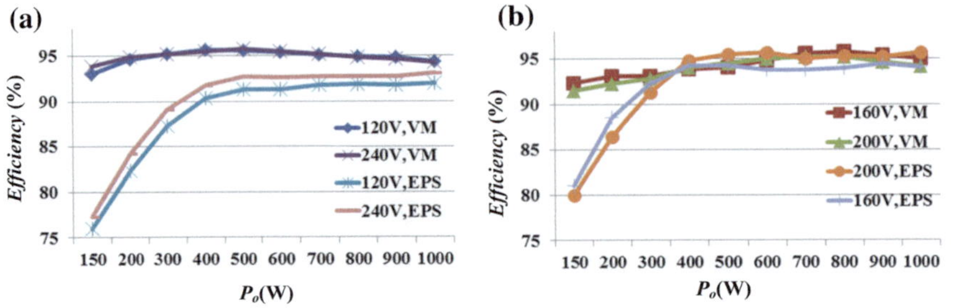

Fig. 3.16 Efficiency comparison between VM control and EPS control. **a** V_1 = 120 V and V_1 = 240 V. **b** V_1 = 160 V and V_1 = 200 V

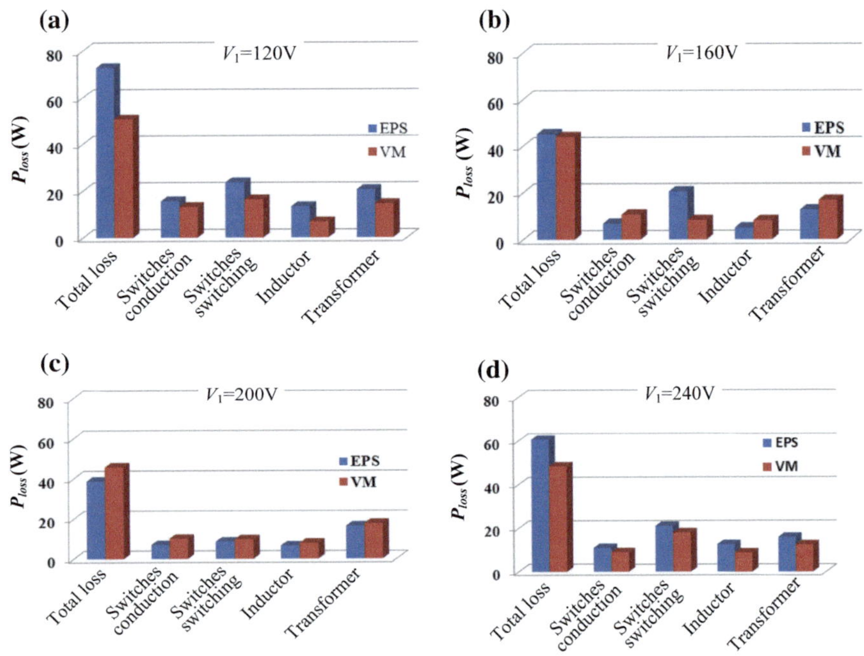

Fig. 3.17 Loss breakdown between VM control and EPS control under rated load. **a** V_1 = 120 V. **b** V_1 = 160 V. **c** V_1 = 200 V. **d** V_1 = 240 V

optimal operating voltage 180–120 and 240 V. Therefore, the proposed VM control maintains high efficiency over the whole operating voltage and load range, while EPS control can only ensure a high efficiency at heavy load and a limited voltage range.

Figure 3.17 shows the loss breakdown comparison for the hybrid-bridge-based converter with proposed VM control and conventional full-bridge DAB converter with EPS control. Four types of losses are compared: the switches conduction loss, switching loss, inductor loss, and transformer loss. The parameters for the two topologies are also the same as those listed in Table 3.7. As shown in Fig. 3.17a, the total loss is much larger for EPS control when V_1 = 120 V, since the four types of loss are all larger. In Fig. 3.17b, when V_1 voltage increases to 160 V, the total loss is similar for both of the two control methods. In this case, the conduction loss is smaller using EPS control, but the switch switching loss is higher. When V_1 = 200 V as seen in Fig. 3.17c, the loss using EPS control becomes a little bit lower. While, EPS control has larger loss when V_1 voltage is moved away from the optimal voltage (180 V) to 240 V compared with VM control as shown in Fig. 3.17d.

3.7 Discussion and Future Work

Like the ZVS discussion of many previous works in DAB converters, the charging/discharging energy is ignored in this chapter since the leakage inductance is relatively large. If the charging/discharging energy is considered, full ZVS performance can be affected slightly in real applications. Therefore, further improvements can be made to achieve accurate ZVS conditions taking into account the small charging/discharging current for drain-to-source capacitors of the power switches. Besides, the optimal target in this chapter is to widen ZVS range and simplify the controller. Hence, only four voltage levels are used at one port of the transformer resulting in that the control has only two degrees of freedom. Future work can be implemented to generate five voltage levels at one port of the transformer. Better performance may be obtained at the cost of increased control degree of freedom.

3.8 Conclusion

A VM control has been proposed for hybrid-bridge-based DAB converter to suit wide voltage conversion gain applications. The topology is an integration of a half-bridge and a full-bridge DAB converter, which utilizes four-level voltage waveforms at one side of the transformer. Based on operation modes and ZVS conditions for the working patterns, the VM control is employed to achieve wide ZVS range and simple controller. This control is suited especially for the applications where the voltage gain is twice of its minimum voltage gain. With VM control, theoretically full load ZVS range can be obtained for the six main switches, and the two auxiliary switches can also operate in a wide ZVS range. Also, the power control is very simple and does not need any complicated calculations.

Hence, it is easy to implement digital control. A 1 kW converter prototype is built up to verify the effectiveness of the VM control for hybrid-bridge-based converters. The results show that it is a good candidate for bidirectional power flow applications needing wide voltage conversion gains.

References

1. Choi W, Rho KM, Cho BH (2016) Fundamental duty modulation of dual-active-bridge converter for wide-range operation. IEEE Trans Power Electron 31(6):4048–4064
2. Zhao B, Song Q, Liu W, Liu G, Zhao Y (2015) Universal high-frequency link characterization and practical fundamental-optimal strategy for dual-active-bridge DC–DC converter under PWM plus phase-shift control. IEEE Trans Power Electron 30(12): 6488–6494
3. Shen Y, Sun X, Li W, Wu X, Wang B (2016) A modified dual active bridge converter with hybrid phase-shift control for wide input voltage range. IEEE Trans Power Electron 31 (10):6884–6990
4. Song W, Lehman B (2004) Dual-bridge DC–DC converter: a new topology characterized with no deadtime operation. IEEE Trans Power Electron 19(1):94–103
5. Sun X, Li X, Shen Y, Wang B, Guo X (2017) A dual-bridge LLC resonant converter with fixed-frequency PWM control for wide input applications. IEEE Trans Power Electron 32 (1):69–80
6. Bezerra PAM, Krismer F, Burkart RM, Kolar JW (2014) Bidirectional isolated non-resonant DAB DC–DC converter for ultra-wide input voltage range applications. In: 2014 International Power Electronics and Application Conference and Exposition (PEAC), pp 1038–1044
7. Inoue S, Akagi H (2007) A bidirectional DC–DC converter for an energy storage system with galvanic isolation. IEEE Trans Power Electron 22(6):2299–2306
8. Oggier GG, Garcia GO, Oliva AR (2009) Switching control strategy to minimize dual active bridge converter losses. IEEE Trans Power Electron 24(7):1826–1838
9. Wu K, Silva CW, Dunford WG (2012) Stability analysis of isolated bidirectional dual active full-bridge DC–DC converter with triple phase-shift control. IEEE Trans Power Electron 27 (4):2007–2017
10. Jain AK, Ayyanar R (2011) PWM control of dual active bridge: comprehensive analysis and experimental verification. IEEE Trans Power Electron 26(4):1215–1227

Chapter 4
Dual-Transformer-Based DAB Converter with Wide ZVS Range for Wide Voltage Gain Application

Abstract Dual-transformer-based dual active bridge (DAB) converter with wide ZVS range for wide voltage conversion gain application is derived and studied in this chapter. The phase shift control is adopted for this converter with minimum power switches employing half-bridge output, and a control law is proposed to achieve wide ZVS range. With the proposed method, four switches of the converter can achieve full range of ZVS. The other two switches can achieve full-range ZVS under positive power flow, while slightly reduced ZVS region under reverse power flow. Unlike the methods employing three control degrees of freedom, the proposed method only utilizes two decoupled control variables, making the controller easy to be implemented. In addition, the design of turns ratios for the two transformers under the proposed control is also optimized based on the gain ranges and conduction loss. The effectiveness of the converter with the proposed control is verified by experimental results from a 1 kW prototype.

Keywords Dual-transformer-based · Dual active bridge converter
Wide ZVS range · Wide conversion gain

4.1 Converter Topology and Operation Principle

The concept of dual-transformer-based converter is used in [1] to achieve full ZVS range and reduced filter size for conventional full bridge with plus width modulation (PWM) control. Also, a dual-transformer-based full-bridge converter with DCM operation [2] is studied to minimize the switching number, reduce filter size, and improve the efficiency. In [3], the dual-transformer-based concept is extended to converters with phase shift control to achieve wide voltage gain. While the power flow is unidirectional with the using of diode rectifier, ZVS range of the converter is still limited. Wide ZVS range and wide conversion gain can be both achieved for dual-transformer-based converter in [4], but the converter has many power devices, and the power is also unidirectional.

D. Sha and G. Xu, *High-Frequency Isolated Bidirectional Dual Active Bridge DC–DC Converters with Wide Voltage Gain*, CPSS Power Electronics Series,
https://doi.org/10.1007/978-981-13-0259-6_4

In this chapter, to achieve wide ZVS range and wide voltage gain for bidirectional power applications, a simple control law is presented for the dual-transformer-based DAB converters with minimum power switches. The topology has been inspired by the PWM converter in [1].

4.1.1 *Topology and Modulation Schedule Using Phase Shift Control*

The dual-transformer-based DAB converter topology is shown in Fig. 4.1. It only has six active switches. In the primary side, the circuit can be divided as two parts including a conventional full-bridge circuit and an auxiliary half-bridge circuit. The two bridges share the same leg consisting of S_1 and S_2. For the output side, it is a half bridge. L_k is the leakage inductance at the secondary side acting as a power link for this DAB converter. The typical operation waveforms with phase shift control are shown in Fig. 4.2. It should be noted that only the case when $0 < \phi < D_1$ is taken as an example to illustrate the operation principle, where D_1 is the duty cycle of primary side, and ϕ is the phase shift ratio between v_{DE} and v_{DE}. Other cases will be analyzed in the next section. All of the six switches are driven with 50% duty cycle. For the primary side, the four switches work the same as conventional phase shift full-bridge converter. The other two capacitors: C_{11}, C_{12}, together with the leg of S_1, S_2, work as an additional half bridge with fixed 50% duty cycle. The primary sides of the two transformers are in parallel, and the secondary sides are in series. With this configuration, four voltage levels can be generated in the v_{DE} waveform and are controlled with the duty ratio D_1. In the secondary side, the leakage L_k provides the power link between the high-frequency transformer output voltage v_{DE} and the output voltage v_{DE} of the half bridge in secondary side. The bidirectional power transfer is controlled by the phase shift ratio ϕ between v_{DE} and v_{DE}.

In view of the high-frequency voltage waveform induced on the leakage inductor (V_{DE}), it actually consists of four voltage levels unlike three voltage levels for the other DAB converter. Through proper method to generate the four voltage levels, the peak current and reactive current of the converter can be tightly controlled when the voltage V_1 or V_2 changes. This characteristic provides the possibility for the converter to have a better performance under wide voltage gain range application.

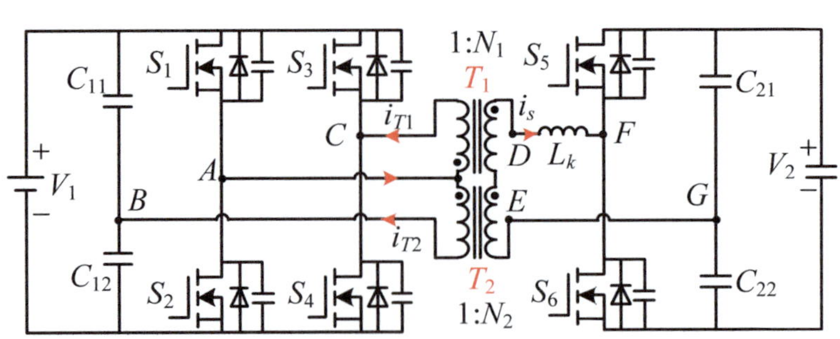

Fig. 4.1 Topology of dual-transformer-based DAB converter

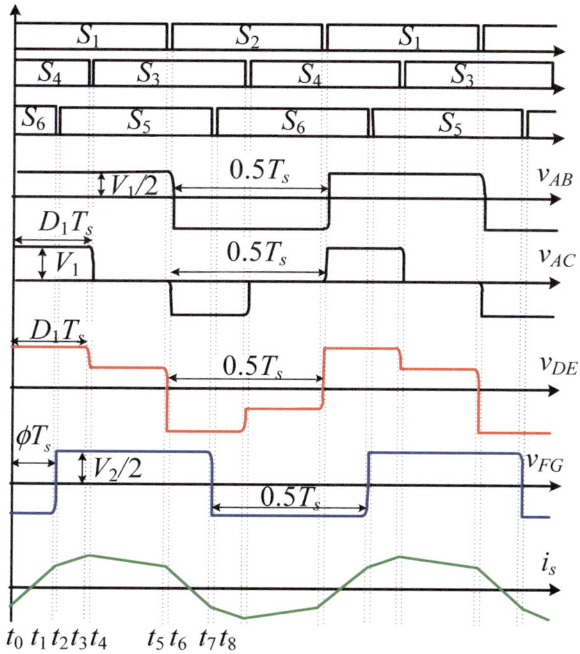

Fig. 4.2 Typical waveforms for dual-transformer-based DAB converter when $0 < \phi < D_1$

4.1.2 *Working Stages of the Converter*

Since the PWM signals are symmetrical during two half switching cycles, the mode of operation is analyzed only within one half cycle from t_0 to t_6 as shown in Fig. 4.2. The corresponding six working modes are illustrated in Fig. 4.3.

Stage 1 (t_0, t_1) [Fig. 4.3a]: Prior to t_0, S_4 and S_6 are on, and S_1, S_2, S_3, and S_5 are off. At t_0, S_1 is turned on. During this stage, v_{AC} is equal to V_1, and v_{AB} is equal to $V_1/2$. Consequently, the total secondary side voltage of the two transformers v_{DE} is $N_1V_1 + N_2V_1/2$, and the high-frequency voltage v_{FG} generated by the half bridge in V_2 side is $-V_2/2$. The transformer current starts to increase with a constant slew rate and the following relationship is obtained,

$$\begin{cases} i_s(t - t_0) = i_s(t_0) + (v_{DE} - v_{FG})(t - t_0)/L_k \\ v_{DE} = N_1V_1 + N_2V_1/2,\ v_{FG} = -V_2/2 \end{cases} \tag{4.1}$$

where 1: N_1 and 1: N_2 are the turns ratios for transformer T_1 and T_2, respectively.

Stage 2 (t_1, t_2) [Fig. 4.3b]: At time t_1, S_6 is turned off. The transformer current starts to discharge the junction capacitor of S_5 and to charge the junction capacitor of S_6. When the drain–source voltage of S_6 is charged to V_2, the drain–source voltage of S_5 is zero and the body diode will be conducted to create ZVS turn-on condition for S_5 in the next stage. Because the leakage inductance is relatively large for DAB converters, the current required to charge/discharge the two drain–source

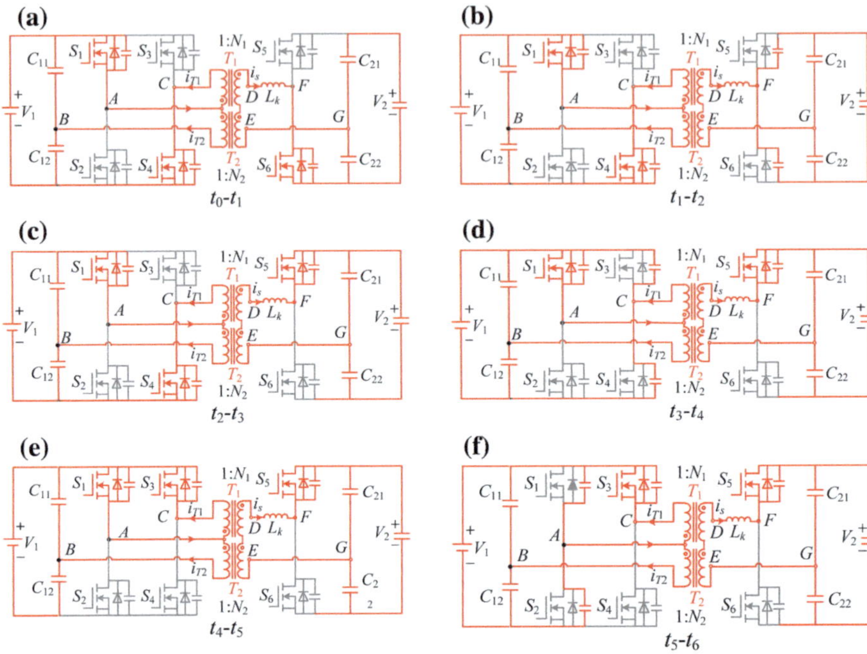

Fig. 4.3 Working stages. **a** Stage 1, **b** Stage 2, **c** Stage 3, **d** Stage 4, **e** Stage 5, and **f** Stage 6

capacitors is small and can be neglected to simplify ZVS analysis. Hence, ZVS condition for S_5 can be obtained as,

$$i_s(t_1) < 0 \tag{4.2}$$

Stage 3 (t_2, t_3) [Fig. 4.3c]: At t_2, S_5 is turned on under ZVS. In this stage, the total secondary side voltage of the two transformers v_{DE} is the same as that of Stage 1. v_{FG} is changed to be $V_2/2$. In this stage, the transformer leakage current is changing with a relatively lower constant slew rate and its value can be expressed as,

$$\begin{cases} i_s(t-t_2) = i_s(t_2) + (v_{DE} - v_{FG})(t-t_2)/L_k \\ v_{DE} = N_1 V_1 + N_2 V_1/2, \; v_{FG} = V_2/2 \end{cases} \tag{4.3}$$

Stage 4 (t_3, t_4) [Fig. 4.3d]: At t_3, S_4 is turned off. i_{T1} starts to discharge the junction capacitor of S_3 and charge the junction capacitor of S_4. At the end of this charging process, the body diode conducts to create ZVS for S_3 in the next stage. The charging energy stored in these capacitors is ignored, and ZVS condition for S_3 is,

$$i_{T1}(t_3) = i_s(t_3)N_1 < 0 \tag{4.4}$$

Stage 5 (t_4, t_5) [Fig. 4.3e]: At t_4, S_3 is turned on under ZVS. During this stage, v_{AC} changes to be zero. v_{DE} becomes $N_2V_1/2$. The relationship between transformer current and voltages can be expressed as,

$$\begin{cases} i_s(t - t_3) = i_s(t_3) + (v_{DE} - v_{FG})(t - t_3)/L_k \\ v_{DE} = N_2V_1/2, v_{FG} = V_2/2 \end{cases} \tag{4.5}$$

Stage 6 (t_5, t_6) [Fig. 4.3f]: S_1 is turned off at t_5. During this interval, the junction capacitor of S_2 is discharged, and its drain–source voltage decreases. At the end of this interval, the drain–source voltage of S_2 is zero, and the body diode is conducted, leading to ZVS turned on for next stage. The ZVS condition for S_2 can be shown as,

$$i_{T1}(t_3) + i_{T2}(t_3) = i_s(t_3)(N_1 + N_2) < 0 \tag{4.6}$$

4.2 ZVS Constraints and Control

4.2.1 Current Range for ZVS

Similar to many other DAB converters, the converter performance is dependent on more than one control variable. It is controlled by the full-bridge duty cycle D_1 and the phase shift ratio ϕ between v_{DE} and v_{FG}. Different combinations of them can have different working patterns, leading to different ZVS performances. In the previous section, the case of $0 < \phi < D_1$ is taken as an example to study the converter working principles. If all the combinations of D_1 and ϕ are taken into consideration, three more cases should be considered: $D_1 < \phi < 0.5$, $D_1 - 0.5 < \phi < 0$, and $-0.5 < \phi < D_1 - 0.5$. The theoretical waveforms for all the four cases are illustrated, respectively, in Fig. 4.4. The cases when $\phi > 0$ are shown in Fig. 4.4a, b, and the cases when $\phi < 0$ are shown in Fig. 4.4c, d. As discussed from working Stages 2, 4, and 6 for the pattern A which is shown in Fig. 4.3, ZVS is only determined by the polarity of transformer current at the instant when switch is turned on. According to this principle, current constraints of ZVS can be summed up in Table 4.1. The expressions for these currents in Table 4.1 can be calculated based on the voltage induced across the leakage inductance during each working stage. Take pattern A for instance, the relationships between the current and the voltages are shown in (4.1), (4.3), and (4.5). For the transformer current, the DC current bias is zero, and the two half cycle waveforms are symmetric. Therefore, it can be obtained,

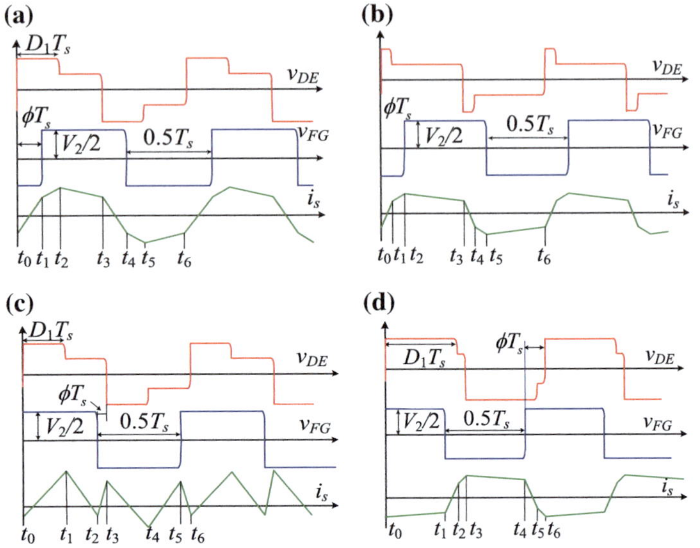

Fig. 4.4 Four working patterns. **a** Pattern A: $0 < \phi < D_1$. **b** Pattern B: $D_1 < \phi < 0.5$. **c** Pattern C: $D_1 - 0.5 < \phi < 0$. **d** Pattern D: $-0.5 < \phi < D_1 - 0.5$

Table 4.1 ZVS current limit for different working patterns

Working patterns	S_1, S_2 (ZVS)	S_3, S_4 (ZVS)	S_5, S_6 (ZVS)
A: $0 < \phi < D_1$	$i_s(t_0) < 0$	$i_s(t_2) > 0$	$i_s(t_1) > 0$
B: $D_1 < \phi < 0.5$	$i_s(t_0) < 0$	$i_s(t_1) > 0$	$i_s(t_2) > 0$
C: $D_1 - 0.5 < \phi < 0$	$i_s(t_0) < 0$	$i_s(t_1) > 0$	$i_s(t_2) < 0$
D: $-0.5 < \phi < D_1 - 0.5$	$i_s(t_0) < 0$	$i_s(t_2) > 0$	$i_s(t_1) < 0$

$$\begin{cases} i_s(t_1) + i_s(t_4) = i_s(t_2) + i_s(t_5) = i_s(t_3) + i_s(t_6) = 0 \\ 1/T_s \int\limits_{t_0}^{t_6} i_s(t_1) d_t = 0 \end{cases} \tag{4.7}$$

Then, based on (4.1), (4.3), (4.5), and (4.7), the current expressions in Table 4.1 for pattern A can be obtained. In addition, the current expressions for other three working patterns can also be calculated using the same method. Substitution of all current expressions back to Table 4.1, the relationship between ZVS and D_1 and ϕ can be summed up in Table 4.2.

Table 4.2 ZVS constraints under different working patterns

	S_1, S_2 (ZVS)	S_3, S_4 (ZVS)	S_5, S_6 (ZVS)
A: $0 < \Phi < D_1$	$-4V_2\Phi < E$	$4[\Phi V_2 + D_1(2N_1V_1 + N_2V_1 - V_2)] > E$	$4(2N_1 + N_2)V_1\Phi > E$
B: $D_1 < \Phi < 0.5$	$-4V_2\Phi < E$	$4[V_2(D_1 - \Phi) + D_1V_1(2N_1 + N_2)] > E$	$4V_1(2D_1N_1 + N_2\Phi) > E$
C: $D_1 - 0.5 < \Phi<0$	$-4V_2\|\Phi\| < E$	$4[V_2(D_1 + \|\Phi\|) - D_1V_1(2N_1 + N_2)] > E$	$-4N_2V_1\|\Phi\| < -E$
D: $-0.5 < \Phi < D_1 - 0.5$	$-4V_2\|\Phi\| < E$	$2[V_2(2D_1 - 1 + 2\|\Phi\|) + N_2V_1(2D_1 - 1)] > -E$	$-4V_1[2N_1(D_1 - 0.5 + \|\Phi\| + N_2\|\Phi\| < -E$

$E = (V_2 - N_1V_1 - 4D_1N_1V_1)$

4.2.2 Proposed Control Law to Achieve Full Range of ZVS for S_1, S_2, S_5 and S_6

Based on Table 4.2, the ZVS performance is affected by D_1, Φ and the two turns ratios: N_1 and N_2. It is complex to select a proper combination of them for achieving wide ZVS range. However, all the expressions in Table 4.2 share the same item: $E = V_2 - N_1V_1 - 4D_1N_1V_1$. If E is set to be zero, the expression for S_1 and S_2 can be easily satisfied regardless of any other variables, implying that ZVS can be achieved independently of the transformer turns ratios, load and two port voltages. Meanwhile, ZVS for S_3 and S_4 can also be achieved independently of the other variables according to Table 4.2. The physical meaning of $E = 0$ is illustrated in Fig. 4.5. As shown, the expression of the voltage second in a half positive cycle for the two ports of the leakage inductor voltages (v_{DE}, v_{FG}) can be calculated as,

$$\begin{cases} S_A = N_2V_1T_s/4 + N_1V_1D_1T_s \\ S_B = V_2T_s/4 \end{cases} \tag{4.8}$$

If $s_A = s_B$, then it can be obtained,

$$D_1 = (V_2 - N_2V_1)/4N_1V_1 \tag{4.9}$$

Substitution of (4.9) into the expression of E leads to $E = 0$, and vice versa. Therefore, $E = 0$ is equivalent to making the half cycle voltage seconds applied to the two ports of the leakage inductor to be equal. Therefore, this control law is aimed at achieving voltage second balance at the two ports of the inductor. With $E = 0$, ZVS results for all the switches can be summarized in Table 4.3. As seen, with the control law, ZVS of S_1, S_2, S_5, and S_6 can always satisfied. ZVS of S_3 and S_4 in pattern B, C, and D is dependent on D_1, ϕ and two transformer turns ratios (N_1, N_2).

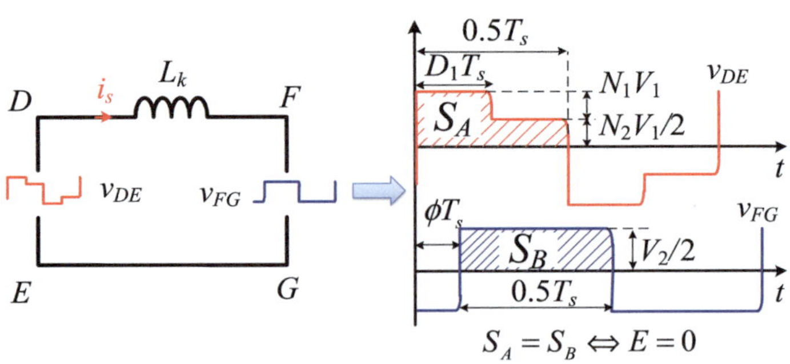

Fig. 4.5 Physical meaning of $E = 0$

Table 4.3 ZVS results when $E = 0$ for four working patterns

Working patterns	S_1, S_2	S_3, S_4	S_5, S_6
A: $0 < \phi < D_1$	AS	AS	AS
B: $D_1 < \phi < 0.5$	AS	$2D_1[N_2 + (2D_1 + 1)N_1] > \phi(N_2 + 4D_1N_1)$	AS
C: $D_1 - 0.5 < \phi < 0$	AS	$4D_1N_1(D_1 + \lvert\phi\rvert - 0.5) < -N_2\lvert\phi\rvert$	AS
D: $-0.5 < \phi < D_1 - 0.5$	AS	$4D_1N_1(D_1 + \lvert\phi\rvert - 0.5) > N_2(1 - 2D_1 - \lvert\phi\rvert)$	AS

AS always satisfied

4.2.3 *Transformer Turns Ratio Consideration and Extension of ZVS Range for S_3 and S_4*

With the control law in (4.9), the relationship between turns ratios and the two port voltages can be obtained as,

$$\begin{cases} 1 = N_2/G_{\max} + 4D_{1\max}N_1/G_{\max} \\ 1 = N_2/G_{\min} + 4D_{1\min}N_1/G_{\min} \end{cases} \tag{4.10}$$

where $G_{\min}$ and $G_{\max}$ are the minimum and maximum value for voltage gain: $V_2 = V_1$, respectively. Also, $D_{1\min}$ and $D_{1\max}$ are the minimum and maximum value for D_1. Based on (4.10), the two turns ratios can be solved as,

$$\begin{cases} N_1 = (G_{\max} - G_{\min})/(4D_{1\max} - 4D_{1\min}) \\ N_2 = (D_{1\max}G_{\min} - D_{1\min}G_{\max})/(D_{1\max} - D_{1\min}) \end{cases} \tag{4.11}$$

As shown in (4.11), the two transformer turns ratios are determined by $D_{1\min}$ and $D_{1\max}$ when the maximum and minimum voltage gain ($G_{\max}$ and $G_{\min}$) of the converter is predetermined. For the proposed converter, D_1 can be in the range of [0, 0.5]. Since phase shift control is applied to the V_1 side full bridge, circulating current for transformer T_1 will exist if D_1 is less than 0.5. The duration of this current circulating stage can be seen during [t_4, t_5] as shown in Fig. 4.2, leading to high conduction loss. In addition, when $D_1 = 0.5$, with the proposed control, the voltages applied to the two ports of the leakage inductance are ensured to be the same except that the phases are shifted. This working condition is actually the optimal operating point for DAB converters. Therefore, $D_{1\max}$ is designed to be 0.5. As a result, the two turns ratios can be designed with a proper selection of $D_{1\min}$ according to (4.11).

Meanwhile, Fig. 4.6a shows ZVS region under whole power transfer range for different $D_{1\min}$. The transferred power P is normalized by $P_{\text{base}} = V_2^2T_s/16L_k$. In Fig. 4.6a, $D_{1\min} = 0$. When the power is positive, ZVS can be obtained almost in full ranges except that when D_1 is close to zero. When the power transfer is negative, ZVS region is limited. Figure 4.6b shows an example of changing the minimum value of duty cycle to enlarge ZVS range, in which $D_{1\min}$ is increased from 0 to a small value 0.05. As shown in Fig. 4.6b, ZVS in positive power flow

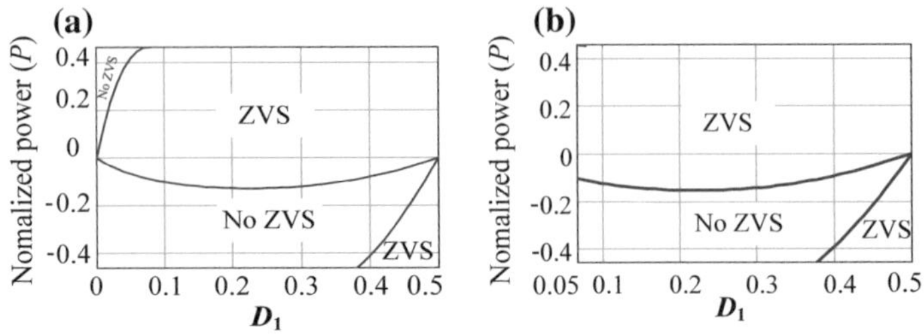

Fig. 4.6 ZVS region for S_3 and S_4 under different $D_{1\min}$ ($G_{\max} = 2G_{\min}$, $D_{1\max} = 0.5$). **a** $D_{1\min} = 0$. **b** $D_{1\min} = 0.05$

can always be satisfied, and ZVS in negative power flow is enlarged also. This implies that through proper design $D_{1\min}$ to determine the two turns ratios, ZVS region of S_3 and S_4 has the possibility to be enlarged. Because the design of two transformers' turns ratios is related to the detail specifications of the converter such as input and output voltage ranges, the detail design will be given in the later section.

4.3 Converter Characteristics with Proposed Control

4.3.1 Power Characteristics Under Proposed Control

The power expression of the converter can be calculated based on the power transferred in a switching cycle, which is shown as,

$$P = \int_{t_0}^{t_6} v_{DE} i_s(t) d_t / T_s \tag{4.12}$$

where t_0 and t_6 are the starting and ending time of the switching cycle, respectively, also denoted in Fig. 4.4. According to Fig. 4.4 and the current expressions for the four working patterns, the transferred power under proposed control can be obtained as below,

$$\begin{cases} P_A = \frac{4P_{\text{base}}[(4D_1\Phi - 2D_1^2 + D_1 - 4\Phi^2)N_1 + (\Phi - 2\Phi^2)N_2]}{N_2 + 4D_1N_1} \\ P_B = \frac{4P_{\text{base}}[(2D_1^2 + D_1 - 4D_1\Phi)N_1 + (\Phi - 2\Phi^2)N_2]}{N_2 + 4D_1N_1} \\ P_C = 4P_{\text{base}}\left(2\Phi^2 + \Phi - \frac{N_1(2D_1^2 + 8D_1\Phi^2 - D_1)}{N_2 + 4D_1N_1}\right) \\ P_D = \frac{4P_{\text{base}}[(2D_1^2 - 4D_1\Phi - 3D_1 + 4\Phi_2 + 4\Phi + 1)N_1 + (\Phi + 2\Phi^2)N_2]}{N_2 + 4D_1N_1} \end{cases} \tag{4.13}$$

where P_A, P_B, P_C and P_D are the power functions for pattern A, pattern B, pattern C, and pattern D, respectively. Then, based on the Φ range for different patterns, the power transfer of the converter with proposed control can be expressed as,

$$P(\Phi, D_1) = \begin{cases} P_A, & 0 < \Phi < D_1 \\ P_B, & D_1 < \Phi < 0.5 \\ P_C, & D_1 - 0.5 < \Phi < 0 \\ P_D, & -0.5 < \Phi < D_1 - 0.5 \end{cases} \tag{4.14}$$

The plot for (4.14) is shown in Fig. 4.7. For different D_1, the maximum power transfer changes. Also, for a particular D_1, $P(\Phi, D_1)$ is similar to a sinusoid waveform. To ensure that the power transfer is monotonously increased with the increasing of Φ, Φ should have both lower and upper limit in the whole operation range. According to (4.13) and (4.14), the operation regions for the four working patterns under different input voltages and different output power can be plotted in Fig. 4.8. As seen in Fig. 4.8a, the converter mainly works in the regions of pattern A and pattern C. Pattern B only occurs when D_1 is relatively small in positive power flow. Pattern D occurs when D_1 is large under reverse power flow. When the maximum voltage gain over minimum voltage gain is increased to 4 as shown in Fig. 4.8b, the regions of pattern B and A become smaller, and the region of pattern C becomes larger.

4.3.2 *Implementation of the Proposed Control*

The diagram for implementation of proposed control is shown in Fig. 4.9. Only one voltage regulator is needed to control the V_2 side voltage. With this method, the two power control variables are decoupled and can be easily obtained based on simple calculation. Phase shift ratio Φ is controlled through a PI regulator, and duty cycle control signal is calculated with the sampling of two voltages according to the

Fig. 4.7 Power transfer characteristic

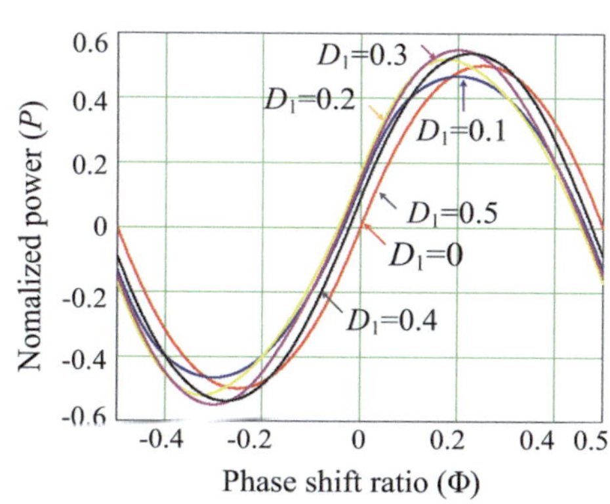

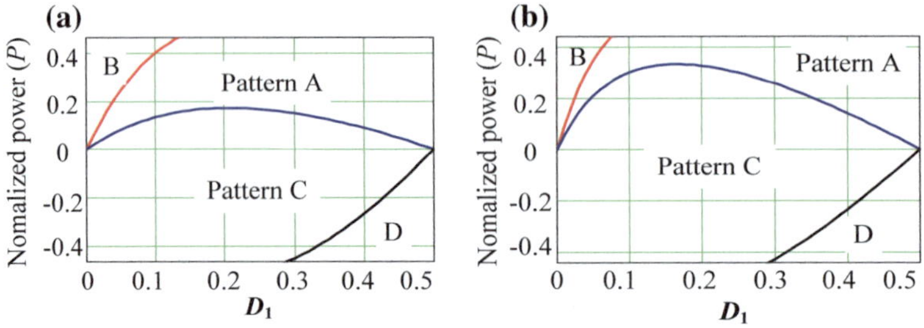

Fig. 4.8 Power transfer characteristic

Fig. 4.9 Implementation of the proposed control

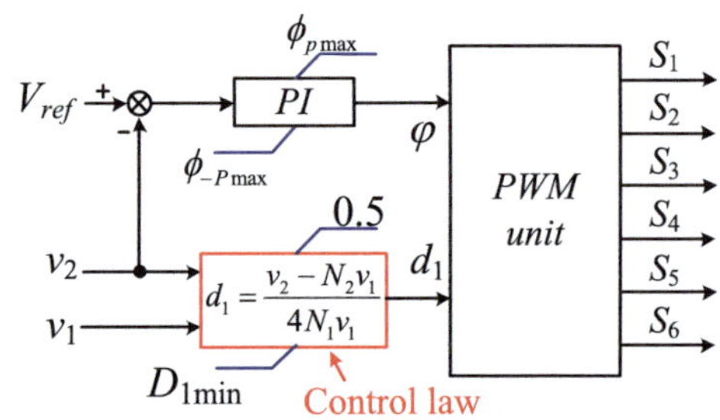

control law shown in (4.9). Based on Φ and D_1, PWM unit generates the corresponding driving signals according to Fig. 4.2. Unlike the other complex control strategies to achieve wide ZVS range under wide voltage gain, this control is quite simple and can be implemented easily with microcontroller.

4.4 Design Consideration and Comparison

In this section, a design example of the proposed converter with V_1 = 120–240 Vψ, V_2 = 96 Vψ, and *Po*ψ = 1 kWψ is introduced. The maximum voltage gain G_{max} = 96/120, and the minimum voltage gain G_{min} = 96/240. The gain range of the converter is wide as G_{max} is twice of G_{min}. A 50 kHz switching frequency is adopted. In the DAB converter design, normally, it is the maximum power transfer instead of voltage gain that matters more. Because the output voltage is fixed in the design case, the converter will work as long as it can have the maximum power transfer ability under the heaviest load. With a particular control, the maximum power transfer of a DAB converter is inversely proportional to the leakage inductance. Therefore, firstly, a leakage inductance can be properly designed to obtain the maximum power transfer. Then, for the turns ratio of the transformer design, the goal is aimed at both achieving wide ZVS range and reducing the reactive current loss.

4.4.1 *Leakage Inductance Design*

The leakage inductance provides a power link between V_1 and V_2, and it should be designed to allow the maximum power transfer. According to the power expression in (4.14), the global maximum power transfer for all the Φ and D_1 is shown as,

$$P_{\max} = 0.42P_{\text{base}} = 0.42V_2^2T_s/(16L_k) \tag{4.15}$$

In practice, 80% margin is given to allow dynamic responses. Hence, the limit for L_k can be shown as,

$$L_k \leq 0.23V_2^2T_s/(16L_k) = 2.65\ \mu\text{H} \tag{4.16}$$

Therefore, L_k is chosen as 2.6 μH.

Equation (4.16) sets up an upper limit to the leakage inductance for the maximum power transfer. One example for the leakage inductor RMS current and peak current when the leakage inductance is smaller than the upper limit is shown in Fig. 4.10. Figure 4.10a, b illustrates the RMS current and peak current of the leakage inductor with different value. When V_1 changes from 140 to 240 V, both the peak current and the RMS current will increase if the leakage inductance becomes smaller. This means high turn-off current and high device RMS current, leading to a higher rating for the device chosen and also higher conduction loss and turn-off loss. In addition, the leakage inductance is also related to the converter dynamic response. In theory, the converter would have faster response if the leakage inductance is lower. In our design, the loss is considered to have a higher priority. Therefore, *Lkψ* is chosen to be close to the upper boundary, which is 2.6 μH.

4.4.2 *Turns Ratios*

With the given specifications, $G_{\min}$ = 96/240 and $G_{\max}$ = 96/120. Substitution of these values along with $D_{1\max}$ = 0.5 into (4.11) leads to,

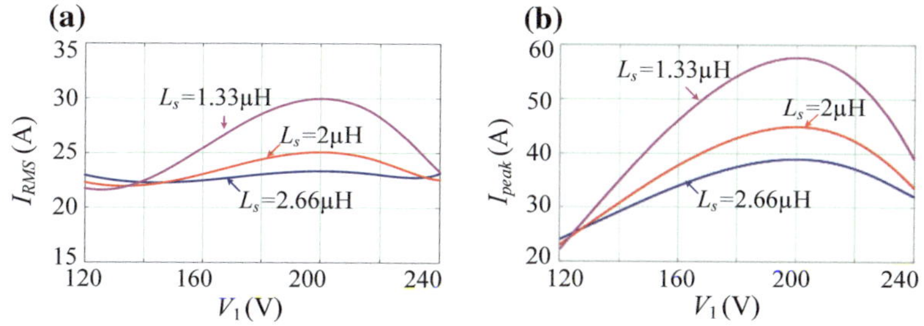

Fig. 4.10 Leakage inductor current comparison when L_k ψ changes under rated load. **a** RMS current. **b** Peak current

$$\begin{cases} N_1 = 1/(5 - 10D_{1\min}) \\ N_2 = (2 - 8D_{1\min})/(5 - 10D_{1\min}) \end{cases} \quad (4.17)$$

According to (4.17), to ensure that N_1 and N_2 are positive values, $D_{1\min}$ should be limited as: $0 \leq D_{1\min} < 0.25$. Besides, based on (4.17), if $D_{1\min}$ ψ is determined, the value for both N_2 and N_1 can be directly solved from (4.17). Therefore, the design of the two turns ratios can be simplified to find an optimal value for $D_{1\ \min}$ ψ which obtains wide ZVS range and low current-related loss.

In Fig. 4.11, RMS currents for the two transformers and the circulating current of transformer T_1 are compared for different $D_{1\min}$ at rated load condition. In Fig. 4.11a, b, the primary currents for transformer T_1 and T_2 are illustrated, respectively. As $D_{1\min}$ decreases from 0.2 to 0, T_1 primary current is decreasing and T_2 primary current is increasing within whole input voltage ranges. If $D_{1\min}$ becomes smaller, the circulating current of T_1 will have less impact of the power transfer because of the reduced total RMS current, which corresponds to Fig. 4.10c. As shown in Fig. 4.11c, the RMS circulating currents is lower if $D_{1\min}$ becomes smaller such as 0 and 0.05. Figure 4.11d shows the RMS current of secondary side. The cases when $D_{1\min} \geq 0.1$ are also not preferred, because the RMS currents are much higher than the other cases when the input voltage is higher than 180 V. The better cases to gain the minimum circulating current and lower RMS current for secondary side should be around 0 or 0.05.

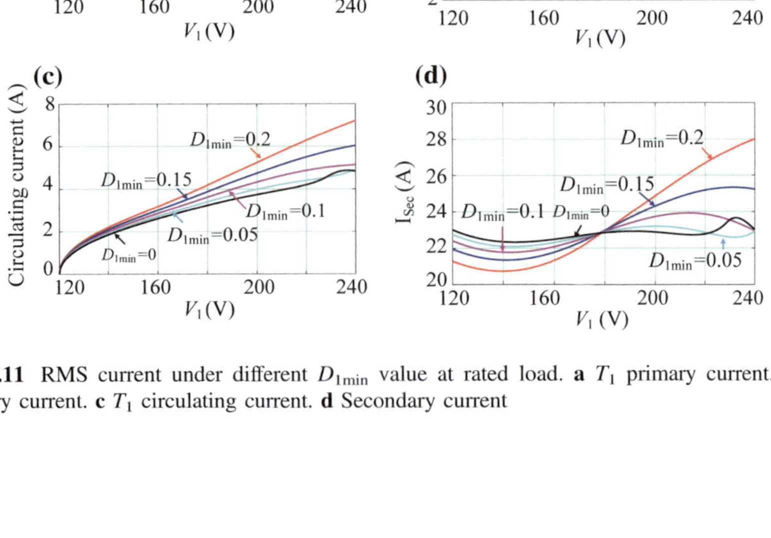

Fig. 4.11 RMS current under different $D_{1\min}$ value at rated load. **a** T_1 primary current. **b** T_2 primary current. **c** T_1 circulating current. **d** Secondary current

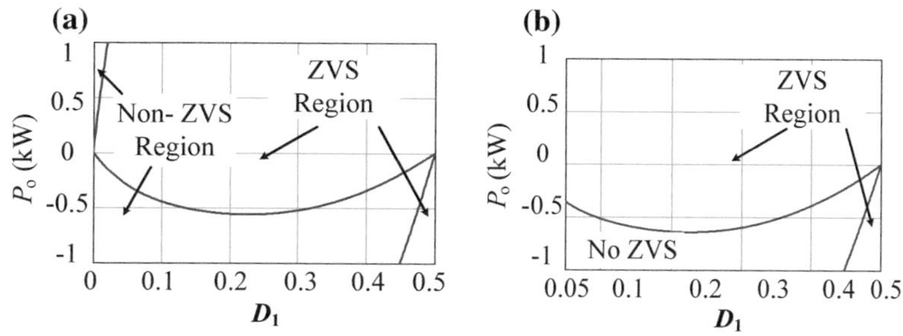

Fig. 4.12 ZVS region for S_3 and S_4 and improvement ($G_{max} = 2G_{min}$, $D_{1max} = 0.5$). **a** $D_{1min} = 0$; **b** $D_{1min} = 0.05$

Meanwhile, as mentioned before in Sect. 4.2, ZVS region of S_3 and S_4 is related to the value of D_{1min}. The cases for ZVS region under different D_{1min} are shown in Fig. 4.12. As shown, if D_{1min} is chosen to be slightly larger than 0, ZVS region can be changed. Unlike the case when $D_{1min} = 0$, ZVS under all the positive power flow conditions can be achieved and ZVS region under reverse power flow is enlarged. Therefore, the minimum duty cycle is designed as $D_{1min} = 0.05$ to obtain wider ZVS ranges and also smaller RMS current performance. Then according to (4.17), the two turns ratios can be obtained as, $N_1 = 2/9$ and $N_2 = 4/11$.

4.5 Comparison

4.5.1 Device RMS and Peak Current Comparison

The RMS currents and turn-off currents of all the devices are presented in Fig. 4.12 for dual-transformer-based DAB converter with proposed control and designed turns ratios under rated load. Also, the comparisons have been made with a conventional half-bridge DAB (CHDAB) converter consisting of a full-bridge at V_1 side and a half bridge at V_2 side. The EPS control in [5] aimed at achieving wide ZVS range is applied to the CHDAB converters, and the turns ratio is designed to be 1:2/5 to make the optimal operating point to be $V_1 = 120$ V to make the voltage matching occur at 120 V. The leakage inductance value is 2.6 μH.

As illustrated in Fig. 4.13a, RMS current of secondary side devices for CHDAB converter will increase greatly as V_1 increases from 120 to 240 V. While the RMS current of secondary side devices (S_5 and S_6 denoted in Fig. 4.1) almost keep the same for dual-bridge-based DAB converter. For the primary side devices, all the four devices have the same RMS current value for CHDAB converter, while two devices (S_1 and S_2) have higher RMS current and the other two (S_3 and S_4) have lower RMS current for dual-transformer-based DAB converter. Therefore, in view of the four switches total conduction loss in primary side, the latter one can have

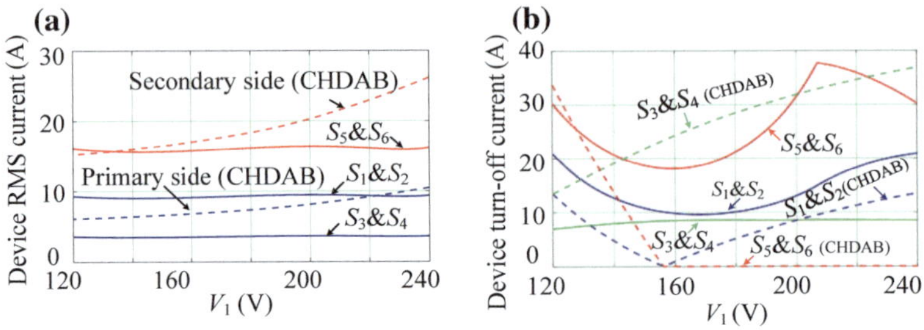

Fig. 4.13 Device RMS and turn-off current comparison under rated power (Po = 1 kW). **a** Device RMS current comparison. **b** Device turn-off current comparison

smaller losses especially when the input voltage becomes high. Figure 4.13b shows the device turn-off current comparison of each device. For the secondary side devices, the device turn-off current of the dual-transformer-based converter is larger. However, ZVS is lost for CHDAB converter when the input voltage is larger than 160 V, because the current is zero and cannot charge/discharge the junction capacitor to satisfy ZVS condition. The device turn-off current of S_1 and S_2 for CHDAB converter is smaller than the dual-transformer-based DAB converter. However, for the switches of S_3 and S_4, the device turn-off current of the CHDAB converter is more than twice larger than the one for dual-transformer-based DAB converter.

4.5.2 ZVS Range Comparison

Figure 4.14a, b shows the ZVS region comparison of the secondary side devices. For the dual-transformer-based converter, ZVS can be achieved for the switches during the whole operating load and voltage ranges. However, for the CHDAB converter, ZVS can only be achieved in a limited region. Figure 4.14c, d shows the ZVS region comparison of switches S_3 and S_4. In both the two figures, ZVS can be achieved when the power flow is positive. When the power is negative, ZVS will be lost in some operation regions. In view of the areas of ZVS region, the one for dual-transformer-based converter is slightly smaller than the CHDAB converter. In addition, full ZVS range for the leading leg (S_1 and S_2) can be obtained for both two topologies.

4.5.3 Transformer Size Comparison

The transferred power for the two transformers with different input voltage is shown in Fig. 4.15. As seen, when the input voltage increases from 120 to 240 V, the

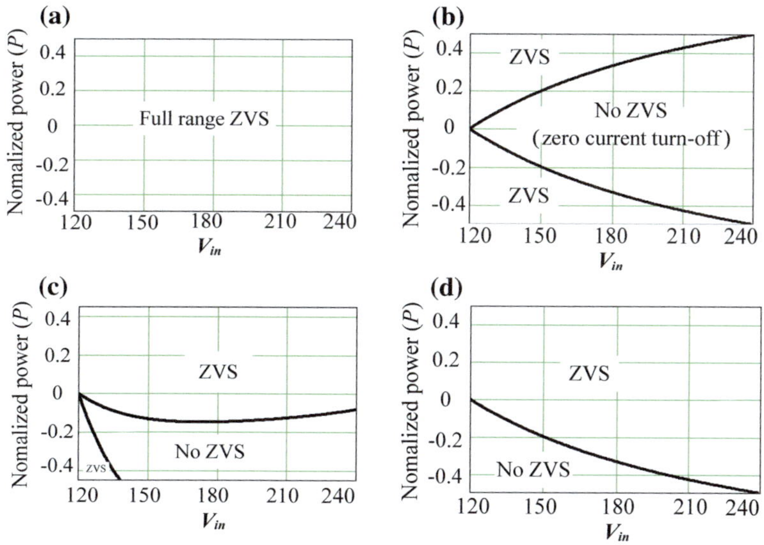

Fig. 4.14 ZVS ranges comparison for switches. **a** S_5 and S_6 for dual-transformer-based DAB converter. **b** S_5 and S_6 for CHDAB converter. **c** S_3 and S_4 for dual-transformer-based DAB converter. **d** S_3 and S_4 for CHDAB converter

Fig. 4.15 Power distribution of the two transformers

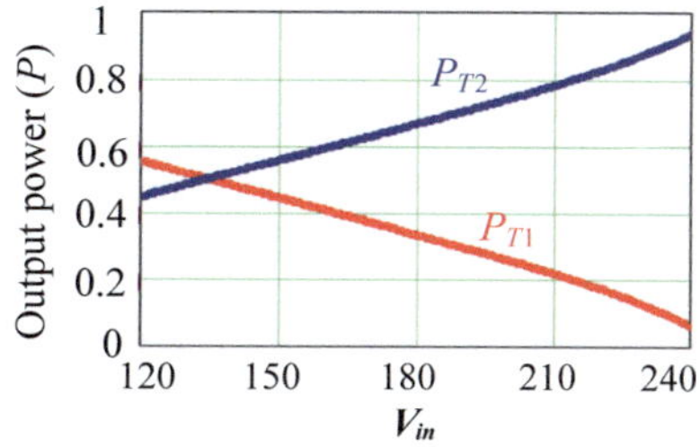

power of transformer T_1 (P_{T1}) decreases, while the power of transformer T_2 (P_{T2}) increases. Under the whole operation range, the maximum power for transformer T_2 is 935 W, and the maximum power for transformer T_1 is 555 W. If the area product (AP) method is used for selection of the core, the AP value of the converter is 50% larger than conventional converter design.

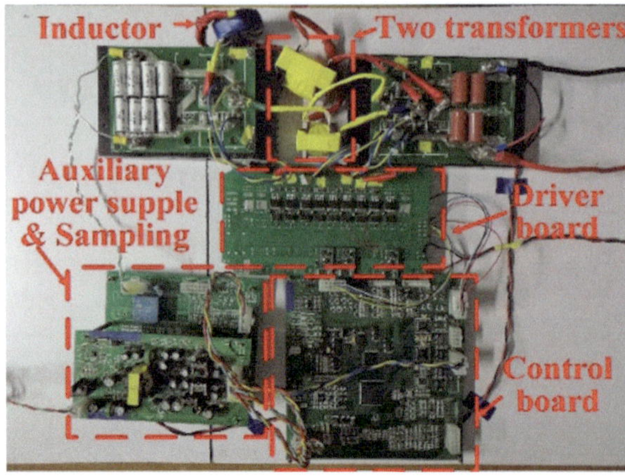

Fig. 4.16 Photograph of the prototype

4.6 Experimental Verification

A laboratory prototype was built to verify the effectiveness of the dual-transformer-based converter and proposed control. The specifications are the same as in the last section. In addition, the capacitance for V_1 side is $C_{11} = C_{12} = 20$ μF, and the capacitance for V_2 side is $C_{21} = C_{22} = 40$ μF. The established experimental prototype is shown in Fig. 4.16. The prototype includes five parts: auxiliary power supply board, sampling board, control board, driver board, and the main circuit. The controller TMS320F28335 from TI Company is used to implement the control. The sampling rate of the control is the same as the switching frequency 50 kHz. FDA38N30 from Fairchild is used for the primary side devices, and FQA90N15 is used for the secondary side devices.

The experimental waveforms under rated power Po = 1 kW are shown in Fig. 4.17. v_{AB} and v_{AC} are the primary side voltages of the two transformers, respectively, and v_{FG} is the voltage generated by the half bridge in V_2 side. Is the current of leakage inductance L_k, whose polarity at switch turn-on instant determines the ZVS performance? As shown in Fig. 4.17a when V_1 = 120 V, v_{AB} and v_{AC} are both square waveforms, and D_1 is at its maximum value 0.5. With the proposed control, the converter works at a condition exactly the same as the optimal operating point for conventional DAB converters with SPS control. Consequently, ZVS is easily achieved as seen from the polarity of the current at the moment when the switches turn on. Lower conduction loss can also be obtained. Meanwhile, when V_1 increases, such as 150, 180, and 210 V in Fig. 4.17b, c, d, respectively, the duty cycle is gradually decreased to obtain the voltage second balance between

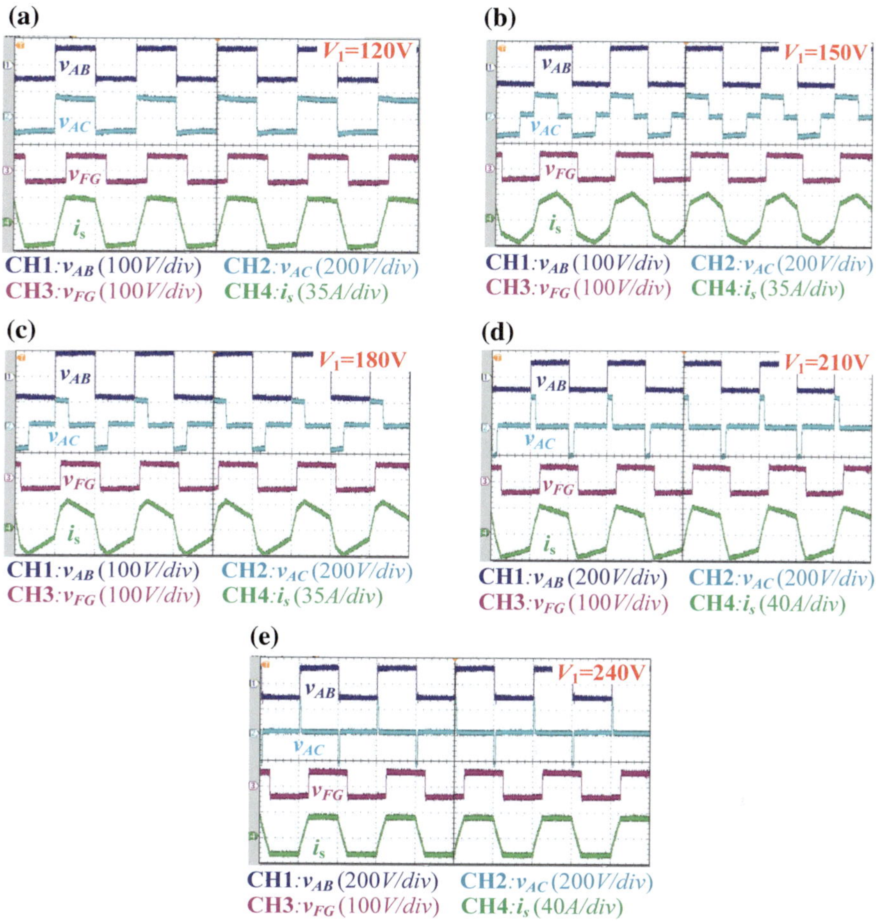

Fig. 4.17 Experimental waveforms under 1 kW. **a** V_1 = 120 V, **b** V_1 = 150 V, **c** V_1 = 180 V, **d** V_1 = 210 V, **e** V_1 = 240 V

the voltages at the two ports of the leakage inductance. Seen from the relationship between the duty cycle and phase shift ratio in these cases, the converter works in pattern A when V_1 = 150 and 180 V, and in pattern B when V_1 = 180 V. These matches with the theoretical waveforms are shown in Fig. 4.4a. In addition, judging from the current polarities of is at the turn-on instants, ZVS is achieved for all the three cases. When V_1 = 240 V as seen from Fig. 4.17e, D_1 is almost zero, therefore, the two voltages applied to L_k is almost the same with the case in Fig. 4.13a, which

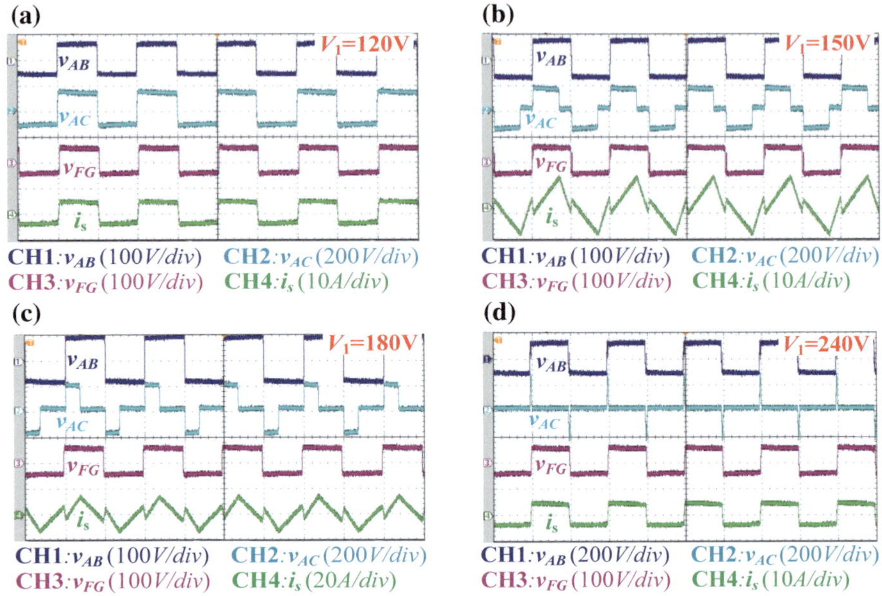

Fig. 4.18 Experimental waveforms under 200 W. **a** V_1 = 120 V, **b** V_1 = 150 V, **c** V_1 = 180 V, **d** V_1 = 210 V, **e** V_1 = 240 V

explains why is similar for the two cases. At the two boundary input voltage values, the converter can automatically shift the working mode to achieve low peak current and circulating current.

Figure 4.18 shows the experimental waveforms at light load conditions (Po = 200 W). As seen in Fig. 4.18a, d, the waveforms are also similar, and both of them are working under the optimal operating condition when full ZVS ranges and low conduction loss can be obtained. As for the cases when V_1 = 150 and 180 V, the converter works in the region of pattern C, which matches with the theoretical waveforms shown in Fig. 4.4c. Also, ZVS for the switches can be obtained judging from the polarity of the current at each turn-on instant.

The soft switching experimental results for different input voltages under different loads are shown in Figs. 4.19 and 4.20. The switch drain-to-source voltage, the switch gate voltage, and the leakage inductor current are captured for each subfigure. As illustrated, ZVS can be achieved both at light load (200 W) and heavy load (1 kW) under different input voltages.

The dynamic response of the converter under load changing is shown in Fig. 4.21. When the load is changing from half load to full load, the voltage recovery time under V_1 = 120 V and V_1 = 240 V is shorter than 6 ms. Under V_1 = 180 V case, the recovery time is a bit longer, but it has lower overshoot voltage. When the load is changing from full load to half load, the recovery time is

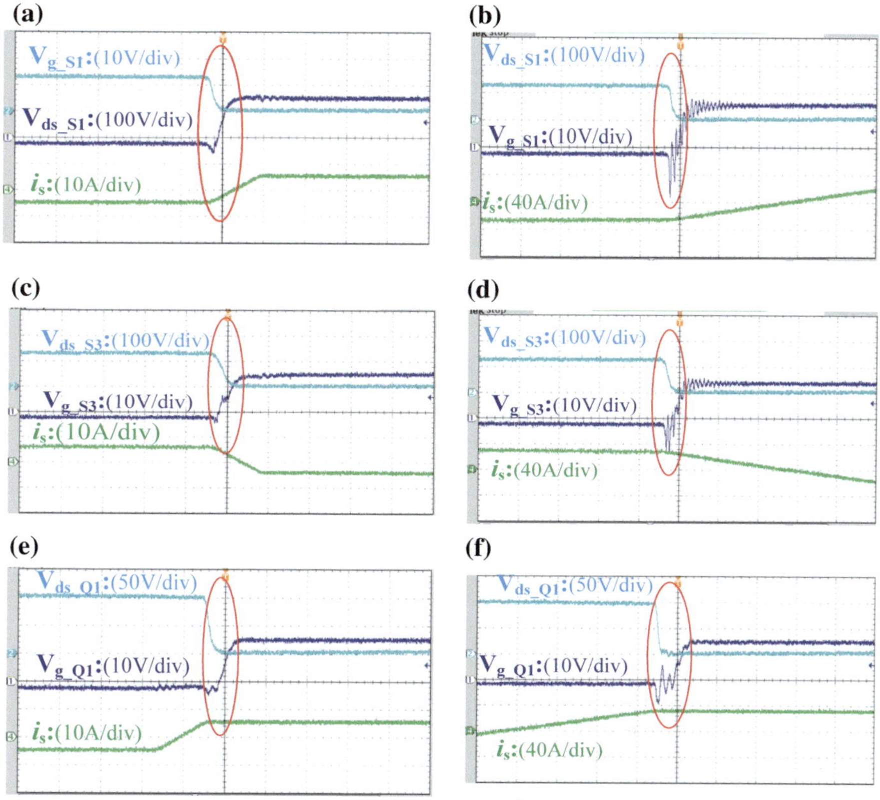

Fig. 4.19 Switches ZVS waveforms under $V_1 = 120$ V for different load. **a** 200 W, S_1. **b** 1000 W, S_1. **c** 200 W, S_3. **d** 1000 W, S_3. **e** 200 W, S_5. **f** 1000 W, S_5

also within 6 ms, which indicates a fast response. The voltage overshooting is also acceptable for many industrial applications.

The efficiency curves of the converter with proposed control under different V_1 voltages are plotted in Fig. 4.22. The efficiency has been compared to the CHDAB converter. When $V_1 = 120$ V, the one with EPS control has similar efficiency with the dual-transformer-based converter. However, when the voltage increases to 180 and 240 V, the efficiency of the converter with EPS control is lower than the dual-transformer-based converter, especially at light load condition. Meanwhile, for the dual-transformer-based converter with proposed control, even though the converter has similar RMS current for each device, the efficiency with $V_1 = 180$ V is higher, because it has lower transformer loss and turn-off loss.

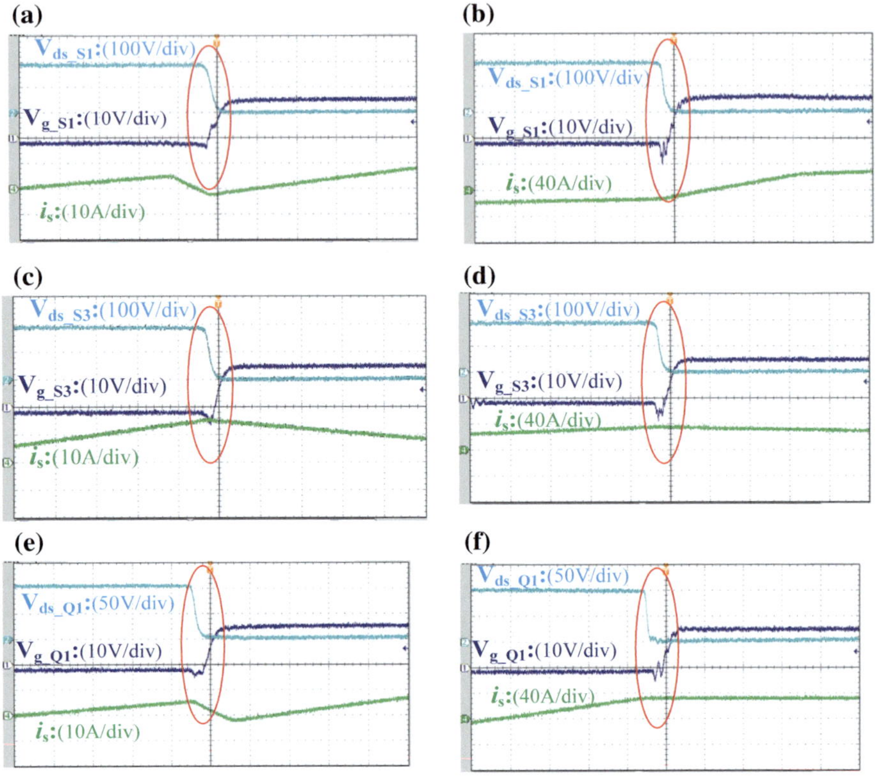

Fig. 4.20 Switches ZVS waveforms under V_1 = 180 V for different load. **a** 200 W, S_1. **b** 1000 W, S_1. **c** 200 W, S_3. **d** 1000 W, S_3. **e** 200 W, S_5. **f** 1000 W, S_5

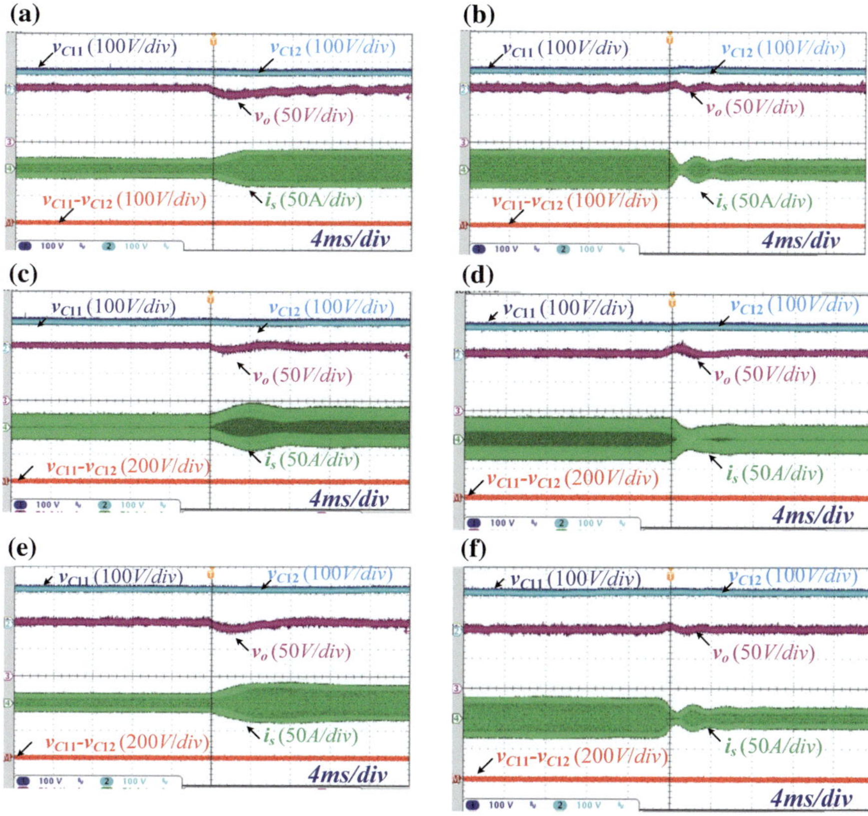

Fig. 4.21 Experimental waveforms under load step changing for different input voltage. V_1 = 120 V: **a** Half to full load. **b** Full to half load; V_1 = 180 V. **c** Half to full load. **d** Full to half load; V_1 = 180 V. **e** Half to full load. **f** Full to half load

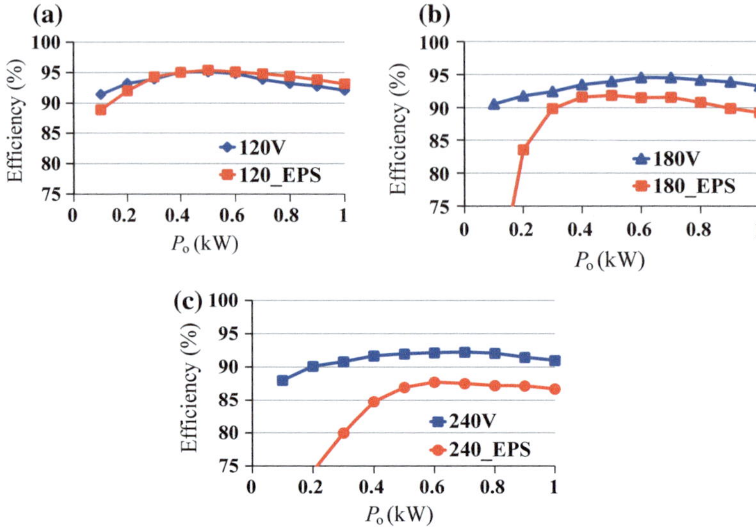

Fig. 4.22 Efficiency comparisons under different voltages

4.7 Conclusion

In this chapter, a control law has been proposed to achieve wide ZVS range for dual-transformer-based DAB converters in wide voltage gain range applications. The characteristics of the dual-transformer-based DAB converters with phase shift control have been analyzed. Based on that, the duty ratio of one bridge can be fixed as 0.5 without losing of ZVS capability. Therefore, half-bridge converter is used at one side of the transformer to reduce the total switching numbers and eliminate one control variable. With the proposed method, four switches of the converter can achieve full range of ZVS. The other two switches can achieve full-range ZVS under positive power flow, while slightly reduced ZVS region under reverse power flow. The power transfer characteristics, ZVS performance, and turns ratios design have been introduced in the chapter. A 1-kW prototype has been built up to verify the effectiveness of the proposed control and the dual-transformer-based converter. The results show that the studied converter with proposed control is an attractive candidate for the applications requiring wide conversion gain range and bidirectional power flow.

References

1. Ayyanar R, Mohan N (2001) Novel soft-switching DC-DC converter with full ZVS-range and reduced filter requirement. i. regulated-output applications. IEEE Trans Power Electron 16 (2):184–192
2. Ning G, Chen W, Shu L, Qu X (2017) A hybrid ZVZCS dual transformer-based full-bridge converter operating in DCM for mvdc grids. IEEE Trans Power Electron 32(7):5162–5170
3. Wu H, Chen L, Xing Y (2015) Secondary-side phase-shift-controlled dual-transformer-based asymmetrical dual-bridge converter with wide voltage gain. IEEE Trans Power Electron 30 (10):5381–5392
4. Shi K, Zhang D, Zhou Z, Zhang M, Gu Y (2016) A novel phaseshift dual full-bridge converter with full soft-switching range and wide conversion range. IEEE Trans Power Electron 31 (11):7747–7760
5. Jain AK, Ayyanar R (2011) PWM control of dual active bridge: Comprehensive analysis and experimental verification. IEEE Trans Power Electron 26(4):1215–1227

Chapter 5
Blocking-Cap-Based DAB Converters

Abstract Blocking-cap-based dual active bridge (DAB) converter with wide voltage conversion gain is studied in this chapter. The DC blocking capacitor, which is used for avoiding the DC flux bias in the transformer of conventional DAB converter, is utilized to achieve better performance under wide voltage gain range. By changing PWM generating method under different voltage input, the converter can work in half-bridge and full-bridge mode correspondingly with presented hybrid control. To simplify the PWM generating method, the duty cycles of primary and secondary bridges are fixed as 50%. The power transfer characteristics, ZVS region, and RMS current are analyzed and are compared with the DAB converter using single-phase shift (SPS) control. The theoretical analysis and conclusion of the converter with the hybrid control are verified by experimental results from a 1 kW prototype.

Keywords Blocking-cap-based dual active bridge · Wide ZVS range
Wide voltage gain · Single-phase shift control

5.1 Topology of the Converter

The topology of blocking-cap-based DAB converter is shown in Fig. 5.1. Compared with the conventional DAB converter, a DC blocking capacitor C_b is added in series with the leakage inductance. This capacitor is not used as a resonated capacitor because the designed value is relatively larger in this case. Actually, in practice, a DC blocking cap is also needed for the conventional DAB converter to avoid the magnetic saturation of the transformer. As a result, in view of the components used in the topology, the blocking-cap-based DAB converter does not need any additional components compared with the conventional one.

5.2 Typical Waveforms of the Converter

The blocking-cap-based DAB converter is aimed at using the capacitor to simplify the control and gain better performance at the same time. Therefore, the duty cycle for the bridge in V_2 side is fixed as 0.5, while the PWM logic for the V_1 side is

D. Sha and G. Xu, *High-Frequency Isolated Bidirectional Dual Active Bridge DC–DC Converters with Wide Voltage Gain*, CPSS Power Electronics Series,
https://doi.org/10.1007/978-981-13-0259-6_5

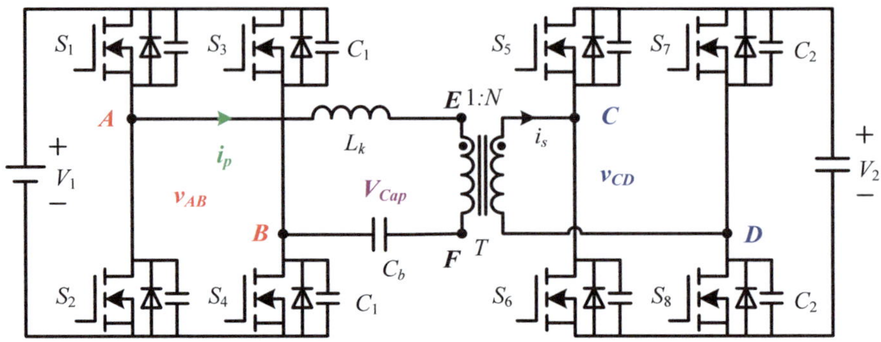

Fig. 5.1 Topology of blocking-cap-based DAB converter

Fig. 5.2 Typical waveforms under full-bridge operation mode ($\phi > 0$)

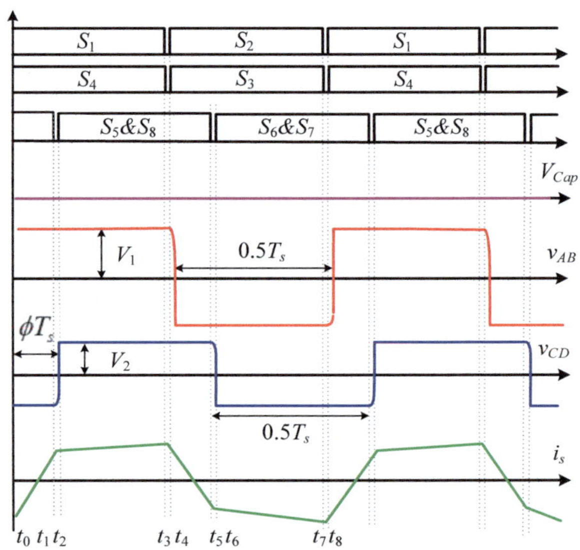

changing in response to different voltages. The converter can shift its working mode under wide voltage range applications. When the input voltage is relatively low, the converter works in full-bridge operation mode whose typical waveforms are shown in Fig. 5.2. When the input voltage is relatively high, the converter will work in half-bridge operating mode whose typical waveforms are shown in Fig. 5.3.

Figure 5.2 shows the waveforms under full-bridge mode. All the PWM switches are turned on with 50% fixed duty cycle. The two high-frequency voltages applied to the two ports of the transformer leakage inductor are also square waveforms with fixed 50% duty cycle. The voltage across the DC blocking capacitor is away zero under full-bridge operation mode. Actually, under this operation mode, the waveforms are the same as conventional SPS control [1–3].

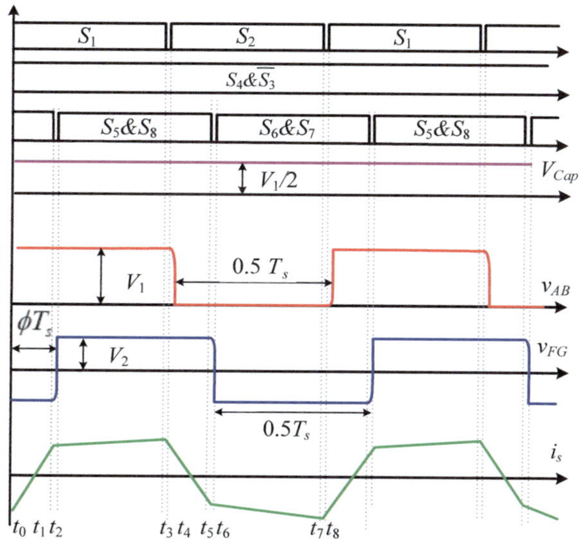

Fig. 5.3 Typical waveforms under half-bridge operation mode ($\phi > 0$)

Figure 5.3 shows the waveforms under half-bridge mode. This mode should be applied when the voltage is relatively high. Under this operating mode, switch S_4 is always on, and switch S_3 is always off. The full bridge in V_1 side generates the high-frequency voltage v_{AB} with a DC bias. Consequently, the blocking cap voltage will exist to cancel that DC bias in order to make the voltage secondary applied to the transformer to be zero. Because of the DC blocking capacitor, the real voltage amplitude that is applied to the transformer will be reduced.

5.3 Working Stages of the Converter

5.3.1 Full-Bridge Operation Mode

The detail modes of full-bridge operation in a switching period from t_0 to t_8 are shown in Fig. 5.4.

Stage 1 (t_0, t_1) (Fig. 5.4a): Prior to t_0, S_6 and S_7 are on, and S_1, S_2, S_3, S_4, S_5, and S_8 are off. At t_0, S_1 and S_4 are turned on. During this stage, v_{AB} is equal to V_1, and v_{CD} is equal to $-V_2$. The voltage of DC blocking capacitor is always zero. The transformer current can be expressed as

$$\begin{cases} i_p(t - t_0) = i_p(t_0) + (v_{AB} - v_{CD}/N)(t - t_0)/L_k \\ v_{AB} = V_1,\ v_{CD} = -V_2 \end{cases} \tag{5.1}$$

where 1:N is the transformer turns ratio from V_1 side to V_2 side.

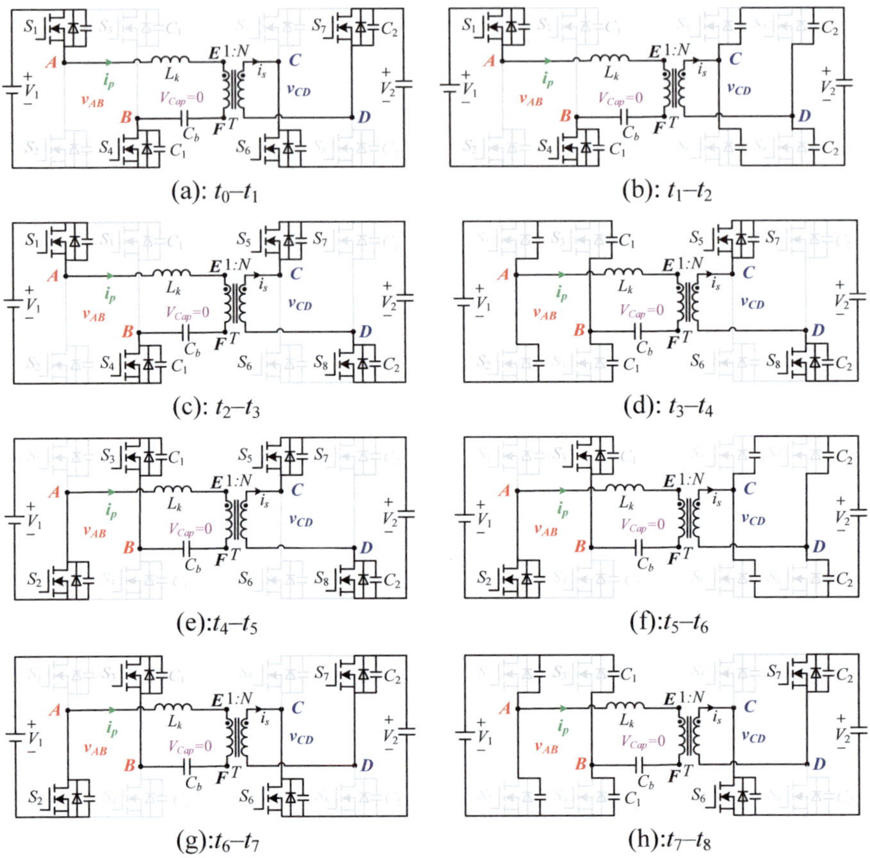

Fig. 5.4 Working stages of full-bridge operation. **a** Stage 1. **b** Stage 2. **c** Stage 3. **d** Stage 4. **e** Stage 5. **f** Stage 6. **g** Stage 7. **h** Stage 8

Stage 2 (t_1, t_2) (Fig. 5.4b): At time t_1, S_6 and S_7 are turned off. The transformer current starts to discharge the junction capacitor of S_5 and S_8 and to charge the junction capacitor of S_6 and S_7. When the drain–source voltage of S_6 and S_7 are charged to V_2, the drain–source voltage of S_5 and S_8 are zero, and the body diode will be conducted to create ZVS turn-on condition for S_5 and S_5 in the next stage. Because the leakage inductance is relatively large for DAB converters, the current required to charge/discharge these drain–source capacitors is small and can be neglected to simplify ZVS analysis. Hence, ZVS condition for S_5 and S_8 can be obtained as

$$i_p(t_1) > 0 \tag{5.2}$$

Stage 3 (t_2, t_3) (Fig. 5.4c): At time t_2, S_5 and S_8 are turned on with ZVS if (5.2) is satisfied. In this stage, v_{AB} is equal to V_1, while v_{CD} is equal to V_2. The current of leakage inductor can be expressed as

$$\begin{cases} i_p(t - t_2) = i_p(t_2) + (v_{AB} - v_{CD}/N)(t - t_2)/L_k \\ v_{AB} = V_1, \; v_{CD} = V_2 \end{cases} \quad (5.3)$$

Stage 4 (t_3, t_4) (Fig. 5.4d): At time t_1, S_1 and S_4 are turned off. The transformer current starts to discharge the junction capacitor of S_2 and S_3 and to charge the junction capacitor of S_1 and S_4. When the drain–source voltage of S_2 and S_3 reach zero, the body diode will be conducted to create ZVS turn-on condition for S_2 and S_3 in the next stage. As a result, ZVS condition for S_2 and S_3 can be obtained as

$$i_p(t_3) > 0 \quad (5.4)$$

The four stages for the other half switching period are shown in Fig. 5.4e, f. Because the PWM signals are symmetrical during two half switching cycles, the mode of operation is similar during the two half switching periods. Consequently, only the current paths shown for the four stages are shown in Fig. 5.4, and they are not discussed in detail.

5.3.2 *Half-Bridge Operation Mode*

When the converter is working in half-bridge mode, the stages in a switching period from t_0 to t_8 are shown in Fig. 5.5. t_0, t_1 … and t_8 are the switching time instant denoted in Fig. 5.3.

Stage 1 (t_0, t_1) (Fig. 5.5a): Prior to t_0, S_6 and S_7 are on, and S_1, S_2, S_3, S_4, S_5, and S_8 are off. At t_0, S_1 and S_4 are turned on. During this stage, v_{AB} is equal to V_1, and v_{CD} is equal to $-V_2$. All these are the same with the Stage 1 under full-bridge operation. However, the voltage across the DC blocking capacitor is $V_1/2$ instead of zero. The transformer current can be expressed as

$$\begin{cases} i_p(t - t_0) = i_p(t_0) + (v_{AB} - V_{\text{cap}} - v_{CD}/N)(t - t_0)/L_k \\ v_{AB} = V_1, \; v_{CD} = -V_2, \; V_{\text{cap}} = V_1/2 \end{cases} \quad (5.5)$$

Stage 2 (t_1, t_2) (Fig. 5.5b): At time t_1, S_6 and S_7 are turned off. The transformer current starts to discharge the junction capacitor of S_5 and S_8 and to charge the junction capacitor of S_6 and S_7. When the drain–source voltage of S_6 and S_7 are charged to V_2, the drain–source voltage of S_5 and S_8 are zero, and the body diode will be conducted to create ZVS turn-on condition for S_5 and S_5 in the next stage. In this stage, ZVS condition for S_5 and S_8 can be obtained as

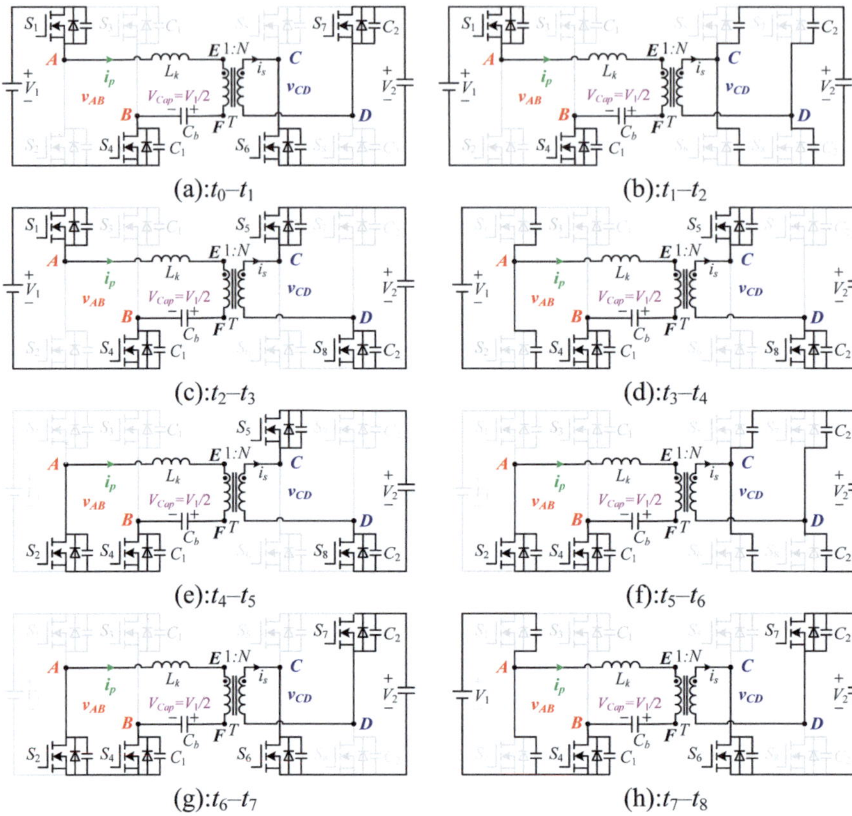

Fig. 5.5 Half-bridge working mode. **a** Stage 1. **b** Stage 2. **c** Stage 3. **d** Stage 4. **e** Stage 5. **f** Stage 6. **g** Stage 7. **h** Stage 8

$$i_p(t_1) > 0 \tag{5.6}$$

Stage 3 (t_2, t_3) (Fig. 5.5c): At time t_2, S_5 and S_8 are turned on with ZVS if (5.6) is satisfied. In this stage, v_{AB} is equal to V_1, while v_{CD} is equal to V_2. The current of leakage inductor can be expressed as

$$\begin{cases} i_p(t - t_2) = i_p(t_2) + (v_{AB} - V_{\text{cap}} - v_{CD}/N)(t - t_2)/L_k \\ v_{AB} = V_1,\ v_{CD} = V_2,\ V_{\text{cap}} = V_1/2 \end{cases} \tag{5.7}$$

Stage 4 (t_3, t_4) (Fig. 5.5d): At time t_3, S_1 is turned off. The transformer current starts to discharge the junction capacitor of S_2 and to charge the junction capacitor of S_1. When the drain–source voltage of S_2 reaches zero, the body diode will be conducted to create ZVS turn-on condition for S_2 and S_3 in the next stage. As a result, ZVS condition for S_2 and S_3 can be obtained as

$$i_p(t_3) > 0 \tag{5.8}$$

Stage 5 (t_4, t_5) (Fig. 5.5e): At time t_4, S_2 is turned on with ZVS if (5.8) is satisfied. During this stage, v_{AB} is equal to 0, and v_{CD} is equal to V_2. The relationship between the leakage inductor current and the voltages can be shown as

$$\begin{cases} i_p(t - t_4) = i_p(t_4) + (v_{AB} - V_{\text{cap}} - v_{CD}/N)(t - t_4)/L_k \\ v_{AB} = 0,\ v_{CD} = V_2,\ V_{\text{cap}} = V_1/2 \end{cases} \tag{5.9}$$

Stage 6 (t_5, t_6) (Fig. 5.5f): At time t_5, S_5 and S_8 are turned off. The junction capacitors of S_6 and S_7 are discharged, and the junction capacitors of S_5 and S_8 are charged. When the drain–source voltage of S_6 and S_7 reach zero, the body diodes will be conducted to create ZVS turn-on condition for S_6 and S_7 in the next stage. As a result, ZVS condition for S_6 and S_7 can be obtained as

$$i_p(t_5) < 0 \tag{5.10}$$

Stage 7 (t_6, t_7) (Fig. 5.5g): At time t_6, S_6 and S_7 are turned on with ZVS if (5.10) is satisfied. During this stage, v_{AB} is equal to 0, and v_{CD} is equal to $-V_2$. The relationship between the leakage inductor current and the voltages can be shown as

$$\begin{cases} i_p(t - t_6) = i_p(t_6) + (v_{AB} - V_{\text{cap}} - v_{CD}/N)(t - t_6)/L_k \\ v_{AB} = 0,\ v_{CD} = -V_2,\ V_{\text{cap}} = V_1/2 \end{cases} \tag{5.11}$$

Stage 8 (t_7, t_8) (Fig. 5.5h): At time t_7, S_2 is turned off. The transformer current starts to discharge the junction capacitor of S_1 and to charge the junction capacitor of S_2. When the drain–source voltage of S_1 reaches zero, the body diode will be conducted to create ZVS turn-on condition for S_1 in the next stage. As a result, ZVS condition for S_1 can be obtained as

$$i_p(t_8) < 0 \tag{5.12}$$

5.4 ZVS Conditions of the Converter

Based on the working stages under different modes of operation, ZVS conditions for all the switches under two working modes can be summarized as shown in Table 5.1.

Because the current waveforms are symmetric during two half switching cycles, the following relationship can be obtained

Table 5.1 ZVS conditions for the two working modes

Working mode	S_1 and S_2 (ZVS)	S_3 and S_4 (ZVS)	S_5 and S_6 (ZVS)	S_7 and S_8 (ZVS)
Full bridge	$i_p(t_0) < 0$	$i_p(t_4) > 0$	$i_p(t_2) > 0$	$i_p(t_6) < 0$
Half bridge	$i_p(t_0) < 0$	–	$i_p(t_2) > 0$	$i_p(t_6) > 0$

$$\begin{cases} i_p(t_0) = -i_p(t_4) \\ i_p(t_2) = -i_p(t_6) \end{cases} \tag{5.13}$$

As a result, the conditions in Table 5.1 can be simplified as given in Table 5.2. To analyze the ZVS region, the expressions of $i_p(t_0)$ and $i_p(t_2)$ need to be derived. The expressions under full-bridge operation mode can be calculated based on the equation in the working stage section.

The current values at the instants when switches are turned on are shown as follows

$$\begin{cases} i_p(t_2) = -i_p(t_6) = i_p(t_0) + (V_1 + V_2/N)(t_2 - t_0)/L_k \\ i_p(t_4) = -i_p(t_0) = i_p(t_2) + (V_1 - V_2/N)(t_4 - t_2)/L_k \end{cases} \tag{5.14}$$

where t_0, t_2, t_4, and t_6 are the switches' turn-on instants denoted in Fig. 5.2.

In addition, for the transformer current, the DC current bias is zero. Therefore, it can be obtained

$$1/T_s \int_{t_0}^{t_7} i_p(t)\mathrm{d}t = 0 \tag{5.15}$$

Combining (5.14) and (5.15), the expressions of $i_p(t_0)$ and $i_p(t_2)$ under full-bridge working mode when $\phi > 0$ can be shown as

$$\begin{cases} i_p(t_0) = \frac{-T_s(NV_1 - V_2 + 4V_2\Phi)}{4L_kN} \\ i_p(t_2) = \frac{T_s(V_2 - NV_1 + 4NV_1\Phi)}{4L_kN} \end{cases} \tag{5.16}$$

Based on the same principle, the expressions of $i_p(t_0)$ and $i_p(t_2)$ under half-bridge working mode when $\phi > 0$ can be shown as

$$\begin{cases} i_p(t_0) = \frac{-T_s(NV_1 - 2V_2 + 8V_2\Phi)}{8L_kN} \\ i_p(t_2) = \frac{T_s(2V_2 - NV_1 + 4NV_1\Phi)}{8L_kN} \end{cases} \tag{5.17}$$

The expressions for the other cases when $\phi < 0$ can also be obtained. Then, combining the ZVS conditions shown in Table 5.2, the expressions for ZVS achievement under different cases can be summarized in Table 5.3.

Table 5.2 ZVS conditions for the two working modes

Working mode	Switches in V_1 side	Switches in V_2 side
Full bridge	$i_p(t_0) < 0$	$i_p(t_2) > 0$
Half bridge	$i_p(t_0) < 0$	$i_p(t_2) > 0$

Table 5.3 Expressions for ZVS achievement

	Working mode	Switches in V_1 side	Switches in V_2 side
$\Phi > 0$	Full bridge	$NV_1 - V_2 + 4V_2\Phi > 0$	$V_2 - NV_1 + 4NV_1\Phi > 0$
	Half bridge	$NV_1 - 2V_2 + 8V_2\Phi > 0$	$2V_2 - NV_1 + 4NV_1\Phi > 0$
$\Phi < 0$	Full bridge	$NV_1 - V_2 - 4V_2\Phi > 0$	$NV_1 - V_2 - 4V_2\Phi > 0$
	Half bridge	$NV_1 - 2V_2 - 8V_2\Phi > 0$	$2V_2 - NV_1 - 4NV_1\Phi > 0$

In addition, the transferred power can also be calculated according to the following expressions.

$$P_o(D_1, \Phi) = \frac{1}{T_s}\int_{t_0}^{t_7} v_{AB}(t)i_p(t)\mathrm{d}t \tag{5.18}$$

Substitution of the expressions for $v_{AB}(t)$ and $i_p(t)$ during each time interval can lead to the power transfer expression under different working mode, which is shown as

$$\begin{cases} P_o = \frac{T_s V_1 V_2 \Phi(1-2\Phi)}{NL_k}, & \text{Full Bridge} \ \& \ \Phi > 0 \\ P_o = \frac{T_s V_1 V_2 \Phi(1-2\Phi)}{2NL_k}, & \text{Half Bridge} \ \& \ \Phi > 0 \\ P_o = \frac{T_s V_1 V_2 \Phi(1+2\Phi)}{NL_k}, & \text{Full Bridge} \ \& \ \Phi < 0 \\ P_o = \frac{T_s V_1 V_2 \Phi(1+2\Phi)}{2NL_k}, & \text{Half Bridge} \ \& \ \Phi < 0 \end{cases} \tag{5.19}$$

5.5 Power Transfer Characteristic and ZVS Region Comparison

5.5.1 Power Transfer Characteristic

In this section, the power transfer characteristics under different V_1 voltage are analyzed. In Fig. 5.6, the characteristics for conventional SPS control are plotted. The conventional SPS control is to operate the converter working in full-bridge mode for all the voltages. In the plots, the transferred power is normalized by $P_{\text{base}} = T_s V_2^2/(4N^2 L_k)$, and the V_1 side voltage is changing from its minimum

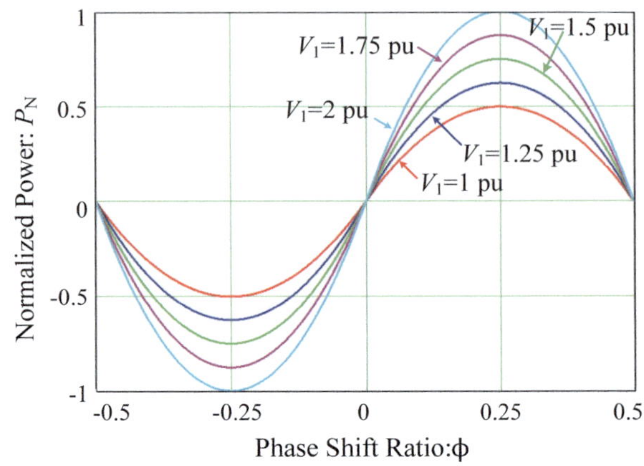

Fig. 5.6 Power transfer characteristics under different V_1 voltage for conventional SPS control

normalized value 1 pu to its maximum normalized value 2 pu. The turns ratio N is designed to make the unity voltage gain occur at the minimal point which is 1.0 pu.

As shown in Fig. 5.6, for a fixed V_1 voltage, the power transfer curve is similar to a sinusoidal wave. In order to make the curve to be monotonously increasing/decreasing, the phase shift ratio ϕ is limited within the range [−1/4, 1/4]. Also, because the duty ratios for the two active bridges are fixed, the power transfer is totally symmetric for the cases between $\phi > 0$ and $\phi < 0$. In addition, the output power is zero when $\phi = 0$. When V_1 is increasing from 1 to 2 pu, the amplitude of the power curve will increase linearly. The global maximum power transfer occurs when $V_1 = 1$ pu and $\phi = 1/4$, and the normalized power value is $P_{N\_max} = 0.5$.

Comparatively, the power transfer characteristics under different V_1 voltage combining full-bridge and half-bridge operation modes (hybrid control) are analyzed in Fig. 5.7. In these plots, the transferred power is normalized by the same value: $P_{\text{base}} = T_s V_2^2/(4N^2 L_k)$, and the V_1 side voltage is also changing from its minimum normalized value 1 pu to its maximum normalized value 2 pu. The unity voltage gain operating point for full-bridge mode is set as 1 pu, and the unity voltage gain operating point for half-bridge mode is set as 2 pu. Consequently, the turns ratio N should be V_2/V_{1_min}. In addition, the shifting point between full-bridge and half-bridge mode is set as $V_1 = 1.5$ pu.

As shown in Fig. 5.7, for a given V_1, the shape of the power curve is the same under two different operation modes, while the amplitude is smaller compared with the full-bridge mode. As V_1 exceeds 1.75 pu, the operation mode will change to half-bridge mode. Consequently, the curve for $V_1 \geq 1.75$ pu will be reduced. Meanwhile, to make a fair comparison with SPS, the leakage inductance under the hybrid control is chosen to be smaller than the one for SPS control to make the normalized power value also $P_{N\_max} = 0.5$ for the global maximum power transfer point. This point occurs when $V_1 = 1.5$ pu and $\phi = 1/4$ as shown in Fig. 5.7.

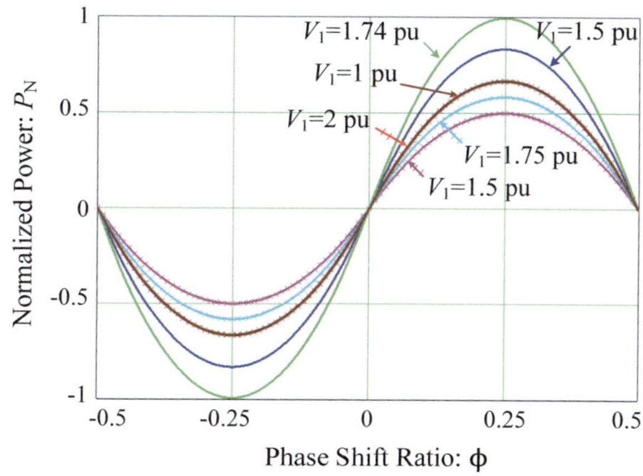

Fig. 5.7 Power transfer characteristics under different V_1 voltage combining full-bridge and half-bridge operation modes

5.5.2 ZVS Region

ZVS region under different power transfer and different V_1 voltage can be obtained based on the power expressions, and the limits for ZVS conditions are shown in Table 5.3. The ZVS region for SPS control is illustrated in Fig. 5.8. The upper and lower lines are the maximum power transfer limits under SPS control. The non-ZVS region occurs in the middle part of the figure. Full ZVS range can be obtained when $V_1 = 1$ pu as it is the designed voltage matching point. When the voltage goes far away from this point, the non-ZVS region will become larger and larger. As a result, the shape of the non-ZVS region is similar to a triangle. As seen, this non-ZVS region occurs in most of the operating region.

In comparison, ZVS region under different power transfer and different V_1 voltage using hybrid control is shown in Fig. 5.9. For the hybrid control, it has two voltage matching points. One occurs at $V_1 = 1$ pu and the other one occurs at

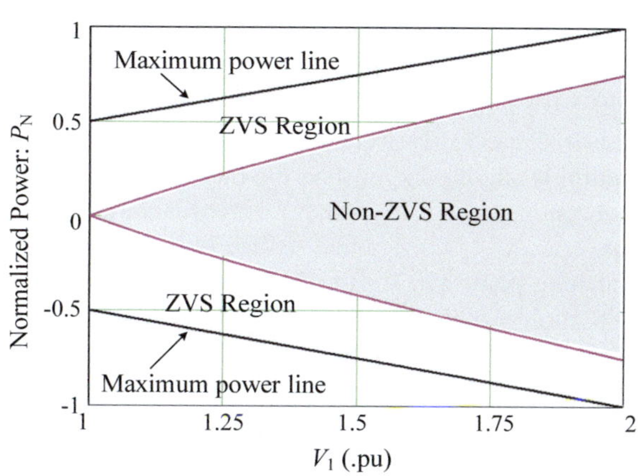

Fig. 5.8 ZVS region for SPS control

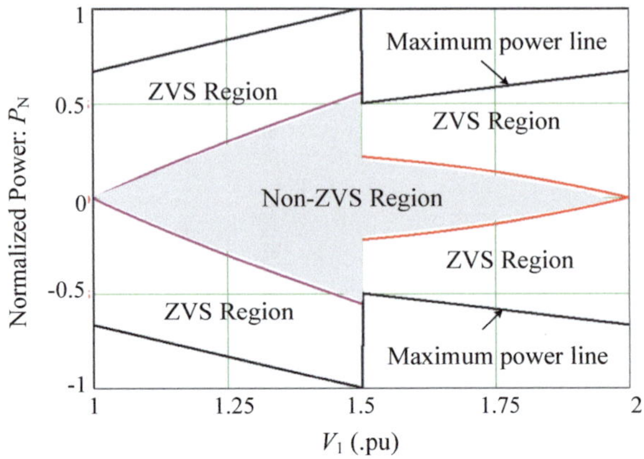

Fig. 5.9 ZVS region combining full-bridge mode and half-bridge mode (hybrid control)

$V_1 = 2$ pu. When V_1 is smaller than 1.5 pu, the converter works in full-bridge mode. Otherwise, the converter will work in half-bridge mode. Consequently, when V_1 is changing from 1 to 1.5 pu, the non-ZVS region will increase, and the non-ZVS region will decrease when V_1 changes from 1.5 to 2 pu. In view of the area of non-ZVS region, the area under SPS control is larger than the one using bridge control.

5.5.3 RMS Current Comparison

The RMS currents for a DAB converter using SPS control and hybrid control are compared in this section. The parameters for the comparison section are listed as follows: (a) minimal $V_{1\min} = 120$ V (1 pu); (b) $f_s = 50$ kHz; (c) $f_s = 50$ kHz; (d) $P_{\text{base}} = 3.6$ kW; (e) $P_{\text{base}} = 3.6$ kW; (f) $P_{\max} = 1.8$ kW; (g) Transformer leakage for SPS control: $L_k = 20$ μH; (h) Transformer leakage for hybrid control: $L_{kH} = 15$ μH.

The RMS current at transformer secondary side (V_2 side) for SPS control and hybrid control under different power transfer and different V_1 voltages is shown in Fig. 5.10. Four cases when output power equals to 1/4, 1/6, 1/8, and 1/12 pu are shown, respectively. When V_1 is lower than 1.5 pu, the RMS current under hybrid control is slightly larger than the one with SPS control. This is because the leakage inductance is relatively small for hybrid control. When V_1 is larger than 1.5 pu, the converter will work under half-bridge mode with hybrid control. The voltage matching point will shift to 2 pu. As V_1 increases from 1.5 pu, the converter under SPS control goes far away from its voltage matching point ($V_1 = 1$ pu), and the converter under hybrid control goes close to its voltage matching point ($V_1 = 2$ pu). Therefore, the current under SPS control will continually increase, while the current under hybrid control will decrease. Note that, under half-bridge mode, only two switches are conducted with current and also with high-frequency turn on/off.

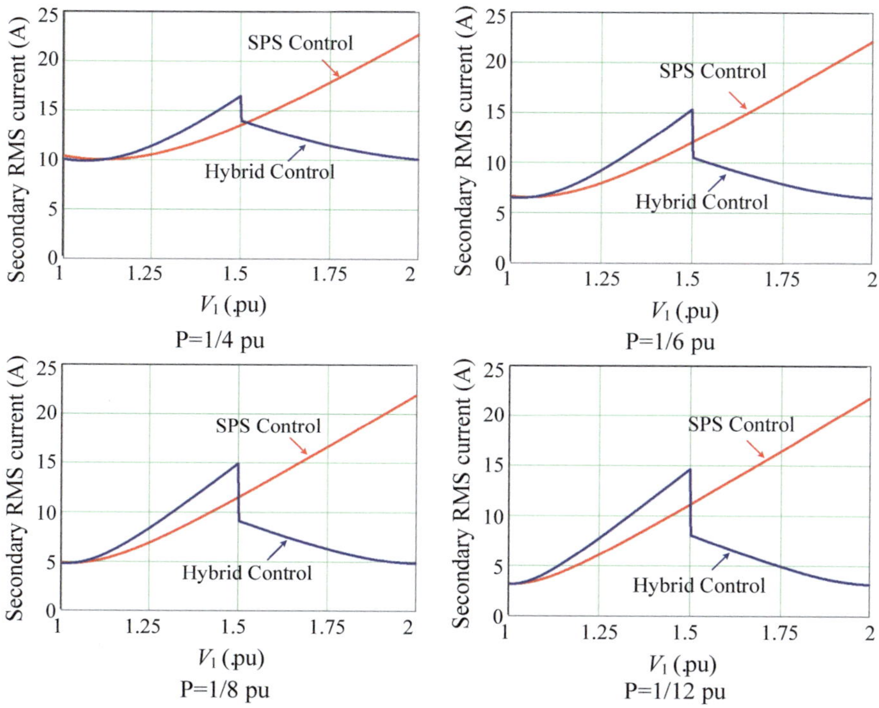

Fig. 5.10 RMS current comparison between SPS control and the control combining full-bridge and half-bridge mode (hybrid control)

Therefore, in view of the conduction loss in primary side, the one with hybrid control can have much smaller conduction loss.

5.6 Experimental Results

A laboratory prototype was built to verify the effectiveness of the dual-transformer-based converter and proposed control. The specifications are the same as these in the RMS comparison section. In addition, DC blocking capacitor is 60 μH. The rated output power is 1 kW, and V_1 side voltage is changing from 120 to 240 V.

5.6.1 Rated Load (1 kW)

In Fig. 5.11, the experimental results under rated load (1 kW) for SPS control and Hybrid control when V_1 = 120 and 150 V are illustrated. Because the converter

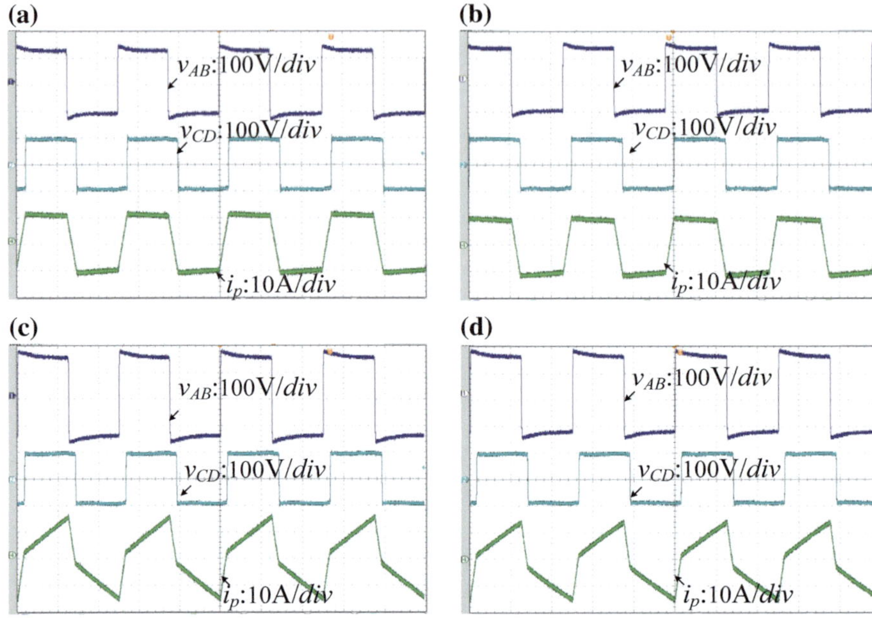

Fig. 5.11 Experimental waveforms (P_o = 1 kW) for SPS control and hybrid control when V_1 < 180 V. **a** 120 V, SPS control. **b** 120 V, hybrid control. **c** 150 V, SPS control. **d** 150 V, hybrid control

works in full-bridge mode when V_1 < 180 V under hybrid control, the working stage is the same as SPS control. As a result, the waveforms are very similar for these two cases. When V_1 = 120 V, the voltage matching point occurs for both SPS control and hybrid control. The transformer leakage current is relatively low, and the slope during the duration when v_{AB} and v_{CD} are both high is almost zero. Under this case, low RMS current can be obtained because of small reactive current. When V_1 increases, the peak current will increase for both of the two control methods.

The experimental results under rated load (1 kW) for SPS control and hybrid control when $V_1 \geq$ 180 V are illustrated in Fig. 5.12. When V_1 reaches 180 V, the transformer peak current i_L will exceed 30 A under SPS control as shown in Fig. 5.12a. In comparison, the transformer current is smaller under hybrid control as shown in Fig. 5.12b, because the converter works at half-bridge mode. As V_1 increases, the current stress will become larger and larger under SPS control, such as the waveforms shown in Fig. 5.12c, e, while the current stress for hybrid control will be reduced as V_1 increases. As shown in Fig. 5.12d, when V_1 increases to be 210 V, the voltage crossing the blocking capacitor is 105 V which makes the transformer RMS current to be smaller than the case for V_1 = 180 V. When V_1 = 240 V, the current slope becomes zero again during the duration when v_{AB} and v_{CD} are both high, which means another voltage matching point occurs under hybrid control. Under this case, minimal current stress can be obtained. Note that,

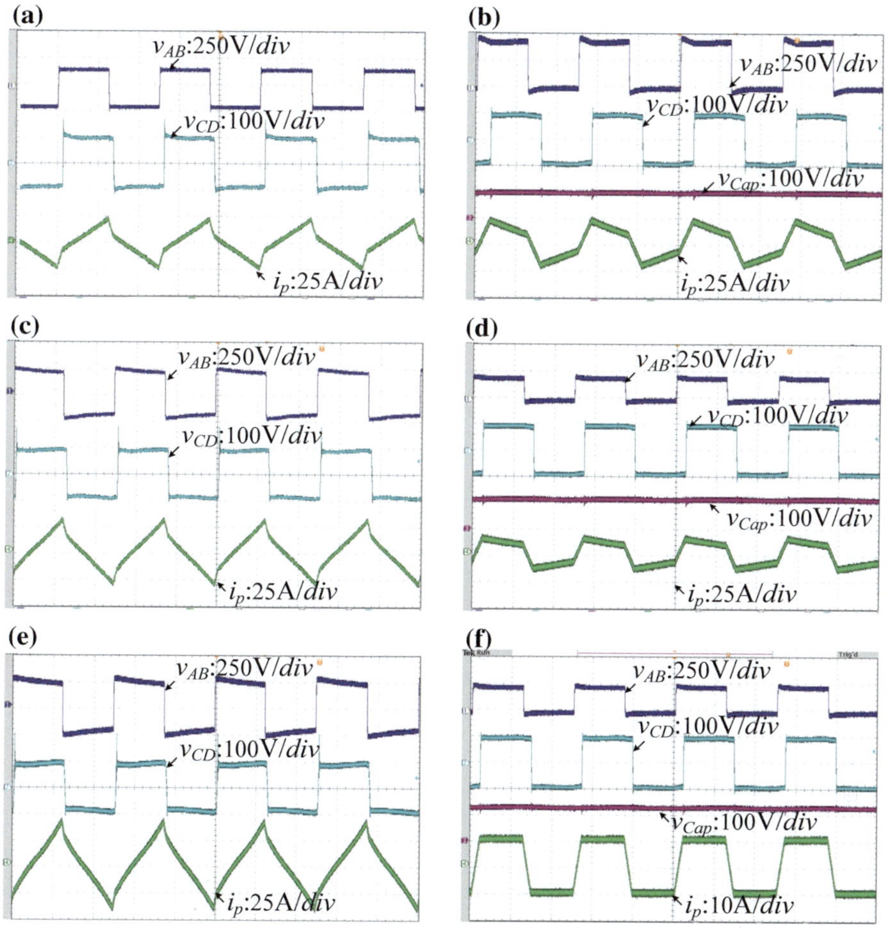

Fig. 5.12 Experimental waveforms (P_o = 1 kW) for SPS control and hybrid control when $V_1 > 180$ V. **a** 180 V, SPS control. **b** 180 V, hybrid control. **c** 210 V, SPS control. **d** 210 V, hybrid control. **e** 240 V, SPS control. **f** 240 V hybrid control

the scale for the current value in Fig. 5.12f is 10 A/div, while the scale for the current value in Fig. 5.12e is 25 A/div.

5.6.2 Light Load (270 W)

In Fig. 5.13, the experimental results are shown under light load P_o = 270 W for SPS control and hybrid control when $V_1 < 180$ V. Similar to the cases under rated load, the waveforms for SPS control and hybrid control are similar. It is because the converter works at full-bridge mode when $V_1 < 180$ V. Under this case, the voltage

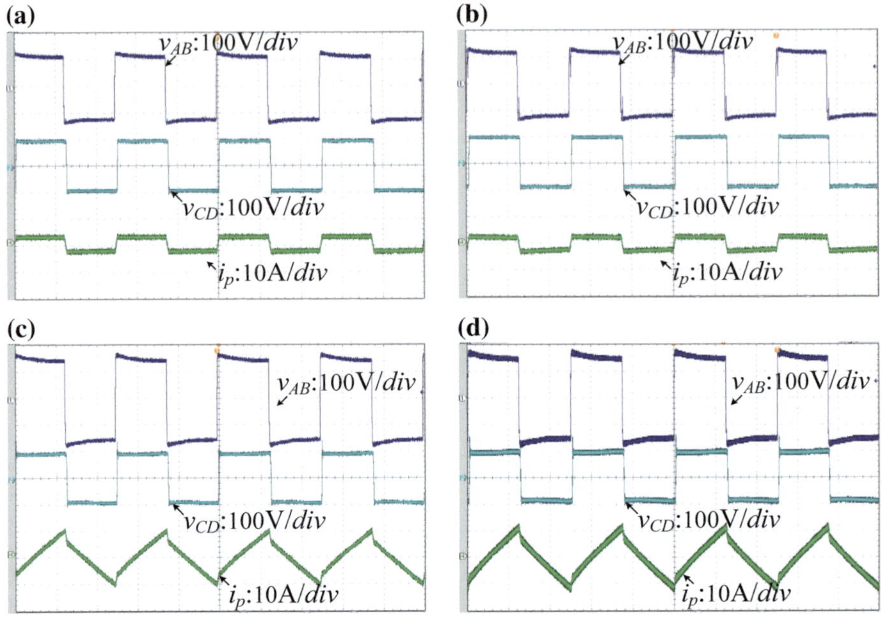

Fig. 5.13 Experimental waveforms (P_o = 270 W) for SPS control and hybrid control when $V_1 < 180$ V. **a** 120 V, SPS control. **b** 120 V, hybrid control. **c** 150 V, SPS control. **d** 150 V, hybrid control

induced across the DC blocking capacitor is zero, and thus the working stage is the same as SPS control. Under light-load conditions, ZVS for the switches in V_2 side is lost judging from the current polarities when the switches turn on as shown in Fig. 5.13c, d. This is because ZVS range under light load will be reduced, which matches with the ZVS region analysis plots shown in Figs. 5.8 and 5.9.

The experimental results under light load (270 W) for SPS control and hybrid control when $V_1 \geq 180$ V are illustrated in Fig. 5.14. As seen from the voltage and current waveforms, large reactive current exists for SPS control compared with the results of hybrid control. For SPS control, ZVS for the secondary side switches cannot be obtained when V_1 changes from 180 to 240 V judging from the transformer leakage current polarity at the instants when switches are turned on. In contrary, when using hybrid control, the DC blocking capacitor will block the DC bias which is half of the DC bus voltage. Judging from the current polarities under hybrid control, ZVS for all the switches can be obtained. This indicates that the ZVS range is enlarged with hybrid control, which matches the ZVS region analysis shown in Figs. 5.8 and 5.9.

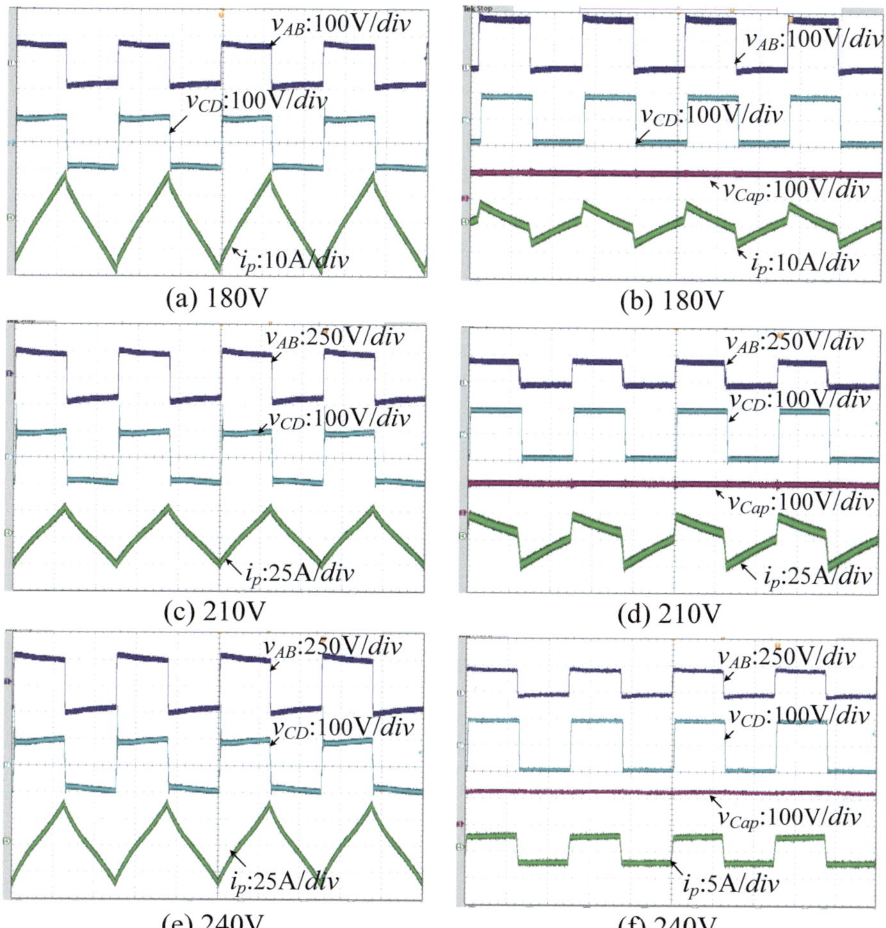

Fig. 5.14 Experimental waveforms (P_o = 270 W) for SPS control and hybrid control when V_1 > 180 V. **a** 180 V, SPS control. **b** 180 V, hybrid control. **c** 210 V, SPS control. **d** 210 V, hybrid control. **e** 240 V, SPS control. **f** 240 V hybrid control

5.7 Summary

The blocking-cap-based dual active bridge (DAB) converter is presented in this chapter. In practice, a capacitor is usually series connected with the transformer leakage inductance to avoid the DC flux bias of the high-frequency transformer. For the presented converter, the DC blocking capacitor is not only limited to avoiding flux bias, but also used to achieve the performance improvement under different input voltages. With changing of the PWM for the switches in one bridge, the converter can work in full-bridge mode when the voltage is relatively low and work in half-bridge mode when the voltage is high. The characteristics of DAB converter with presented hybrid control are analyzed and made comparison with the converter using conventional SPS control. It has been shown that the blocking-cap-based DAB converter can achieve low wide ZVS range and low RMS current for the switches. In addition, the converter has less total switching time when operating at half-bridge mode. The theoretical analysis and conclusion of the converter with the hybrid control are verified by experimental results from a 1 kW prototype.

References

1. Inoue S, Akagi H (2007) A bidirectional DC-DC converter for an energy storage system with galvanic isolation. IEEE Trans Power Electron 22(6):2299–2306
2. Bai H, Mi C, Gargies S (2008) The short-time-scale transient processes in high-voltage and high-power isolated bidirectional dc-dc converters. IEEE Trans Power Electron 23(6):2648–2656
3. Li X, Li YF (2014) An optimized phase-shift modulation for fast transient response in a dual-active-bridge converter. IEEE Trans Power Electron 29(6):2661–2665

Chapter 6
Three-Level Bidirectional DC–DC Converter with an Auxiliary Inductor in Adaptive Working Mode for Full-Operation Zero-Voltage Switching

Abstract This chapter proposes a three-level bidirectional DC–DC converter with an auxiliary inductor for full-operation zero-voltage switching (ZVS) in high-output voltage applications. The auxiliary inductor is connected across the middle node of the split flying capacitors and the center tap of the secondary winding in the transformer. In this topology, the outer and inner switches in the three-level stage can generate two independent 50% duty-cycle square waveforms, which is used to control the current in the auxiliary inductor to extend ZVS range from no loads to full loads condition. Considering the phase shift angle in three-level stage, the ZVS range of the converter is analyzed, and the modulation trajectory to maintain the full-operation ZVS range with low conduction loss is proposed. A flowchart implementation can guarantee the seamless transfer in different working modes. Then, the conduction loss in the proposed converter is compared with the previous three-level bidirectional solution, which illustrates that the conduction loss in the proposed converter only increases in light loads. Finally, the experimental results verify the theoretical analyses and ZVS performance across the whole power and voltage range, and the efficiency curves demonstrate the efficiency improvement.

Keywords Three-level · Dual active bridge · Auxiliary inductor
Zero-voltage switching

6.1 Three-Level Bidirectional DAB Converter Full-Operation Zero-Voltage Switching

The voltage level in the DC microgrid has reached 600–800 V or higher. Si-based MOSFETs in this voltage level have large turn-on resistance and output capacitance, which causes large conduction loss and switching loss. IGBTs can meet this voltage level, but it does not meet high-frequency requirement due to the tail-current effect. SiC-based MOSFETs can meet high switching frequency and high-voltage applications, but the cost is still very high. To meet high-DC voltage

D. Sha and G. Xu, *High-Frequency Isolated Bidirectional Dual Active Bridge DC–DC Converters with Wide Voltage Gain*, CPSS Power Electronics Series,
https://doi.org/10.1007/978-981-13-0259-6_6

applications with low cost, a three-level bidirectional DC–DC converter was proposed [1], as shown in Fig. 6.1. The three-level structure has been extended to current-fed bidirectional DC–DC converters [2]. By controlling the slew rate of the current in the transformer, the conduction loss is reduced. By replacing some switches with diodes, a semiactive rectifier is derived from the three-level structure for high-voltage applications [3]. Splitting the flying capacitor into two series-connected capacitors, the rectifier can achieve quadruple step-up voltage by using phase shift control. By using split flying capacitors in the three-level stage, the outer and inner switches can generate two 50% duty-cycle square waveforms [4] in unidirectional DC–DC converters, which can reduce the circulating current.

In this chapter, a three-level bidirectional DC–DC converter with an auxiliary inductor is presented for full-operation ZVS from no loads to full loads. The auxiliary inductor is connected across the middle node of the split flying capacitors in the three-level stage and the center tap of the secondary winding in the transformer. With this topology, the outer and inner switches can generate two independent 50% duty-cycle square waveforms, which are used to control the current in the auxiliary inductor.

Figure 6.2 shows the circuit of the three-level bidirectional DAB converter. V_{bat} is the battery voltage, and V_{o} is the output voltage. On the battery side, the power stage is a full bridge. To meet high output voltage, the output-side power stage is a three-level structure. There are two series-connected windings in the secondary side of the transformer. The turns ratio of the three windings N_1:N_2:N_3 is 2:n:n. The auxiliary inductor is connected across the middle node of the split flying capacitors and the center tap of the secondary winding. L_r is the series inductor, and L_a is the auxiliary inductor for full-operation ZVS. C_{d1} and C_{d2} are the split output filter capacitors. The output filter capacitance is large enough to regard v_{Cd1} and v_{Cd2} as

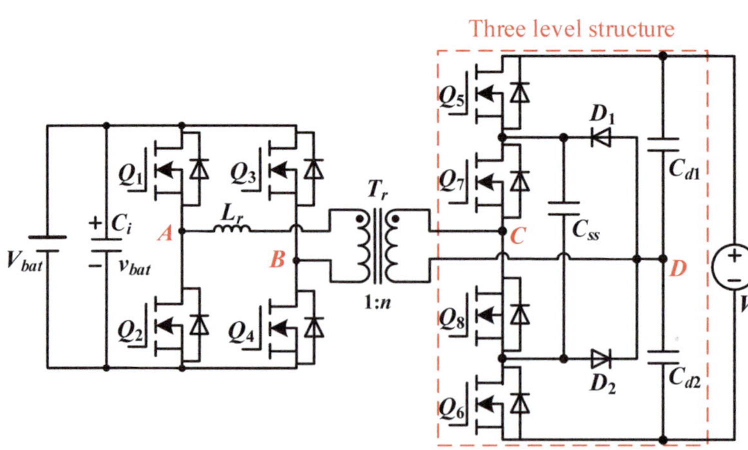

Fig. 6.1 Three-level bidirectional DC–DC converter

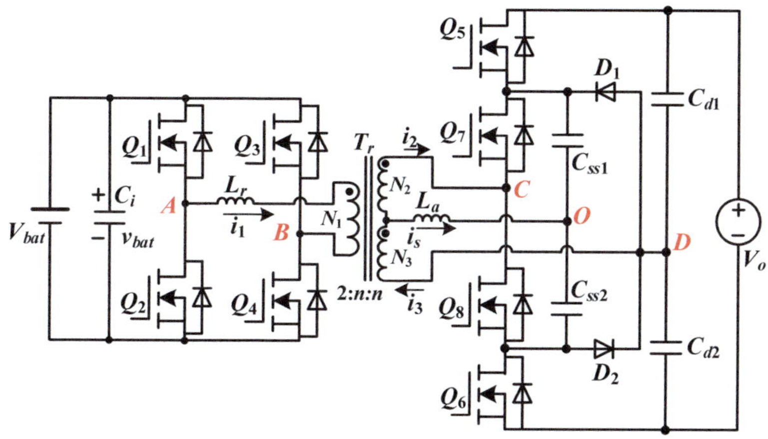

Fig. 6.2 Three-level bidirectional DAB converter with an auxiliary inductor

constant voltage sources. The power transmitted from the battery to the output is defined as the forward power flow, and the power from the output to the battery is defined as the reverse power flow.

Taking the forward power flow into consideration, the converter can work in four modes, which are shown in Fig. 6.3. The gate signals for Q_1–Q_8 are 50% duty cycle. v_{AB} is modulated by the phase shift angle between the two switching legs of the full bridge, and D_1 is the effective duty cycle. T_s is the switching period. φ_a is the phase shift between v_{AB} and v_{CO}, and φ_b is the phase shift angle between v_{AB} and v_{OD}. The working stages in two halves of switching period are symmetrical. One of the split output filter capacitors works in the half of the switching period, and the other one works in the other half of the switching period. The output voltage of the split output filter capacitors can be naturally balanced. Therefore, the voltage across the output filter capacitors is $V_o/2$. In the three-level stage, the phase shift angle between v_{CO} and v_{OD} is very small. Otherwise, there is large current in the auxiliary inductor. The converter cannot work normally. The full bridge in the battery side must work in PWM in this case. The effective voltage amplitude in the battery side of the full bridge is expressed as nV_{bat}. It must be larger than the voltage amplitude across v_{CD} in the three-level stage in EPPS modulation, which is expressed as $V_o/2$. Therefore, the converter must meet the conditions: $nV_{\text{bat}} \geq V_o/2$.

The boundary conditions of each mode are shown in Table 6.1. Because of the symmetry of the working mode, the same conclusions for reversed power flow can be drawn. i_{12} is assigned as the current component in winding N_2 interacted by winding N_1. i_{13} is assigned as the current component in winding N_3 interacted by winding N_1. i_{23} is assigned as the current component in winding N_3 interacted by winding N_2. Therefore, the winding current meets the following condition.

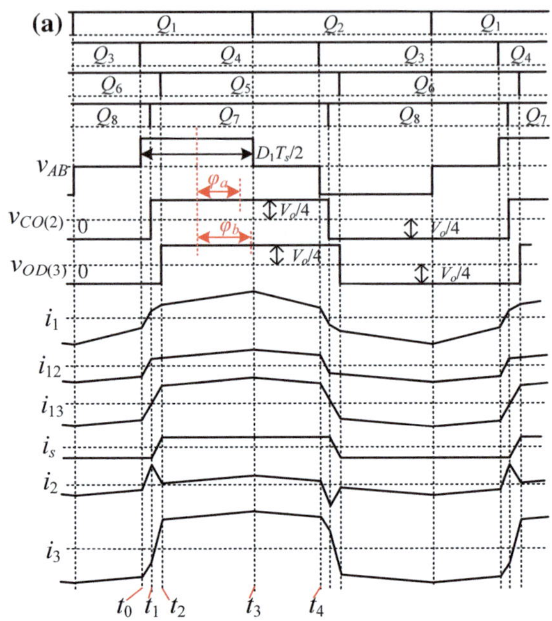

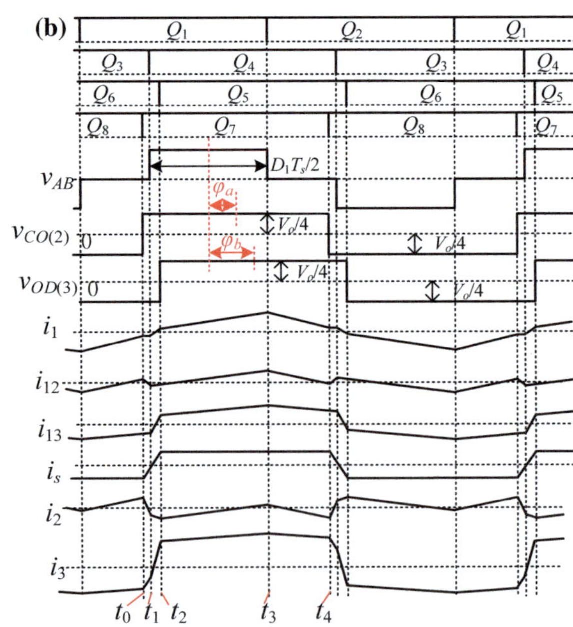

Fig. 6.3 Working mode in forward power flow: **a** Mode 1a; **b** Mode 2a; **c** Mode 3; **d** Mode 4

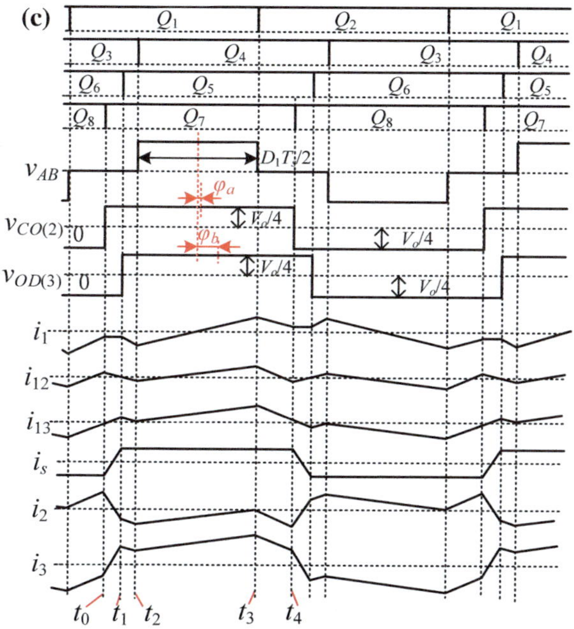

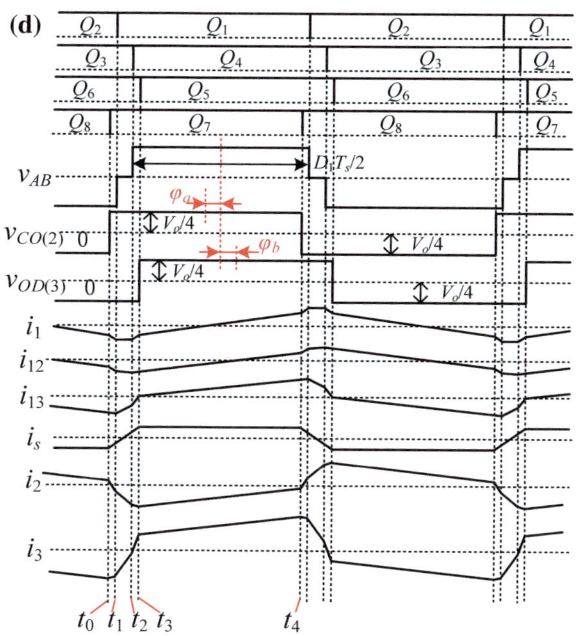

Fig. 6.3 (continued)

Table 6.1 Boundary conditions of each mode

Mode 1a	Mode 2a	Mode 3	Mode 4
$\frac{1}{2}-\frac{D_1}{2}\le\frac{\varphi_a}{\pi}\le\frac{\varphi_b}{\pi}\le\frac{1}{2}$	$-\left(\frac{1}{2}-\frac{D_1}{2}\right)\le\frac{\varphi_a}{\pi}\le\left(\frac{1}{2}-\frac{D_1}{2}\right)\le\frac{\varphi_b}{\pi}$	$-\left(\frac{1}{2}-\frac{D_1}{2}\right)T_s\le\frac{\varphi_a}{\pi}$ $\le\frac{\varphi_b}{\pi}\le\left(\frac{1}{2}-\frac{D_1}{2}\right)T_s$	$\frac{\varphi_a}{\pi}\le-\left(\frac{1}{2}-\frac{D_1}{2}\right)T_s\le 0$ $\le\left(\frac{1}{2}-\frac{D_1}{2}\right)T_s\le\frac{\varphi_b}{\pi}$
$t_0=0$ $t_1=\left(\frac{D_1}{4}+\frac{\varphi_a}{2\pi}-\frac{1}{4}\right)T_s$ $t_2=\left(\frac{D_1}{4}+\frac{\varphi_b}{2\pi}-\frac{1}{4}\right)T_s$ $t_3=\frac{D_1T_s}{2}$ $t_4=\frac{T_s}{2}$	$t_0=0$ $t_1=\left(\frac{1}{4}-\frac{D_1}{4}-\frac{\varphi_a}{2\pi}\right)T_s$ $t_2=\left(\frac{\varphi_b}{2\pi}-\frac{\varphi_a}{2\pi}\right)T_s$ $t_3=\left(\frac{1}{4}+\frac{D_1}{4}-\frac{\varphi_a}{2\pi}\right)T_s$ $t_4=\frac{T_s}{2}$	$t_0=0$ $t_1=\left(\frac{\varphi_b}{2\pi}-\frac{\varphi_a}{2\pi}\right)T_s$ $t_2=\left(\frac{1}{4}-\frac{\varphi_a}{2\pi}-\frac{D_1}{4}\right)T_s$ $t_3=\left(\frac{1}{4}-\frac{\varphi_a}{2\pi}+\frac{D_1}{4}\right)T_s$ $t_4=\frac{T_s}{2}$	$t_0=0$ $t_1=\left(\frac{D_1}{4}-\frac{\varphi_a}{2\pi}-\frac{1}{4}\right)T_s$ $t_2=\left(\frac{1}{4}-\frac{\varphi_a}{2\pi}-\frac{D_1}{4}\right)T_s$ $t_3=\left(\frac{\varphi_b}{2\pi}-\frac{\varphi_a}{2\pi}\right)T_s$ $t_4=\frac{T_s}{2}$

$$\begin{cases} i_s = 2i_{23} \\ i_2 = i_{12} - i_{23} \\ i_3 = i_{13} + i_{23} \end{cases} \tag{6.1}$$

Ignoring the dead time of the gate signals, there are four working stages for each working mode in half of the switching period. To illustrate, the working stages in Mode 1a are shown in Fig. 6.4.

Stage 1([t_0, t_1]) (Fig. 6.4a): Prior to t_0, Q_1, Q_3, Q_6, and Q_8 are on, and Q_2, Q_4, Q_5, and Q_7 are off. At t_0, Q_3 is switched off and i_1 is negative. The junction capacitor of Q_4 is discharged, and Q_4 is turned on with ZVS. i_1, i_2, and i_3 start to increase. Clamping diode D_1 is forward-biased. The energy stored in auxiliary inductor discharges capacitor C_{ss2} and charges capacitor C_{ss1} until i_s becomes positive. When i_2 is positive, the body diode of Q_7 is forward-biased. The currents in each winding in this stage are expressed as

$$\begin{cases} i_1(t) = i_1(t_0) + \frac{V_{\text{bat}} + V_o/(2n)}{L_r}(t - t_0) \\ i_{12}(t) = i_{12}(t_0) + \frac{nV_{\text{bat}}/2 + V_o/4}{n^2L_r/2}(t - t_0) \\ i_{13}(t) = i_{13}(t_0) + \frac{nV_{\text{bat}}/2 + V_o/4}{n^2L_r/2}(t - t_0) \\ i_{23}(t) = i_{23}(t_0) \end{cases} \tag{6.2}$$

Stage 2([t_1, t_2]) (Fig. 6.4b): At t_1, Q_8 is switched off and Q_7 is turned on with ZVS. When i_3 becomes positive, i_3 flows through the body diode of Q_5. When i_s is positive, clamping diode D_2 starts to become forward-biased in this stage. The energy stored in auxiliary discharges capacitor C_{ss1} and charges capacitor C_{ss2}. The current in each winding in this stage is expressed as

$$\begin{cases} i_1(t) = i_1(t_1) + \frac{V_{\text{bat}}}{L_r}(t - t_1) \\ i_{12}(t) = i_{12}(t_1) + \frac{nV_{\text{bat}}/2 - V_o/4}{n^2L_r/2}(t - t_1) \\ i_{13}(t) = i_{13}(t_1) + \frac{nV_{\text{bat}}/2 + V_o/4}{n^2L_r/2}(t - t_1) \\ i_{23}(t) = i_{23}(t_1) + \frac{2V_o/4}{4L_a}(t - t_1) \end{cases} \tag{6.3}$$

Stage 3([t_2, t_3]) (Fig. 6.4c): At t_2, Q_6 is switched off and Q_5 is turned on with ZVS. In this stage, the power is transmitted from the battery to the output. i_s still discharges C_{ss1} and charges capacitor C_{ss2}. The voltage $v_{Css1} + v_{Css2}$ is clamped to the half of the output voltage C_{d1} through D_2 and Q_5. The current in each winding in this stage is expressed as

$$\begin{cases} i_1(t) = i_1(t_2) + \frac{V_{\text{bat}} - V_o/(2n)}{L_r}(t - t_2) \\ i_{12}(t) = i_{12}(t_2) + \frac{nV_{\text{bat}}/2 - V_o/4}{n^2L_r/2}(t - t_2) \\ i_{13}(t) = i_{13}(t_2) + \frac{V_{\text{bat}}/2 - V_o/4}{n^2L_r/2}(t - t_2) \\ i_{23}(t) = i_{23}(t_2) \end{cases} \tag{6.4}$$

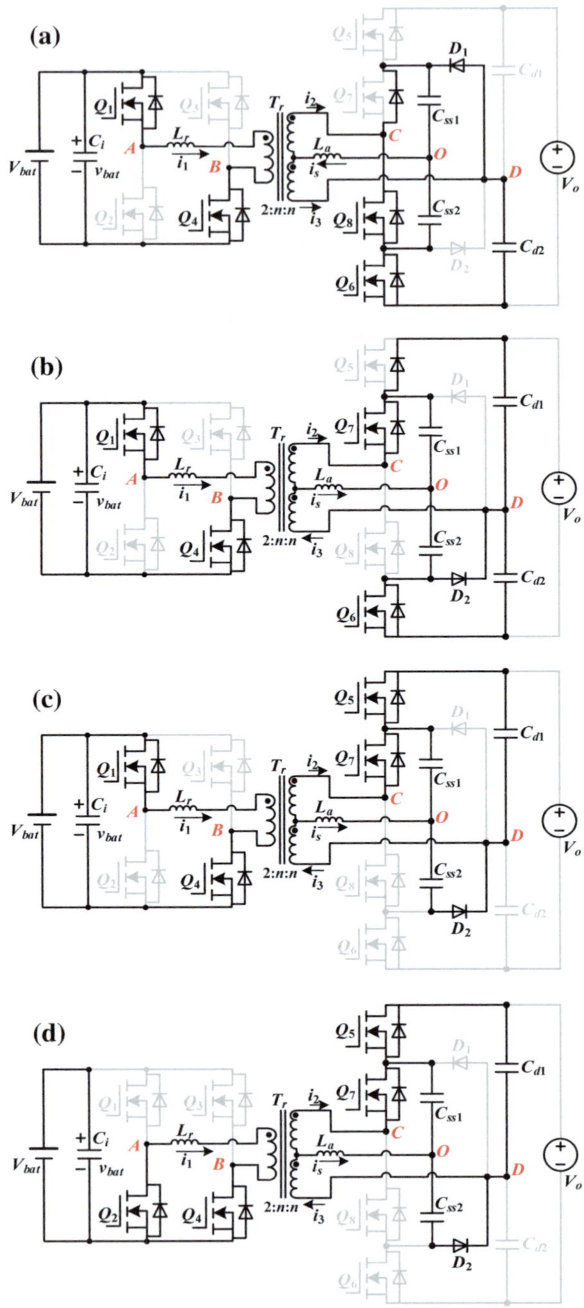

Fig. 6.4 Topological stages with half-switching period for Mode 1a: **a** $[t_0, t_1]$, **b** $[t_1, t_2]$, **c** $[t_2, t_3]$, **d** $[t_3, t_4]$

Stage 4($[t_3, t_4]$) (Fig. 6.4d): At t_3, Q_1 is switched off, and i_1 starts to charge and discharge the junction capacitor of Q_1 and Q_2. When i_1 flows through the body diode of Q_2, Q_2 is turned on with ZVS. The status of the switches in the three-level stage is the same as Stage 3. The current in each winding in this stage is expressed as

$$\begin{cases} i_1(t) = i_1(t_3) - \frac{V_o/(2n)}{L_r}(t - t_3) \\ i_{12}(t) = i_{12}(t_3) - \frac{V_o/4}{n^2 L_r/2}(t - t_3) \\ i_{13}(t) = i_{13}(t_3) - \frac{V_o/4}{n^2 L_r/2}(t - t_3) \\ i_{23}(t) = i_{23}(t_3) \end{cases} \quad (6.5)$$

Based on (6.1)–(6.5), current i_1, i_2, and i_3 at the time of t_0, t_1, t_2, and t_3 can be derived. The working stages in Mode 2a, Mode 3, and Mode 4 can be analyzed in the same manner. Currents i_1, i_2, and i_3 for different working modes at the time of t_0, t_1, t_2, and t_3 are also shown in appendix.

6.2 Key Feature and Modulation Scheme of the Converter

6.2.1 Voltage Balance of the Flying Capacitor

The key feature for three-level converter is that the voltage stress of the switches is reduced to half of the DC-link voltage. Therefore, the challenge for the three-level converter is how to maintain the voltage stress of the outer and inner switches. In view of the modulation scheme, whether for conventional three-level DC-AC inverter or three-level DC-DC converter [5–7], the outer switch should be turned off earlier than the inner switch in order to clamp the voltage stress to be half of the DC-link voltage. As seen the conventional three-level stage in Fig. 6.5a, when Q_5 is turned off before Q_7 is turned off, node *M* is clamped to the node *D*. The voltage across Q_5 is equal to the voltage across capacitor C_{d1}, which is half of the DC-link voltage. As seen in Fig. 6.5b, when Q_8 and Q_6 are turned on, the voltage across Q_5 still can be clamped to half of the DC-link voltage. The flying capacitor is clamped to C_{d2} through D_1 and Q_6, and the voltage across the flying capacitor maintains $V_o/2$.

Figure 6.6 shows the equivalent circuit of the three-level stage in the proposed converter. When Q_5 is turned off before Q_7 is turned off, the equivalent circuit is shown in Fig. 6.6a. v_{OD} leads v_{CO}, and i_s flows out of node o. i_2 is larger than i_3. Therefore, the energy stored in the L_a will charge capacitor C_{ss1}. The voltage $v_{Css1} + v_{Css2}$ will become larger than $V_o/2$, and D_2 will be reverse-biased. The voltage stresses of the inner switches Q_7 and Q_8 will be larger than half of the DC-link voltage, so the switches may be damaged. When Q_7 is turned off before Q_5 is turned off, i.e., v_{CO} leads v_{OD}, the equivalent circuit is shown in Fig. 6.6a. In this case, i_s flows into node o. i_3 is larger than i_2, so the energy stored in the auxiliary

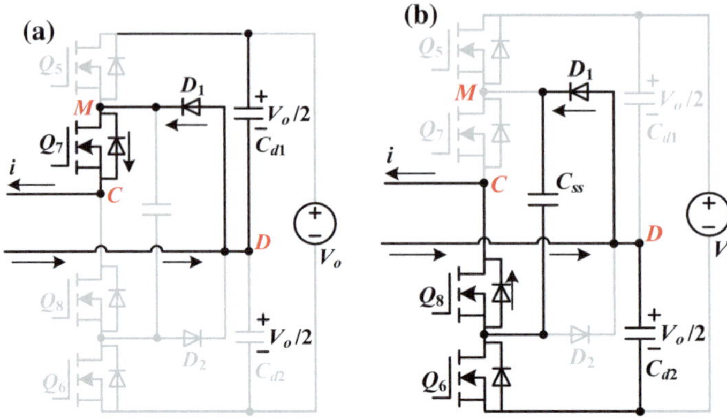

Fig. 6.5 Equivalent circuit in conventional three-level stage: **a** Q_5, Q_6, and Q_8 are turned off, and Q_7 is turned on. **b** Q_5 and Q_7 are turned off, and Q_6 and Q_8 are turned on

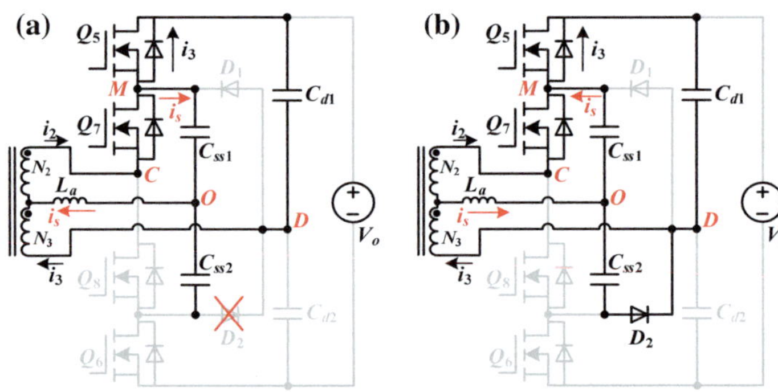

Fig. 6.6 Equivalent circuit in proposed three-level stage: **a** Q_5 is turned off before Q_7 is turned off. **b** Q_7 is turned off before Q_5 is turned off

inductor will discharge the capacitor C_{ss1}. The voltage $v_{Css1} + v_{Css2}$ will be less than $V_o/2$. In this case, the voltages of C_{ss1} and C_{ss2} will be clamped by C_{d1} through D_2 and Q_5. Therefore, the flying capacitor can maintain half of the DC-link voltage. When Q_6 and Q_8 are switched on, the same conclusion can be drawn.

In above, the constraint for conventional three-level converters is not suitable for the proposed three-level structure with auxiliary inductor. In order to keep the voltage stress of the switches to be half of the DC-link voltage, v_{CO} should not lag behind v_{OD}; i.e., φ_b should be no less than φ_a.

6.2.2 ZVS Analyses for Q_1–Q_4

Because of the symmetry in two halves of the switching period, the ZVS conditions for Q_1 and Q_2 are the same, and the ZVS conditions for Q_3 and Q_4 are also the same. Therefore, the conditions for Q_1 and Q_4 are analyzed in half of the switching period. The commutation during the dead time for Q_1 and Q_4 is shown in Fig. 6.7. To achieve ZVS of Q_1 and Q_4, i_1 should be negative and large enough, i.e., $i_1 < -I_{ZVS}$, where I_{ZVS} is the minimum current to charge and discharge the junction capacitors of the switches during the dead time. Therefore, the ZVS constraint for Q_1 and Q_4 is shown in Tables 6.2 and 6.3.

Considering the voltage-second balance of the primary and secondary windings in the transformer, $nV_{\text{bat}}D_1$ should equal $\frac{V_o}{2}\left(1-\frac{\varphi_b}{\pi}+\frac{\varphi_a}{\pi}\right)$. If the effective duty cycle D_1 is designed as

$$D_1=\frac{V_o}{2nV_{\text{bat}}}\left(1-\frac{\varphi_b}{\pi}+\frac{\varphi_a}{\pi}\right)+D_m \tag{6.6}$$

Tables 6.2 and 6.3 can be rewritten to Tables 6.4 and 6.5.

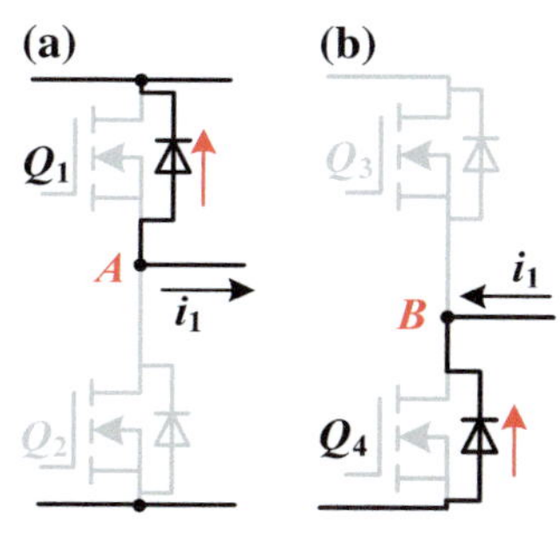

Fig. 6.7 Commutation during the dead time for Q_1 and Q_4: **a** Q_1, **b** Q_4

Table 6.2 ZVS conditions for Q_1 and Q_4—Part A

	Mode 1a $\frac{1}{2}-\frac{D_1}{2}\le\frac{\varphi_a}{\pi}\le\frac{\varphi_b}{\pi}\le\frac{1}{2}$	Mode 2a $-\left(\frac{1}{2}-\frac{D_1}{2}\right)\le\frac{\varphi_a}{\pi}\le\left(\frac{1}{2}-\frac{D_1}{2}\right)\le\frac{\varphi_b}{\pi}$
ZVS for Q_1	$-i_1(t_3)=-\left[\left(\frac{\varphi_a}{\pi}+\frac{\varphi_b}{\pi}-D_1\right)\frac{V_oT_s}{8nL_r}+\frac{V_{\text{bat}}D_1T_s}{4L_r}\right] < -I_{ZVS}$	$-i_1(t_3)=-\left[\frac{V_{\text{bat}}D_1T_s}{4L_r}-\left(D_1-\frac{\varphi_a}{\pi}-\frac{\varphi_b}{\pi}\right)\frac{V_oT_s}{8nL_r}\right] < -I_{ZVS}$
ZVS for Q_4	$i_1(t_0)=\left(2-\frac{\varphi_a}{\pi}-\frac{\varphi_b}{\pi}-D_1\right)\frac{V_oT_s}{8nL_r}-\frac{V_{\text{bat}}D_1T_s}{4L_r} < -I_{ZVS}$	$i_1(t_1)=\left(1+\frac{\varphi_a}{\pi}-\frac{\varphi_b}{\pi}\right)\frac{V_oT_s}{8nL_r}-\frac{V_{\text{bat}}D_1T_s}{4L_r} < -I_{ZVS}$

Table 6.3 ZVS conditions for Q_1 and Q_4—Part B

	Mode 3 $-\left(\frac{1}{2}-\frac{D_1}{2}\right)T_s \le \frac{\varphi_a}{\pi} \le \frac{\varphi_b}{\pi} \le \left(\frac{1}{2}-\frac{D_1}{2}\right)T_s$	Mode 4 $\frac{\varphi_a}{\pi} \le -\left(\frac{1}{2}-\frac{D_1}{2}\right)T_s \le 0$ $\le 0\left(\frac{1}{2}-\frac{D_1}{2}\right)T_s \le \frac{\varphi_b}{\pi}$
ZVS for Q_1	$-i_1(t_3) = -\left[\frac{V_{\text{bat}}D_1T_s}{4L_r} - \left(D_1 - \frac{\varphi_a}{\pi} - \frac{\varphi_b}{\pi}\right)\frac{V_oT_s}{8nL_r}\right]$ $< -I_{\text{ZVS}}$	$i_1(t_1) = \left(1 + \frac{\varphi_a}{\pi} - \frac{\varphi_b}{\pi}\right)\frac{V_oT_s}{8nL_r} - \frac{V_{\text{bat}}D_1T_s}{4L_r}$ $< -I_{\text{ZVS}}$
ZVS for Q_4	$i_1(t_2) = \left(D_1 + \frac{\varphi_b}{\pi} + \frac{\varphi_a}{\pi}\right)\frac{V_oT_s}{8nL_r} - \frac{V_{\text{bat}}D_1T_s}{4L_r}$ $< -I_{\text{ZVS}}$	$i_1(t_2) = \left(1 + \frac{\varphi_a}{\pi} - \frac{\varphi_b}{\pi}\right)\frac{V_oT_s}{8nL_r} - \frac{V_{\text{bat}}D_1T_s}{4L_r}$ $< -I_{\text{ZVS}}$

Summarizing these two tables, as long as $\frac{V_{\text{in}}D_mT_s}{4L_r} \ge I_{\text{ZVS}}$, i.e., $D_m \ge \frac{4L_rI_{\text{ZVS}}}{V_{\text{in}}T_s}$, then Q_1–Q_4 in every mode can achieve ZVS. Therefore, D_m is selected as $\frac{4L_rI_{\text{ZVS}}}{V_{\text{in}}T_s}$ in the following analyses and prototype design.

6.2.3 ZVS Analyses for Q_5–Q_8

Because of the symmetry in two halves of the switching period, the ZVS conditions for Q_5 and Q_6 are the same, and the ZVS conditions for Q_7 and Q_8 are also the same. Therefore, the conditions for Q_5 and Q_7 are analyzed in half of the switching period. The commutation during the dead time for Q_5 and Q_7 is shown in Fig. 6.8. To achieve ZVS of Q_5 and Q_7, i_2 and i_3 should be positive and large enough, i.e., $i_2 > I_{\text{ZVS}}$ and $i_3 > I_{\text{ZVS}}$. The angle error between φ_a and φ_b is denoted as $\Delta\varphi$. Defining $\varphi_a = \varphi - \frac{\Delta\varphi}{2}$ and $\varphi_b = \varphi + \frac{\Delta\varphi}{2}$, i.e., $\varphi = \frac{\varphi_a + \varphi_b}{2}$ and $\Delta\varphi = \varphi_b - \varphi_a$, the ZVS constraints for Q_5–Q_8 are shown in Tables 6.6 and 6.7.

To achieve ZVS for Q_5 and Q_7 in all the operation modes, the following conditions should be satisfied.

From Mode 1a in Table 6.6, if the phase shift angle φ is large enough, the ZVS for Q_5 and Q_7 is easy to achieve ZVS. When $\Delta\varphi = 0$, the following condition should be satisfied for the soft switching of Q_5 and Q_7:

$$\frac{(V_o - 2nV_{\text{bat}})T_s}{8n^2L_r} + \frac{nV_{\text{bat}}T_s}{4n^2L_r}\frac{2\varphi}{\pi} \ge I_{\text{ZVS}} \tag{6.7}$$

Rewriting (6.7), the phase shift angle φ should meet the following condition in this case:

$$\varphi \ge \varphi_{m1} \tag{6.8}$$

Table 6.4 Derived ZVS conditions for Q_1 and Q_4—Part A

	Mode 1a $\frac{1}{2}-\frac{D_1}{2} \leq \frac{\varphi_a}{\pi} \leq \frac{\varphi_b}{\pi} \leq \frac{1}{2}$	Mode 2a $-\left(\frac{1}{2}-\frac{D_1}{2}\right) \leq \frac{\varphi_a}{\pi} \leq \left(\frac{1}{2}-\frac{D_1}{2}\right) \leq \frac{\varphi_b}{\pi}$
ZVS for Q_1	$-i_1(t_3) = -\left(\frac{2\varphi_a}{\pi} + 1 - D_1\right)\frac{V_o T_s}{8nL_r} - \frac{V_{\text{bat}} D_m T_s}{4L_r} - \frac{V_{\text{bat}} D_m T_s}{4L_r} \leq -I_{\text{ZVS}}$	$-i_1(t_3) = -\left(\frac{2\varphi_a}{\pi} + 1 - D_1\right)\frac{V_o T_s}{8nL_r} - \frac{V_{\text{bat}} D_m T_s}{4L_r} \leq -\frac{V_{\text{bat}} D_m T_s}{4L_r} \leq -I_{\text{ZVS}}$
ZVS for Q_4	$i_1(t_0) = \left(1 - D_1 - \frac{2\varphi_a}{\pi}\right)\frac{V_o T_s}{8nL_r} - \frac{V_{\text{bat}} D_m T_s}{4L_r} \leq -\frac{V_{\text{bat}} D_m T_s}{4L_r} \leq -I_{\text{ZVS}}$	$i_1(t_1) = -\frac{V_{\text{bat}} D_m T_s}{4L_r} \leq -I_{\text{ZVS}}$

Table 6.5 Derived ZVS conditions for Q_1 and Q_4—Part B

	Mode 3 $-\left(\frac{1}{2}-\frac{D_1}{2}\right)T_s \le \frac{\varphi_a}{\pi} \le \frac{\varphi_b}{\pi} \le \left(\frac{1}{2}-\frac{D_1}{2}\right)T_s$	Mode 4 $\frac{\varphi_a}{\pi} \le -\left(\frac{1}{2}-\frac{D_1}{2}\right)T_s \le 0 \le \left(\frac{1}{2}-\frac{D_1}{2}\right)T_s \le \frac{\varphi_b}{\pi}$
ZVS for Q_1	$-i_1(t_3) = -\left(\frac{2\varphi_a}{\pi}+1-D_1\right)\frac{V_oT_s}{8nL_r} - \frac{V_{\text{bat}}D_mT_s}{4L_r} \le -\frac{V_{\text{bat}}D_mT_s}{4L_r} \le -I_{\text{ZVS}}$	$i_1(t_1) = -\frac{V_{\text{bat}}D_mT_s}{4L_r} \le -I_{\text{ZVS}}$
ZVS for Q_4	$i_1(t_2) = -\left(1-D_1-\frac{2\varphi_b}{\pi}\right)\frac{V_oT_s}{8nL_r} - \frac{V_{\text{bat}}D_mT_s}{4L_r} \le -\frac{V_{\text{bat}}D_mT_s}{4L_r} \le -I_{\text{ZVS}}$	$i_1(t_1) = -\frac{V_{\text{bat}}D_mT_s}{4L_r} \le -I_{\text{ZVS}}$

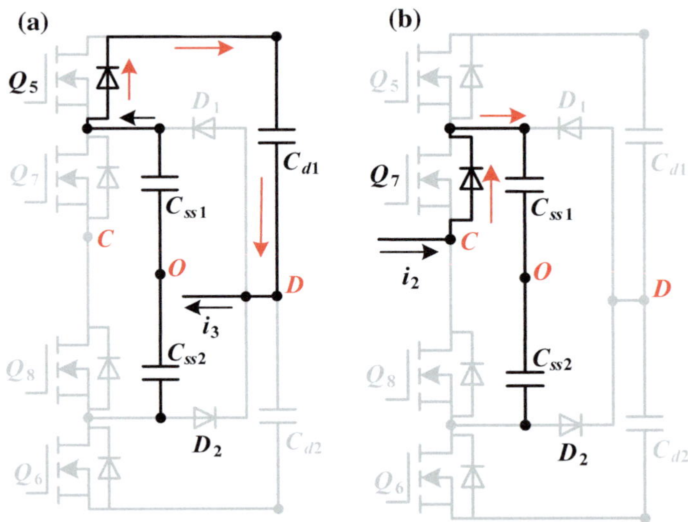

Fig. 6.8 Commutation during the dead time for Q_5 and Q_7: **a** Q_5, **b** Q_7

where $\varphi_{m1} = \frac{2\pi I_{\text{ZVS}} nL_r}{V_{\text{bat}}T_s} + \left(2 - \frac{V_o}{nV_{\text{bat}}}\right)\frac{\pi}{4}$. If $\varphi \ge \varphi_{m1}$ and $\Delta\varphi \ge 0$, then Q_5 and Q_7 in Mode 1a can achieve ZVS.

When $\varphi < \varphi_{m1}$, the ZVS of Q_5–Q_8 is obtained with the aid of auxiliary inductor L_a. As seen for Mode 1a in Table 6.6, the ZVS condition for Q_5 is expressed as

$$i_3(t_2) = \frac{(V_o - 2nV_{\text{bat}})T_s}{8n^2L_r} + \frac{nV_{\text{bat}}T_s}{4n^2L_r}\frac{2\varphi}{\pi} + \left(\frac{nV_{\text{bat}}T_s}{4n^2L_r} + \frac{V_oT_s}{32L_a}\right)\frac{\Delta\varphi}{\pi} \ge I_{\text{ZVS}} \quad (6.9)$$

Table 6.6 ZVS conditions for Q_5 and Q_7 for Mode 1a and Mode 2a

	Mode 1a $\frac{1}{2}-\frac{D_1}{2}\le\frac{\varphi_a}{\pi}\le\frac{\varphi_b}{\pi}\le\frac{1}{2}$	Mode 2a $-\left(\frac{1}{2}-\frac{D_1}{2}\right)\le\frac{\varphi_a}{\pi}\le\left(\frac{1}{2}-\frac{D_1}{2}\right)\le\frac{\varphi_b}{\pi}$
ZVS for Q_5	$i_3(t_2)=\frac{(V_o-2nV_{\text{bat}})T_s}{8n^2L_r}+\frac{nV_{\text{bat}}T_s}{4n^2L_r}\frac{2\varphi}{\pi}+\left(\frac{nV_{\text{bat}}T_s}{4n^2L_r}+\frac{V_oT_s}{32L_a}\right)\frac{\Delta\varphi}{\pi}\ge I_{\text{ZVS}}$	$i_3(t_2)=\frac{(V_o-2nV_{\text{bat}})T_s}{8n^2L_r}+\frac{nV_{\text{bat}}T_s}{4n^2L_r}\frac{2\varphi}{\pi}+\left(\frac{nV_{\text{bat}}T_s}{4n^2L_r}+\frac{V_oT_s}{32L_a}\right)\frac{\Delta\varphi}{\pi}\ge I_{\text{ZVS}}$
ZVS for Q_7	$i_2(t_1)=\frac{(V_o-2nV_{\text{bat}})T_s}{8n^2L_r}+\frac{nV_{\text{bat}}T_s}{4n^2L_r}\frac{2\varphi}{\pi}+\left(\frac{V_oT_s}{32L_a}-\frac{nV_{\text{bat}}T_s}{4n^2L_r}\right)\frac{\Delta\varphi}{\pi}\ge I_{\text{ZVS}}$	$i_2(t_0)=\left(\frac{V_oT_s}{8n^2L_r}+\frac{V_oT_s}{32L_a}\right)\frac{\Delta\varphi}{\pi}-\frac{nV_{\text{bat}}D_mT_s}{4L_r}\ge I_{\text{ZVS}}$

Rearranging (6.9), $\Delta\varphi$ is expressed as

$$\Delta\varphi\ge\frac{\pi}{\frac{nV_{\text{bat}}T_s}{4n^2L_r}+\frac{V_oT_s}{32L_a}}\left(I_{\text{ZVS}}-\frac{(V_o-2nV_{\text{bat}})T_s}{8n^2L_r}-\frac{nV_{\text{bat}}T_s}{4n^2L_r}\frac{2\varphi}{\pi}\right) \tag{6.10}$$

As seen for Mode 1a in Table 6.6, the ZVS condition for Q_7 is expressed as

$$i_2(t_1)=\frac{(V_o-2nV_{\text{bat}})T_s}{8n^2L_r}+\frac{nV_{\text{bat}}T_s}{4n^2L_r}\frac{2\varphi}{\pi}+\left(\frac{V_oT_s}{32L_a}-\frac{nV_{\text{bat}}T_s}{4n^2L_r}\right)\frac{\Delta\varphi}{\pi}>I_{\text{ZVS}} \tag{6.11}$$

Rearranging inequality (6.11), $\Delta\varphi$ is expressed as

$$\Delta\varphi\ge\frac{\pi}{\frac{V_oT_s}{32L_a}-\frac{V_{\text{bat}}T_s}{4nL_r}}\left(I_{\text{ZVS}}-\frac{(V_o-2nV_{\text{bat}})T_s}{8n^2L_r}-\frac{nV_{\text{bat}}T_s}{4n^2L_r}\frac{2\varphi}{\pi}\right) \tag{6.12}$$

If (6.12) is satisfied, then (6.13) is satisfied. Moreover, $\frac{V_oT_s}{32L_a}-\frac{V_{\text{bat}}T_s}{4nL_r}$ should be greater than zero. Therefore, the auxiliary inductor should meet the following condition.

$$L_a<\frac{nV_oL_r}{8V_{\text{bat}}} \tag{6.13}$$

As seen in Table 6.6, the ZVS condition for Q_5 in Mode 2a is the same as that in Mode 1a, so it is also expressed in (6.10).

The ZVS condition for Q_7 in Mode 2a is expressed as

$$i_2(t_0)=\left(\frac{V_oT_s}{8L_r}+\frac{V_oT_s}{32L_a}\right)\frac{\Delta\varphi}{\pi}-\frac{nV_{\text{bat}}D_mT_s}{4L_r}\ge I_{\text{ZVS}} \tag{6.14}$$

It is independent of φ. Rearranging (6.14), $\Delta\varphi$ is expressed as

Table 6.7 ZVS conditions for Q_5 and Q_7 for Mode 3 and Mode 4

	Mode 3 $-\left(\frac{1}{2}-\frac{D_1}{2}\right) \leq \frac{\varphi_a}{\pi} \leq \frac{\varphi_b}{\pi} \leq \frac{1}{2}-\frac{D_1}{2}$	Mode 4 $\frac{\varphi_a}{\pi} \leq -\left(\frac{1}{2}-\frac{D_1}{2}\right) \leq 0 \leq \frac{1}{2}-\frac{D_1}{2} \leq \frac{\varphi_b}{\pi}$
ZVS for Q_5	$i_3(t_1) = \left(\frac{V_o T_s}{8n^2 L_r} + \frac{V_o T_s}{32 L_a}\right)\frac{\Delta\varphi}{\pi} - \frac{n V_{\text{bat}} D_m T_s}{4n^2 L_r} \geq I_{\text{ZVS}}$	$i_3(t_3) > \frac{(V_o - 2nV_{\text{bat}})T_s}{8n^2 L_r} + \frac{n V_{\text{bat}} T_s}{4n^2 L_r}\frac{2\varphi}{\pi} + \left(\frac{n V_{\text{bat}} T_s}{4n^2 L_r} + \frac{V_o T_s}{32 L_a}\right)\frac{\Delta\varphi}{\pi} \geq I_{\text{ZVS}}$
ZVS for Q_7	$i_2(t_0) = \left(\frac{V_o T_s}{8n^2 L_r} + \frac{V_o T_s}{32 L_a}\right)\frac{\Delta\varphi}{\pi} - \frac{n V_{\text{bat}} D_m T_s}{4n^2 L_r} \geq I_{\text{ZVS}}$	$i_2(t_0) > \frac{(V_o - 2nV_{\text{bat}})T_s}{8n^2 L_r} - \frac{n V_{\text{bat}} T_s}{4n^2 L_r}\frac{2\varphi}{\pi} + \left(\frac{n V_{\text{bat}} T_s}{4n^2 L_r} + \frac{V_o T_s}{32 L_a}\right)\frac{\Delta\varphi}{\pi} \geq I_{\text{ZVS}}$

$$\Delta\varphi \geq \Delta\varphi_m \tag{6.15}$$

where $\Delta\varphi_m = \frac{3\pi I_{ZVS}}{\frac{V_o T_s}{4L_r} + \frac{V_o T_s}{16L_s}}$.

As seen in Table 6.7, the ZVS condition for Q_5 and Q_7 in Mode 3 is the same as the condition for Q_7 in Mode 2a, which is expressed in (6.15).

As seen in Table 6.7, the ZVS condition for Q_5 in Mode 4 is the same as in Mode 1a, which is expressed in (6.10).

The ZVS condition for Q_7 in Mode 4 is expressed as

$$i_2(t_0) > \frac{(V_o - 2nV_{\text{bat}})T_s}{8n^2 L_r} - \frac{nV_{\text{bat}}T_s}{4n^2 L_r}\frac{2\varphi}{\pi} + \left(\frac{V_o T_s}{32L_a} + \frac{nV_{\text{bat}}T_s}{4n^2 L_r}\right)\frac{\Delta\varphi}{\pi} \geq I_{ZVS} \tag{6.16}$$

Rewriting (6.16), $\Delta\varphi$ is expressed as

$$\Delta\varphi \geq \frac{\pi}{\frac{V_o T_s}{32L_a} + \frac{V_{\text{bat}}T_s}{4nL_r}}\left[I_{ZVS} - \frac{(V_o - 2nV_{\text{bat}})T_s}{8n^2 L_r} + \frac{V_{\text{bat}}T_s}{4nL_r}\frac{2\varphi}{\pi}\right] \tag{6.17}$$

Above all, the ZVS condition for Q_5–Q_8 can be synthesized to (6.12) in Mode 1a, (6.10) and (6.15) in Mode 2a, (6.15) in Mode 3, and (6.10) and (6.17) in Mode 4.

6.2.4 Modulation Trajectory

According to the ZVS conditions for Q_5–Q_8 in different working modes, Fig. 6.9 shows the ZVS range of Q_5–Q_8 for different voltage conversion ratios, where $L_r = 24$ μH, $T_s = 12.5$ μs, $L_a = 12$ μH, $n = 1.5$, $I_{zvs} = 1$ A, and $V_o = 600$ V. The ZVS area is in the shadow region of the figures. As seen, with the increase of the conversion ratio, the ZVS area becomes smaller. As seen in Fig. 6.5a, the conversion ratio is 2, so the effective voltage amplitudes across the primary and secondary windings of the transformer are the same. In this case, the converter only loses ZVS in light loads. When the output voltage is constant and the battery voltage increases, the conversion ratio decreases simultaneously. It is more difficult for the converter to achieve ZVS when the battery voltage becomes higher. Therefore, if the effective voltages across the primary and secondary windings of the transformer do not coincide, the ZVS range will dramatically decrease.

With the help of the auxiliary inductor, the converter can maintain ZVS performance within the whole power and voltage range. However, it may cause more conduction loss. The modulation strategy for ZVS should be designed based on the boundary of the ZVS range at the cost of low conduction loss. As seen in Fig. 6.9, the coordinates of point E are $(\varphi_{m2}, \Delta\varphi_m)$, where φ_{m2} is equal to $\frac{\pi(1-D_1)+\Delta\varphi_m}{2}$. The coordinates of point F are $(\varphi_{m1}, 0)$. The line EF is expressed as

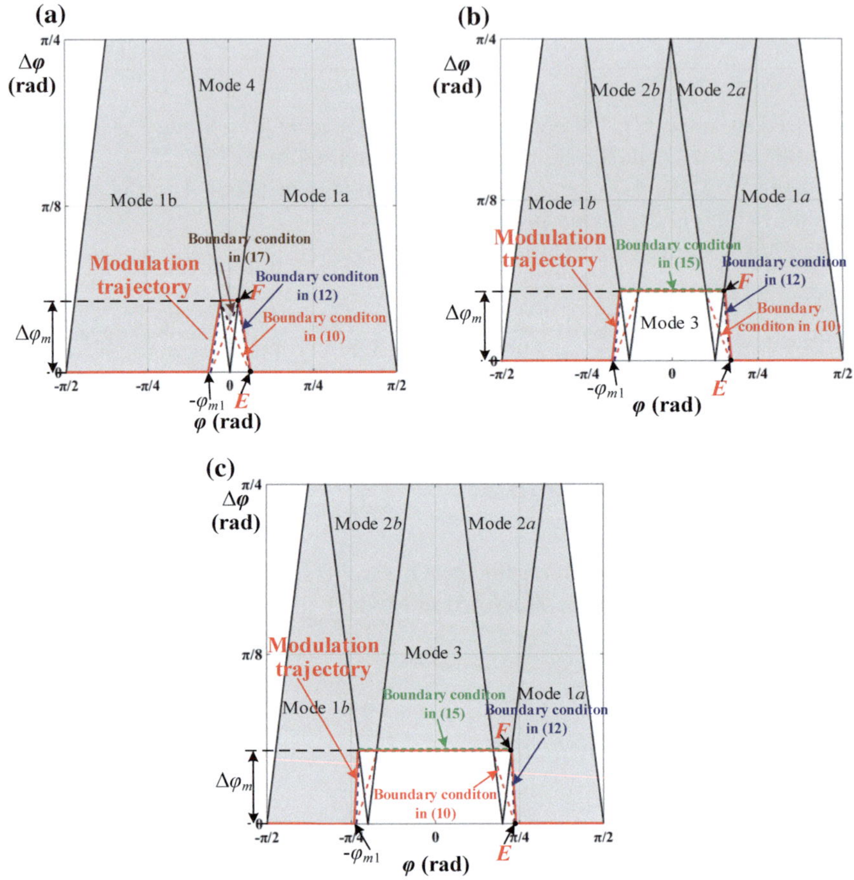

Fig. 6.9 ZVS range and modulation trajectory for different voltage conversion ratios: **a** $V_o/(nV_{bat}) = 2$; **b** $V_o/(nV_{bat}) = 1.5$; **c** $V_o/(nV_{bat}) = 1.2$

$$\Delta\varphi = \frac{\Delta\varphi_m}{\varphi_{m2} - \varphi_{m1}}(\varphi - \varphi_{m1}) \tag{6.18}$$

The modulation trajectories are shown in the solid line in Fig. 6.9, which is based on the boundary condition analyzed in Sect. 6.2.3. The control diagram is shown in Fig. 6.10. The output voltage reference V_{o_ref} is compared with the output voltage. The voltage error is the input of the voltage controller $G_v(s)$. The output of the voltage controller is the phase shift angle φ. The flowchart of the modulation trajectory is shown in Fig. 6.11. According to the flowchart in Fig. 6.11, D_1, φ_a, and φ_b are calculated based on φ. Finally, the PWM signals of Q_1–Q_8 are obtained.

Based on the flowchart of the modulation trajectory in Fig. 6.11, the output power versus the phase shift angle φ in per unit (p.u.) for different conversion ratios

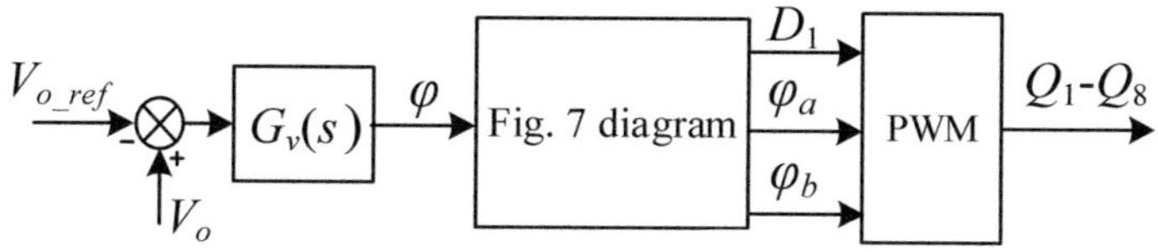

Fig. 6.10 Control diagram of the converter

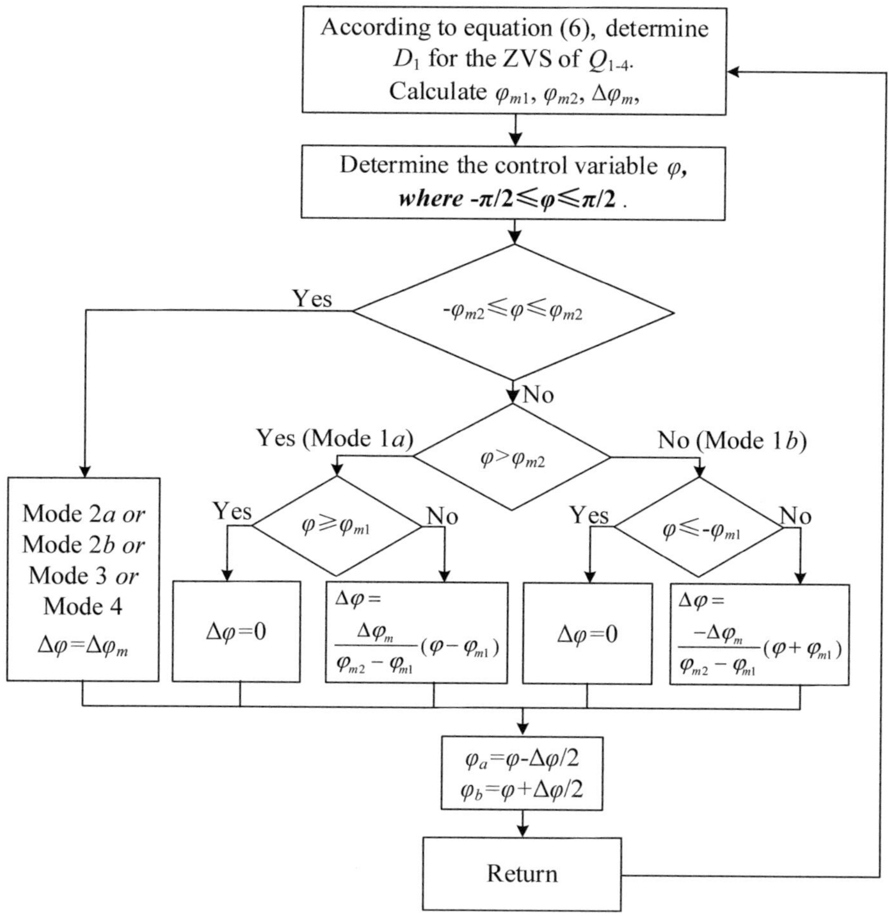

Fig. 6.11 Flowchart of the modulation trajectory

is shown in Fig. 6.12. For different conversion ratios, the modulation trajectories are seamless transfers in different operation modes. The output power increases monotonically with the phase shift angle φ, when $-\pi/2 \leq \varphi \leq \pi/2$. In the proposed modulation trajectory, the modulation trajectories for φ_a and φ_b for different

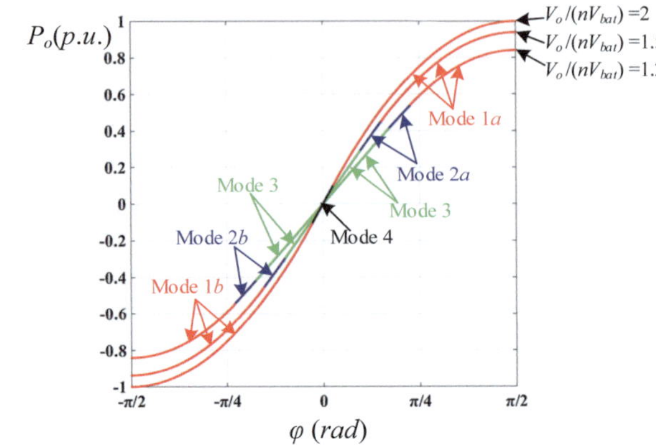

Fig. 6.12 Output power versus phase shift angle φ for different voltage conversion ratios

conversion ratios are shown in Fig. 6.13. It illustrates that φ_a equals φ_b when the phase angle is large enough. Otherwise, $\Delta\varphi$ is greater than zero for the ZVS of Q_5–Q_8. As illustrated in Fig. 6.13, $\Delta\varphi$ equals $\Delta\varphi_m$ in Mode 3, Mode 4, Mode 2a, and Mode 2b.

Within the whole load range, the proposed converter can achieve ZVS. Actually, if $\Delta\varphi$ equals zero, the working modes in the proposed converter are the same as the three-level bidirectional converter with EPPS, whose output power versus the phase shift angle φ in per unit (p.u.) is shown in Fig. 6.14. The shadow area is the hard-switching range. With the decrease of the voltage conversion ratios, it becomes harder for the converter to achieve soft switching.

6.2.5 Conduction Loss Comparison

To study the conduction loss, the root-mean-square (RMS) of the current in the transformer should be calculated. The RMS of i_1 reflects the conduction loss on the battery side of the switches, and the RMS of i_2 and i_3 reflects the conduction loss on the output side of the switches. Therefore, the RMS current in the converter is defined as

$$I_{\mathrm{RMS}} = \frac{1}{n}\sqrt{\frac{1}{T_s}\int_0^{Ts} i_1^2(t)\,\mathrm{d}t} + \sqrt{\frac{1}{T_s}\int_0^{Ts} i_2^2(t)\,\mathrm{d}t} + \sqrt{\frac{1}{T_s}\int_0^{Ts} i_3^2(t)\,\mathrm{d}t} \tag{6.19}$$

With the help of MATLAB, the RMS currents in the converter for different conversion ratios are shown in Fig. 6.15. The solid lines are the RMS current in the proposed converter, and the dashed lines are the RMS current in the three-level

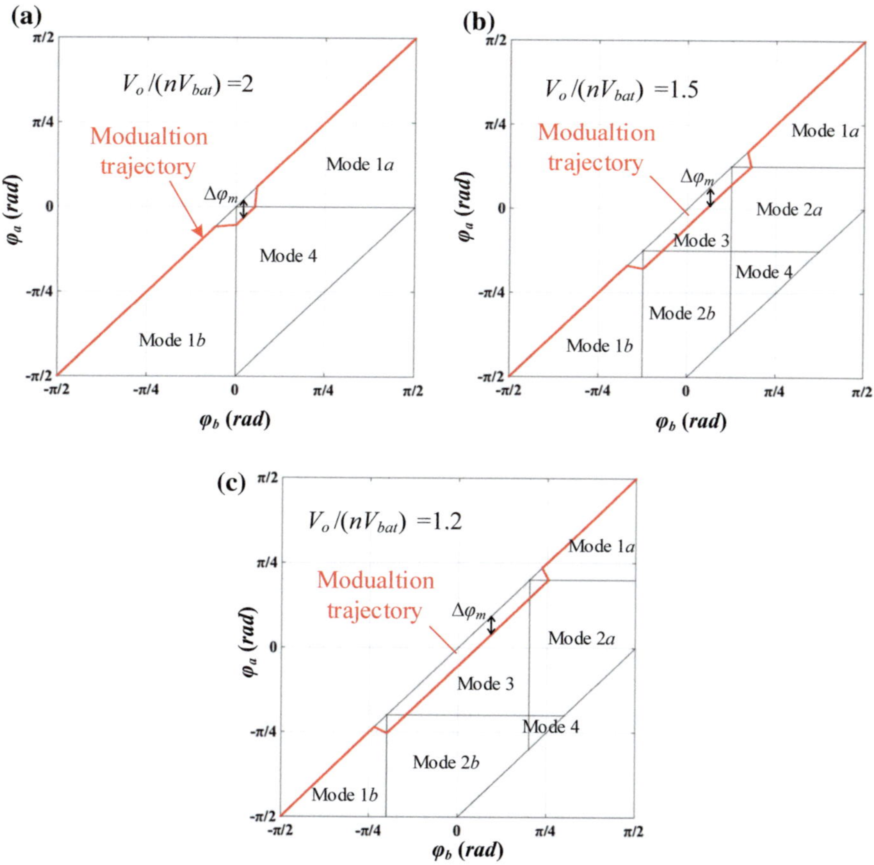

Fig. 6.13 Modulation trajectories for φ_a and φ_b for different voltage conversion ratios: **a** $V_o/(nV_{bat}) = 2$; **b** $V_o/(nV_{bat}) = 1.5$; **c** $V_o/(nV_{bat}) = 1.2$

Fig. 6.14 Output power versus phase shift angle φ in per unit for extended phase shift control

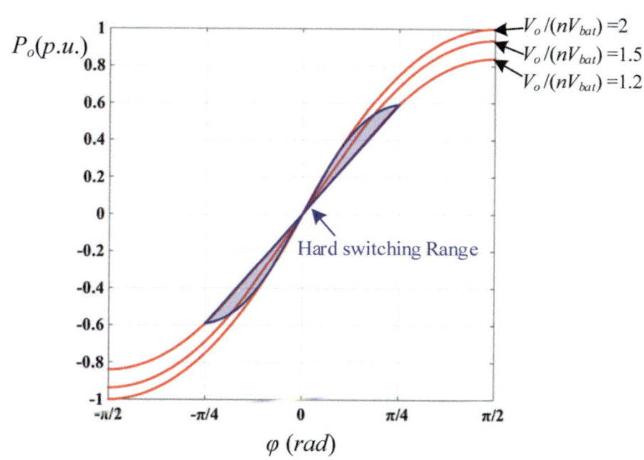

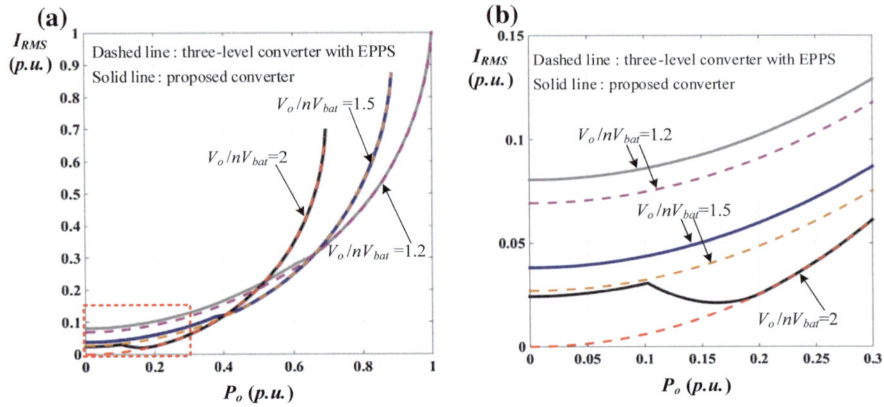

Fig. 6.15 **a** RMS current of the proposed converter and three-level converter with EPPS; **b** Zoom-in figure of the dashed box in Fig. 5.12a

converter shown in Fig. 6.1 with EPPS. Figure 6.15b is the zoom-in figure of the dashed box in Fig. 6.15a. As seen, the conduction loss in the two converters is the same under heavy loads. When the effective conversion ratio is 2, the three-level converter with EPPS only loses ZVS status with light loads. To guarantee the ZVS within whole power range, the proposed converter has more conduction loss. However, the switching loss is the dominant losses with light loads, and the conduction loss does not have a strong impact on the efficiency in this case. Moreover, the conduction loss with heavy loads in the proposed converter is the same as the three-level converter with EPPS. Therefore, hard switching in the three-level converter with EPPS may degrade the efficiency. With the decrease of the conversion ratio, i.e., with the increase of the battery voltage for the constant output voltage, it is much harder for the three-level converter with EPPS to achieve ZVS of all the switches. According to the modulation trajectory, the proposed converter can maintain ZVS of all the switches within the whole power range and over a wide voltage range with low conduction loss, so the efficiency can be improved.

6.3 Experimental Verifications

A 1200 W experimental prototype with a 200–300 V battery voltage and 600 V output voltage was built to verify the proposed converter. Figure 6.16 shows the prototype used for test. The detailed specifications are shown in Table 6.8.

Figure 6.17 shows key waveforms for the forward power flow with light loads. To achieve ZVS of all the switches with light loads, the current in the auxiliary inductor should be used, and the angle error $\Delta\varphi$ is greater than zero. Figure 6.17a shows the key waveforms when V_{bat} = 200 V and P_o = 280 W. In this case, the

Fig. 6.16 1.2 kW proposed converter

Table 6.8 Detailed specifications

Items	Symbol	Parameter
Battery voltage	V_{bat}	200–300 V
Output voltage	V_{o}	600 V
Turns ratio of the transformers	n	1.5
Switching frequency	f_s (T_s)	80 kHz (12.5 μs)
Primary switches	Q_{1-4}	FQL40N50
Secondary switches	Q_{5-8}	FQL40N50
Clamping diodes	D_1, D_2	DSEP12-12A
Series inductor	L_r	24 μH
auxiliary inductor	L_a	12 μH
Output filter capacitance	C_{d1}, C_{d2}	10 μF
Flying capacitance	C_{ss1}, C_{ss2}	4.7 μF

converter works in Mode 4. As seen in Fig. 6.17b, c, with the increase of the battery voltage, D_1 is decreased. In these cases, the converter transits into Mode 3.

Figure 6.18 shows the key waveforms for the forward power flow with heavy loads. In Fig. 6.18a, when the battery voltage is 200 V, the converter works in Mode 1a for a 1200 W load and the angle error $\Delta\varphi$ becomes zero. In Fig. 6.18b, when the battery voltage increases to 240 V, the converter still works in Mode 1a. The currents in the transformer are large enough for ZVS during the commutation, so the angle error $\Delta\varphi$ is still zero. In Fig. 6.18c, when the input voltage is 280 V, not all the switches can achieve ZVS for a conventional three-level converter with

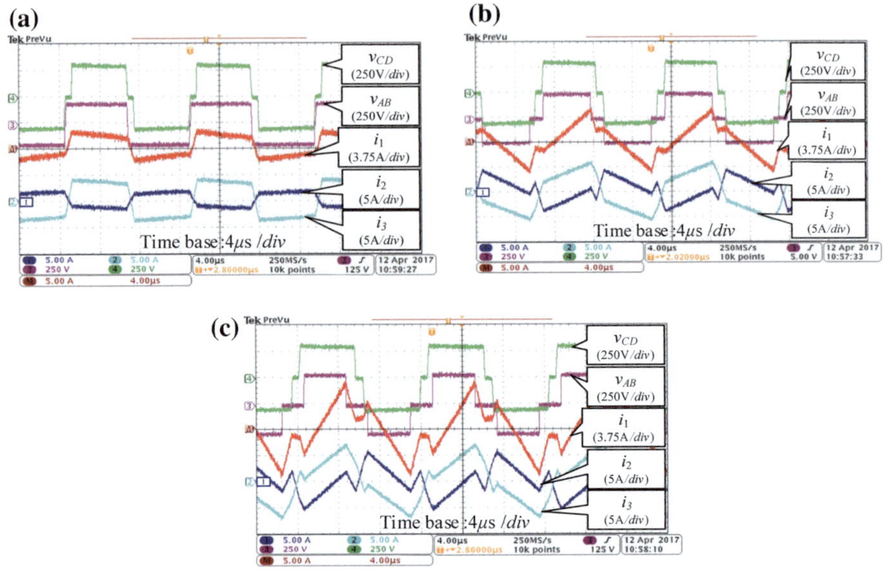

Fig. 6.17 Measured key waveforms for the forward power flow in light loads: **a** V_{bat} = 200 V, P_o = 280 W, **b** V_{bat} = 240 V, P_o = 280 W, and **c** V_{bat} = 280 V, P_o = 280 W

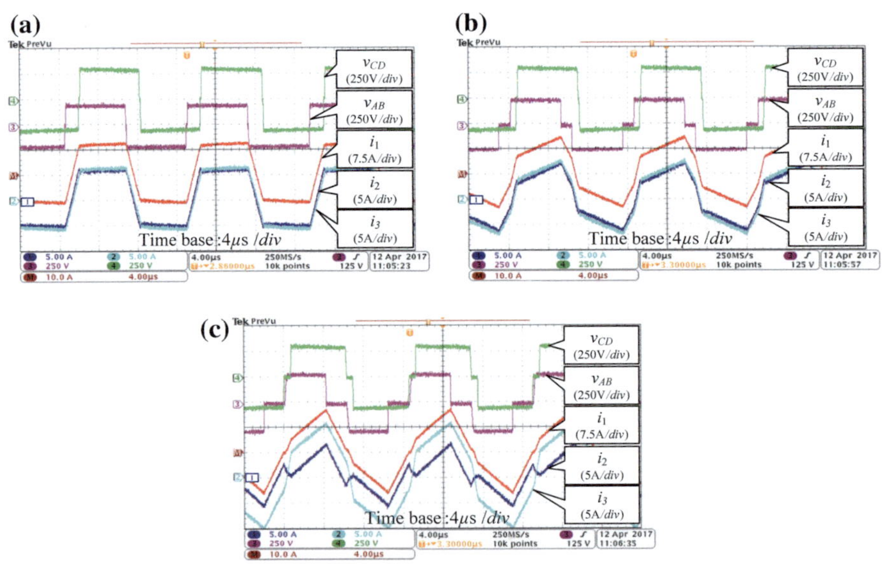

Fig. 6.18 Measured key waveforms for the forward power flow in heavy loads: **a** V_{bat} = 200 V, P_o = 1200 W, **b** V_{bat} = 240 V, P_o = 1200 W, and **c** V_{bat} = 280 V, P_o = 1200 W

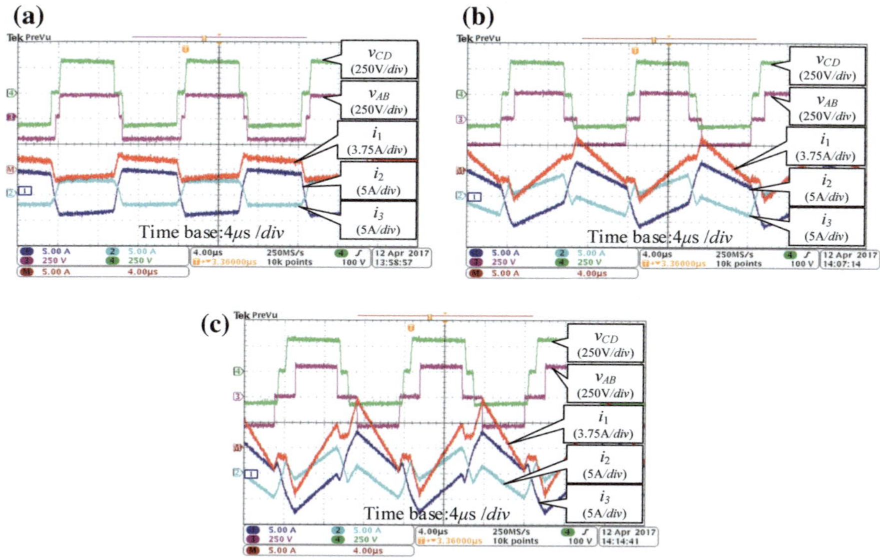

Fig. 6.19 Measured key waveforms for the reverse power flow in light loads: **a** V_{bat} = 200 V, P_{o} = −240 W, **b** V_{bat} = 240 V, P_{o} = −280 W, and **c** V_{bat} = 280 V, P_{o} = −330 W

EPPS, so the angle error $\Delta\varphi$ is greater than zero. As seen in Fig. 6.18c, the converter works in Mode 2a in this case.

Figure 6.19 shows the key waveforms for the reverse power flow with light loads. In Fig. 6.19a, the converter works in Mode 4. In Fig. 6.19b, the converter works in Mode 2b. In Fig. 6.19c, the converter works in Mode 3. In these cases, $\Delta\varphi$ is greater than zero for the ZVS of Q_5–Q_8. The working modes in Fig. 6.19 are nearly symmetrical with those in Fig. 6.17.

Figure 6.20 shows the key waveforms for the reverse power flow with heavy loads. In Fig. 6.20a, the converter works in Mode 1b. The current in the transformer is large enough to achieve the ZVS of all the switches, so $\Delta\varphi$ is equal to zero. In Fig. 6.20b, the battery voltage is 240 V, and the power is −1000 W. The converter works in Mode 1b, but $\Delta\varphi$ is greater than zero. As illustrated in Fig. 6.20c, with the increase of the battery voltage, the phase angle is decreased and the converter switches to Mode 2b.

Figure 6.21 shows the gate signal and drain–source voltage of switch Q_1. In different working states, i_1 is less than zero during this commutation, illustrating that Q_1 works in ZVS in both forward and reverse power flows. Q_2 operates in the same working mode in the other half of the switching period, so it also can achieve ZVS.

Figure 6.22 shows the gate signal and drain–source voltage of switch Q_3. Before Q_3 is turned on, i_1 is greater than zero in different working modes, which highlights the ZVS of Q_3 in both forward and reverse power flows. Q_4 operates in the same

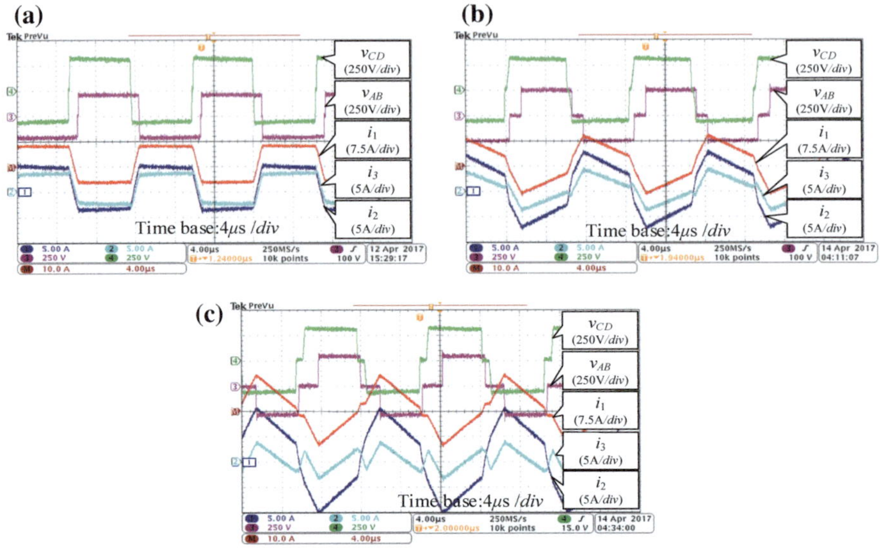

Fig. 6.20 Measured key waveforms for the reverse power flow in heavy loads: **a** $V_{bat} = 200$ V, $P_o = -1000$ W, **b** $V_{bat} = 240$ V, $P_o = -1000$ W, and **c** $V_{bat} = 280$ V, $P_o = -1000$ W

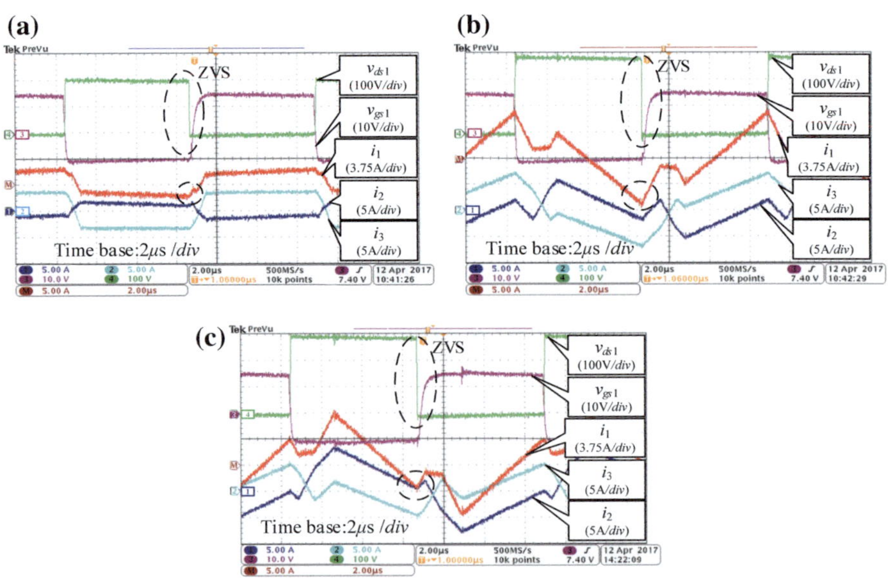

Fig. 6.21 Gate signal and drain–source voltage of switch Q_1: **a** $V_{bat} = 200$ V, $P_o = 280$ W, **b** $V_{bat} = 280$ V, $P_o = 280$ W, **c** $V_{bat} = 280$ V, $P_o = -334$ W

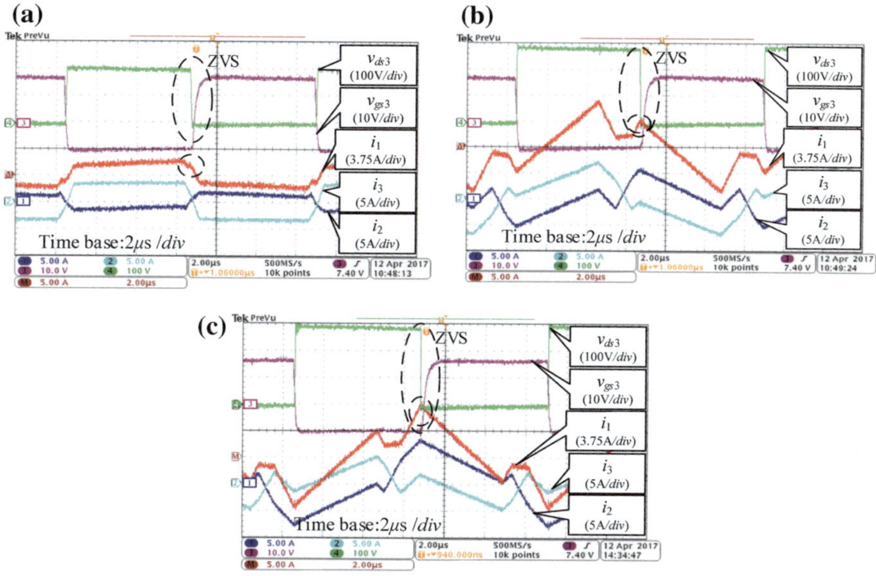

Fig. 6.22 Gate signal and drain–source voltage of switch Q_3: **a** $V_{bat} = 200$ V, $P_o = 280$ W, **b** $V_{bat} = 280$ V, $P_o = 280$ W, **c** $V_{bat} = 280$ V, $P_o = -334$ W

working mode in the other half of the switching period, so it is also easy to achieve ZVS.

Figure 6.23 shows the gate signal and drain–source voltage of switch Q_5. As seen in different working states, before Q_5 is turned on, i_3 has a large current to charge and discharge the junction capacitors of Q_6 and Q_5. Therefore, Q_5 can work in ZVS in both forward and reverse power flows. Q_6 operates in the same working mode in the other half of the switching period, so it is also easy to achieve ZVS.

Figure 6.24 shows the gate signal and drain–source voltage of switch Q_7. As seen in different working states, i_2 has large enough value to charge and discharge the junction capacitors of Q_8 and Q_7 before Q_7 is turned on. It demonstrates the ZVS performance for Q_7 in both forward and reverse power flows. Q_8 operates in the same working mode in the other half of the switching period, so it also achieves ZVS.

The measured efficiency curves for the proposed converter and the conventional three-level bidirectional converter shown in Fig. 6.1 with EPPS are illustrated in Fig. 6.25. Figure 6.25a shows the efficiency curve for the 200 V battery voltage. In this case, the effective conversion ratio is 2, while the slew rate of the current in the transformer during the power transmission interval is zero. Therefore, the three-level bidirectional converter with EPPS has a wide ZVS range. The converter loses ZVS status only in light loads. The proposed converter can achieve full-operation ZVS, so the efficiency in light loads is higher than that of the three-level converter with EPPS. As seen in Fig. 6.25b and c, with the increase of

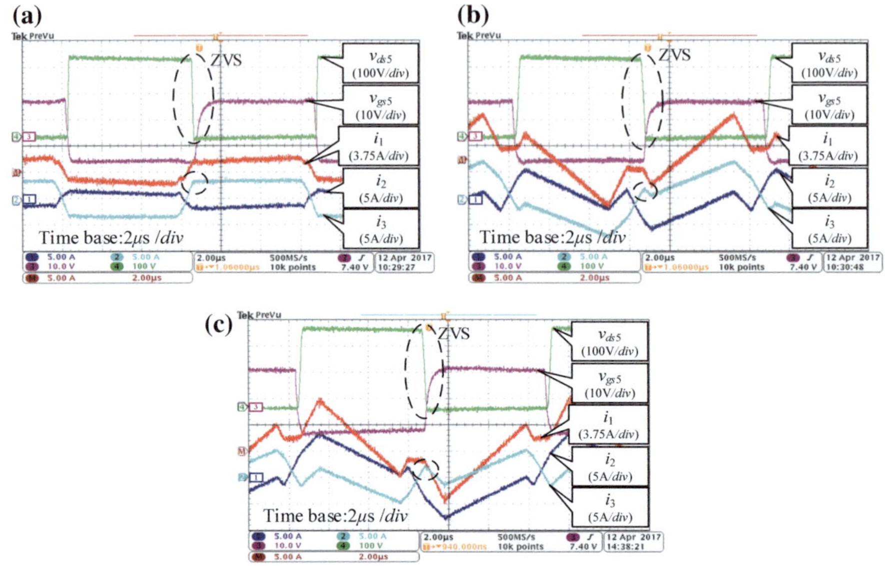

Fig. 6.23 Gate signal and drain–source voltage of switch Q_5: **a** $V_{bat} = 200$ V, $P_o = 280$ W, **b** $V_{bat} = 280$ V, $P_o = 280$ W, **c** $V_{bat} = 280$ V, $P_o = -334$ W

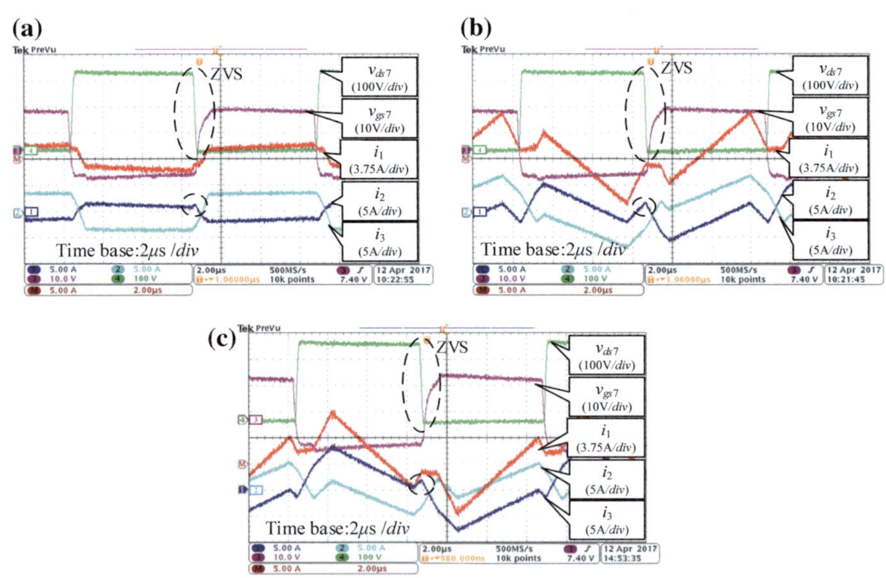

Fig. 6.24 Gate signal and drain–source voltage of switch Q_7: **a** $V_{bat} = 200$ V, $P_o = 280$ W, **b** $V_{bat} = 280$ V, $P_o = 280$ W, **c** $V_{bat} = 280$ V, $P_o = -334$ W

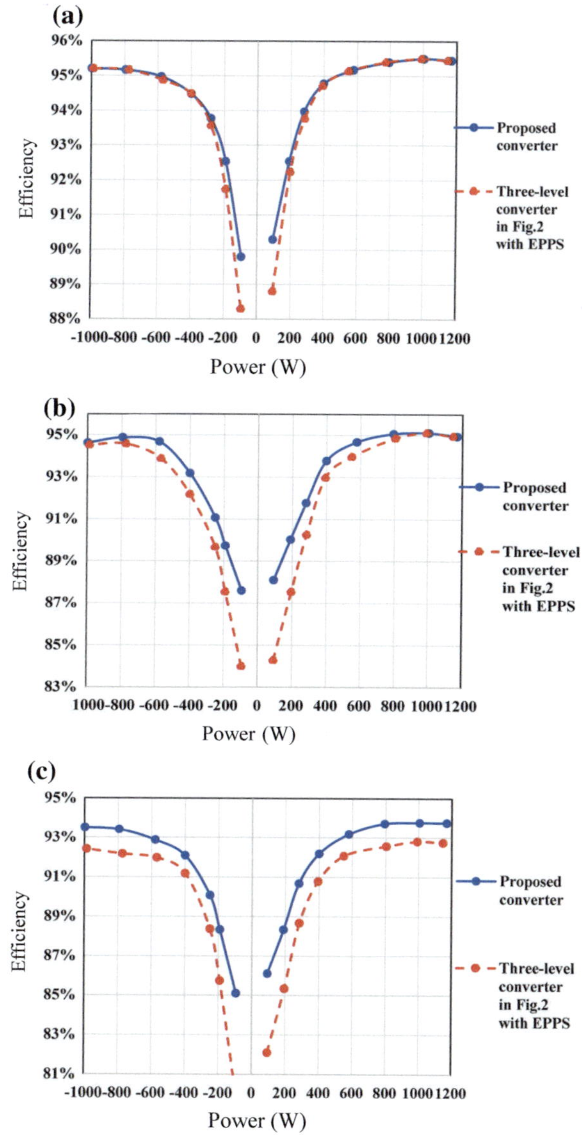

Fig. 6.25 Efficiency curves for the proposed converter and the three-level bidirectional converter with EPPS at **a** $V_{bat} = 200$ V, **b** $V_{bat} = 240$ V, and **c** $V_{bat} = 280$ V

the battery voltage, it becomes harder for the three-level converter with EPPS to achieve ZVS of all the switches. However, the proposed converter still can maintain high efficiency because of the ZVS performance within the whole power and voltage range. Especially at the 280 V battery voltage and a 1200 W load, an efficiency improvement of more than 1% is achieved.

The calculated loss distribution under rated load condition is shown in Fig. 6.26. With increase of the battery voltage, the loss is increased. The conduction loss in

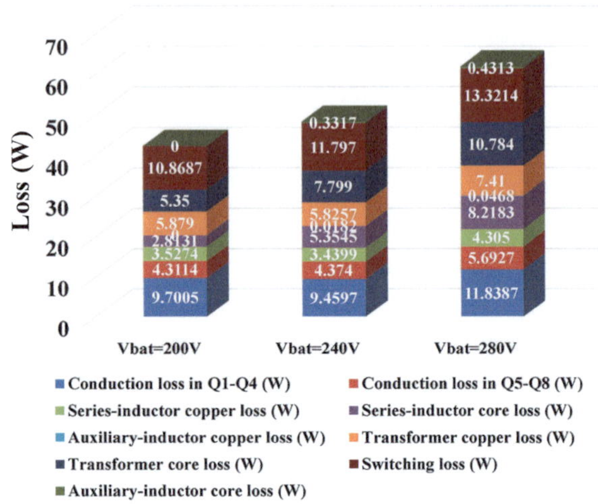

Fig. 6.26 Calculated power loss distribution under the rated load condition

the three-level stage and loss in the auxiliary inductor are only in a small portion. Therefore, the loss caused by the auxiliary components is very low. However, because of the ZVS of all the switching, the efficiency can be improved.

6.4 Conclusion

In this chapter, a three-level bidirectional DC–DC converter with an auxiliary inductor is proposed. The converter is composed of a full bridge on the low-voltage side and a three-level stage on the high-voltage side, linked with a transformer and a series inductor. The flying capacitor in the three-level stage is split into two series-connected capacitors. The auxiliary inductor is connected across the middle node of the split flying capacitors and the center tap of the secondary winding of the transformer. The outer and inner switches in the three-level stage can generate two independent 50% square waveforms, which can be used to control the current in the auxiliary inductor. According to the different working modes of the converter, the ZVS range is analyzed, which is associated with not only the loads but also the voltage conversion ratio. When the effective voltage conversion ratio is 2, the converter only loses ZVS status under light loads. With the decrease of the conversion ratio, the converter encounters difficulty for achieving ZVS. Therefore, the current in the auxiliary inductor is used to maintain the ZVS within the whole power and over a wide voltage range. To reduce the conduction loss in the auxiliary inductor, the boundary condition of the ZVS range is used to design the modulation trajectory. The flowchart of the modulation trajectory can guarantee the ZVS of all the switches with low conduction loss. Finally, the experimental results verified the ZVS performance in the proposed converter. The efficiency curves demonstrate the efficiency improvement for the conventional three-level bidirectional DC–DC converter with EPPS.

Appendix

The currents i_1, i_2, and i_3 at the time of t_0, t_1, t_2, and t_3 in Mode 1a are expressed as follows:

$$\begin{cases} i_1(t_0) = \left(2 - \frac{\varphi_a}{\pi} - \frac{\varphi_b}{\pi} - D_1\right)\frac{V_oT_s}{8nL_r} - \frac{V_{\text{bat}}D_1T_s}{4L_r} \\ i_1(t_1) = \left(1 + \frac{\varphi_a}{\pi} - \frac{\varphi_b}{\pi}\right)\frac{V_oT_s}{8nL_r} + \left(\frac{2\varphi_a}{\pi} - 1\right)\frac{V_{\text{bat}}T_s}{4L_r} \\ i_1(t_2) = \left(1 + \frac{\varphi_a}{\pi} - \frac{\varphi_b}{\pi}\right)\frac{V_oT_s}{8nL_r} + \left(\frac{2\varphi_b}{\pi} - 1\right)\frac{V_{\text{bat}}T_s}{4L_r} \\ i_1(t_3) = \left(\frac{\varphi_a}{\pi} + \frac{\varphi_b}{\pi} - D_1\right)\frac{V_oT_s}{8nL_r} + \frac{V_{\text{bat}}D_1T_s}{4L_r} \\ i_1(t_4) = \frac{V_{\text{bat}}D_1T_s}{4L_r} - \left(2 - \frac{\varphi_a}{\pi} - \frac{\varphi_b}{\pi} - D_1\right)\frac{V_oT_s}{8nL_r} \end{cases} \tag{6.20}$$

$$\begin{cases} i_2(t_0) = \left(2 - \frac{2\varphi_a}{\pi} - D_1\right)\frac{V_oT_s}{16n^2L_r} - \frac{V_{\text{bat}}D_1T_s}{8nL_r} - \frac{V_oT_s}{32L_a}\left(\frac{\varphi_a}{\pi} - \frac{\varphi_b}{\pi}\right) \\ i_2(t_1) = \frac{V_oT_s}{8n^2L_r} - \left(1 - \frac{2\varphi_a}{\pi}\right)\frac{V_{\text{bat}}T_s}{4nL_r} - \frac{V_oT_s}{32L_a}\left(\frac{\varphi_a}{\pi} - \frac{\varphi_b}{\pi}\right) \\ i_2(t_2) = \left(1 + \frac{2\varphi_a}{\pi} - \frac{2\varphi_b}{\pi}\right)\frac{V_oT_s}{8n^2L_r} - \left(1 - \frac{2\varphi_b}{\pi}\right)\frac{V_{\text{bat}}T_s}{4nL_r} + \frac{V_oT_s}{32L_a}\left(\frac{\varphi_a}{\pi} - \frac{\varphi_b}{\pi}\right) \\ i_2(t_3) = \frac{V_{\text{bat}}D_1T_s}{4nL_r} - \left(D_1 - \frac{2\varphi_a}{\pi}\right)\frac{V_oT_s}{8n^2L_r} + \frac{V_oT_s}{32L_a}\left(\frac{\varphi_a}{\pi} - \frac{\varphi_b}{\pi}\right) \\ i_2(t_4) = \frac{V_{\text{bat}}D_1T_s}{4nL_r} - \left(2 - \frac{2\varphi_a}{\pi} - D_1\right)\frac{V_oT_s}{8n^2L_r} + \frac{V_oT_s}{32L_a}\left(\frac{\varphi_a}{\pi} - \frac{\varphi_b}{\pi}\right) \end{cases} \tag{6.21}$$

$$\begin{cases} i_3(t_0) = \left(2 - \frac{2\varphi_b}{\pi} - D_1\right)\frac{V_oT_s}{8n^2L_r} - \frac{V_{\text{bat}}D_1T_s}{4nL_r} + \frac{V_oT_s}{32L_a}\left(\frac{\varphi_a}{\pi} - \frac{\varphi_b}{\pi}\right) \\ i_3(t_1) = \left(1 - \frac{2\varphi_b}{\pi} + \frac{2\varphi_a}{\pi}\right)\frac{V_oT_s}{8n^2L_r} - \left(1 - \frac{2\varphi_a}{\pi}\right)\frac{V_{\text{bat}}T_s}{4nL_r} + \frac{V_oT_s}{32L_a}\left(\frac{\varphi_a}{\pi} - \frac{\varphi_b}{\pi}\right) \\ i_3(t_2) = \frac{V_oT_s}{8n^2L_r} - \left(1 - \frac{2\varphi_b}{\pi}\right)\frac{V_{\text{bat}}T_s}{4nL_r} - \frac{V_oT_s}{32L_a}\left(\frac{\varphi_a}{\pi} - \frac{\varphi_b}{\pi}\right) \\ i_3(t_3) = \frac{V_{\text{bat}}D_1T_s}{4nL_r} - \left(D_1 - \frac{2\varphi_b}{\pi}\right)\frac{V_oT_s}{8n^2L_r} - \frac{V_oT_s}{32L_a}\left(\frac{\varphi_a}{\pi} - \frac{\varphi_b}{\pi}\right) \\ i_3(t_4) = \frac{V_{\text{bat}}D_1T_s}{4nL_r} - \left(2 - \frac{2\varphi_b}{\pi} - D_1\right)\frac{V_oT_s}{8n^2L_r} - \frac{V_oT_s}{32L_a}\left(\frac{\varphi_a}{\pi} - \frac{\varphi_b}{\pi}\right) \end{cases} \tag{6.22}$$

The currents i_1, i_2, and i_3 at the time of t_0, t_1, t_2, and t_3 in Mode 2a are expressed as follows:

$$\begin{cases} i_1(t_0) = \left(1 + \frac{\varphi_a}{\pi} - \frac{\varphi_b}{\pi}\right)\frac{V_oT_s}{8nL_r} - \frac{V_{\text{bat}}D_1T_s}{4L_r} \\ i_1(t_1) = \left(1 + \frac{\varphi_a}{\pi} - \frac{\varphi_b}{\pi}\right)\frac{V_oT_s}{8nL_r} - \frac{V_{\text{bat}}D_1T_s}{4L_r} \\ i_1(t_2) = \left(1 + \frac{\varphi_a}{\pi} - \frac{\varphi_b}{\pi}\right)\frac{V_oT_s}{8nL_r} - \left(1 - \frac{2\varphi_b}{\pi}\right)\frac{V_{\text{bat}}T_s}{4L_r} \\ i_1(t_3) = \frac{V_{\text{bat}}D_1T_s}{4L_r} - \left(D_1 - \frac{\varphi_a}{\pi} - \frac{\varphi_b}{\pi}\right)\frac{V_oT_s}{8nL_r} \\ i_1(t_4) = \frac{V_{\text{bat}}D_1T_s}{4L_r} - \left(1 + \frac{\varphi_a}{\pi} - \frac{\varphi_b}{\pi}\right)\frac{V_oT_s}{8nL_r} \end{cases} \tag{6.23}$$

$$\begin{cases} i_2(t_0) = \frac{V_oT_s}{8n^2L_r} - \frac{V_{\text{bat}}D_1T_s}{4nL_r} - \frac{V_oT_s}{32L_a}\left(\frac{\varphi_a}{\pi} - \frac{\varphi_b}{\pi}\right) \\ i_2(t_1) = \left(D_1 + \frac{2\varphi_a}{\pi}\right)\frac{V_oT_s}{8n^2L_r} - \frac{V_{\text{bat}}D_1T_s}{4nL_r} - \frac{V_oT_s}{32L_a}\left(1 - \frac{\varphi_a}{\pi} - \frac{\varphi_b}{\pi} - D_1\right) \\ i_2(t_2) = \left(1 + \frac{2\varphi_a}{\pi} - \frac{2\varphi_b}{\pi}\right)\frac{V_oT_s}{8n^2L_r} - \left(1 - \frac{2\varphi_b}{\pi}\right)\frac{V_{\text{bat}}T_s}{4nL_r} + \frac{V_oT_s}{32L_a}\left(\frac{\varphi_a}{\pi} - \frac{\varphi_b}{\pi}\right) \\ i_2(t_3) = \frac{V_{\text{bat}}D_1T_s}{4nL_r} - \left(D_1 - \frac{2\varphi_a}{\pi}\right)\frac{V_oT_s}{8n^2L_r} + \frac{V_oT_s}{32L_a}\left(\frac{\varphi_a}{\pi} - \frac{\varphi_b}{\pi}\right) \\ i_2(t_4) = \frac{V_{\text{bat}}D_1T_s}{4nL_r} - \frac{V_oT_s}{8n^2L_r} + \frac{V_oT_s}{32L_a}\left(\frac{\varphi_a}{\pi} - \frac{\varphi_b}{\pi}\right) \end{cases} \tag{6.24}$$

$$\begin{cases} i_3(t_0) = \left(1 + \frac{2\varphi_a}{\pi} - \frac{2\varphi_b}{\pi}\right)\frac{V_oT_s}{8n^2L_r} - \frac{V_{\text{bat}}D_1T_s}{4nL_r} + \frac{V_oT_s}{32L_a}\left(\frac{\varphi_a}{\pi} - \frac{\varphi_b}{\pi}\right) \\ i_3(t_1) = \left(2 - D_1 - \frac{2\varphi_b}{\pi}\right)\frac{V_oT_s}{8n^2L_r} - \frac{V_{\text{bat}}D_1T_s}{4nL_r} + \frac{V_oT_s}{32L_a}\left(1 - \frac{\varphi_a}{\pi} - \frac{\varphi_b}{\pi} - D_1\right) \\ i_3(t_2) = \frac{V_oT_s}{8n^2L_r} - \left(1 - \frac{2\varphi_b}{\pi}\right)\frac{V_{\text{bat}}T_s}{4nL_r} - \frac{V_oT_s}{32L_a}\left(\frac{\varphi_a}{\pi} - \frac{\varphi_b}{\pi}\right) \\ i_3(t_3) = \frac{V_{\text{bat}}D_1T_s}{4nL_r} - \left(D_1 - \frac{2\varphi_b}{\pi}\right)\frac{V_oT_s}{8n^2L_r} - \frac{V_oT_s}{32L_a}\left(\frac{\varphi_a}{\pi} - \frac{\varphi_b}{\pi}\right) \\ i_3(t_4) = \frac{V_{\text{bat}}D_1T_s}{4nL_r} - \left(1 + \frac{2\varphi_a}{\pi} - \frac{2\varphi_b}{\pi}\right)\frac{V_oT_s}{8n^2L_r} - \frac{V_oT_s}{32L_a}\left(\frac{\varphi_a}{\pi} - \frac{\varphi_b}{\pi}\right) \end{cases} \tag{6.25}$$

The currents i_1, i_2, and i_3 at the time of t_0, t_1, t_2, and t_3 in Mode 3 are expressed as follows:

$$\begin{cases} i_1(t_0) = \left(1 + \frac{\varphi_a}{\pi} - \frac{\varphi_b}{\pi}\right)\frac{V_oT_s}{8nL_r} - \frac{V_{\text{bat}}D_1T_s}{4L_r} \\ i_1(t_1) = \left(1 + \frac{\varphi_a}{\pi} - \frac{\varphi_b}{\pi}\right)\frac{V_oT_s}{8nL_r} - \frac{V_{\text{bat}}D_1T_s}{4L_r} \\ i_1(t_2) = \left(D_1 + \frac{\varphi_b}{\pi} + \frac{\varphi_a}{\pi}\right)\frac{V_oT_s}{8nL_r} - \frac{V_{\text{bat}}D_1T_s}{4L_r} \\ i_1(t_3) = \frac{V_{\text{bat}}D_1T_s}{4L_r} - \left(D_1 - \frac{\varphi_a}{\pi} - \frac{\varphi_b}{\pi}\right)\frac{V_oT_s}{8nL_r} \\ i_1(t_4) = \frac{V_{\text{bat}}D_1T_s}{4L_r} - \left(1 + \frac{\varphi_a}{\pi} - \frac{\varphi_b}{\pi}\right)\frac{V_oT_s}{8nL_r} \end{cases} \tag{6.26}$$

$$\begin{cases} i_2(t_0) = \frac{V_oT_s}{18n^2L_r} - \frac{V_{\text{bat}}D_1T_s}{4nL_r} - \frac{V_oT_s}{32L_a}\left(\frac{\varphi_a}{\pi} - \frac{\varphi_b}{\pi}\right) \\ i_2(t_1) = \left(1 + \frac{2\varphi_a}{\pi} - \frac{2\varphi_b}{\pi}\right)\frac{V_oT_s}{8n^2L_r} - \frac{V_{\text{bat}}D_1T_s}{4nL_r} + \frac{V_oT_s}{32L_a}\left(\frac{\varphi_a}{\pi} - \frac{\varphi_b}{\pi}\right) \\ i_2(t_2) = \left(D_1 + \frac{2\varphi_a}{\pi}\right)\frac{V_oT_s}{8n^2L_r} - \frac{V_{\text{bat}}D_1T_s}{4nL_r} + \frac{V_oT_s}{32L_a}\left(\frac{\varphi_a}{\pi} - \frac{\varphi_b}{\pi}\right) \\ i_2(t_3) = \frac{V_{\text{bat}}D_1T_s}{4nL_r} - \left(D_1 - \frac{2\varphi_a}{\pi}\right)\frac{V_oT_s}{8n^2L_r} + \frac{V_oT_s}{32L_a}\left(\frac{\varphi_a}{\pi} - \frac{\varphi_b}{\pi}\right) \\ i_2(t_4) = \frac{V_{\text{bat}}D_1T_s}{4nL_r} - \frac{V_oT_s}{8n^2L_r} + \frac{V_oT_s}{32L_a}\left(\frac{\varphi_a}{\pi} - \frac{\varphi_b}{\pi}\right) \end{cases} \tag{6.27}$$

$$\begin{cases} i_3(t_0) = \left(1 + \frac{2\varphi_a}{\pi} - \frac{2\varphi_b}{\pi}\right)\frac{V_oT_s}{8n^2L_r} - \frac{V_{\text{bat}}D_1T_s}{4nL_r} + \frac{V_oT_s}{32L_a}\left(\frac{\varphi_a}{\pi} - \frac{\varphi_b}{\pi}\right) \\ i_3(t_1) = \frac{V_oT_s}{8n^2L_r} - \frac{V_{\text{bat}}D_1T_s}{4nL_r} - \frac{V_oT_s}{32L_a}\left(\frac{\varphi_a}{\pi} - \frac{\varphi_b}{\pi}\right) \\ i_3(t_2) = \left(D_1 + \frac{2\varphi_b}{\pi}\right)\frac{V_oT_s}{8n^2L_r} - \frac{V_{\text{bat}}D_1T_s}{4nL_r} - \frac{V_oT_s}{32L_a}\left(\frac{\varphi_a}{\pi} - \frac{\varphi_b}{\pi}\right) \\ i_3(t_3) = \frac{V_{\text{bat}}D_1T_s}{4nL_r} - \left(D_1 - \frac{2\varphi_b}{\pi}\right)\frac{V_oT_s}{8n^2L_r} - \frac{V_oT_s}{32L_a}\left(\frac{\varphi_a}{\pi} - \frac{\varphi_b}{\pi}\right) \\ i_3(t_4) = \frac{V_{\text{bat}}D_1T_s}{4nL_r} - \left(1 + \frac{2\varphi_a}{\pi} - \frac{2\varphi_b}{\pi}\right)\frac{V_oT_s}{8n^2L_r} - \frac{V_oT_s}{32L_a}\left(\frac{\varphi_a}{\pi} - \frac{\varphi_b}{\pi}\right) \end{cases} \tag{6.28}$$

The currents i_1, i_2, and i_3 at the time of t_0, t_1, t_2, and t_3 in Mode 4 are expressed as follows:

$$\begin{cases} i_1(t_0) = \left(1 + \frac{\varphi_a}{\pi} - \frac{\varphi_b}{\pi}\right)\frac{V_oT_s}{8nL_r} - \left(1 + \frac{2\varphi_a}{\pi}\right)\frac{V_{\text{bat}}T_s}{4L_r} \\ i_1(t_1) = \left(1 + \frac{\varphi_a}{\pi} - \frac{\varphi_b}{\pi}\right)\frac{V_oT_s}{8nL_r} - \frac{V_{\text{bat}}D_1T_s}{4L_r} \\ i_1(t_2) = \left(1 + \frac{\varphi_a}{\pi} - \frac{\varphi_b}{\pi}\right)\frac{V_oT_s}{8nL_r} - \frac{V_{\text{bat}}D_1T_s}{4L_r} \\ i_1(t_3) = \left(1 + \frac{\varphi_a}{\pi} - \frac{\varphi_b}{\pi}\right)\frac{V_oT_s}{8nL_r} - \left(1 - \frac{2\varphi_b}{\pi}\right)\frac{V_{\text{bat}}D_1T_s}{4L_r} \\ i_1(t_4) = \left(1 + \frac{2\varphi_a}{\pi}\right)\frac{V_{\text{bat}}T_s}{4L_r} - \left(1 + \frac{\varphi_a}{\pi} - \frac{\varphi_b}{\pi}\right)\frac{V_oT_s}{8nL_r} \end{cases} \tag{6.29}$$

$$\begin{cases} i_2(t_0) = \frac{V_oT_s}{8n^2L_r} - \left(1 + \frac{2\varphi_a}{\pi}\right)\frac{V_{\text{bat}}T_s}{4nL_r} - \frac{V_oT_s}{32L_s}\left(\frac{\varphi_a}{\pi} - \frac{\varphi_b}{\pi}\right) \\ i_2(t_1) = \left(2 + \frac{2\varphi_a}{\pi} - D_1\right)\frac{V_oT_s}{8n^2L_r} - \frac{V_{\text{bat}}D_1T_s}{4nL_r} - \frac{V_oT_s}{32L_s}\left(D_1 - 1 - \frac{\varphi_a}{\pi} - \frac{\varphi_b}{\pi}\right) \\ i_2(t_2) = \left(D_1 + \frac{2\varphi_a}{\pi}\right)\frac{V_oT_s}{8n^2L_r} - \frac{V_{\text{bat}}D_1T_s}{4nL_r} - \frac{V_oT_s}{32L_s}\left(1 - D_1 - \frac{\varphi_a}{\pi} - \frac{\varphi_b}{\pi}\right) \\ i_2(t_3) = \left(1 + \frac{2\varphi_a}{\pi} - \frac{2\varphi_b}{\pi}\right)\frac{V_oT_s}{8n^2L_r} - \left(1 - \frac{2\varphi_b}{\pi}\right)\frac{V_{\text{bat}}T_s}{4nL_r} + \frac{V_oT_s}{32L_s}\left(\frac{\varphi_a}{\pi} - \frac{\varphi_b}{\pi}\right) \\ i_2(t_4) = \left(1 + \frac{2\varphi_a}{\pi}\right)\frac{V_{\text{bat}}T_s}{4nL_r} - \frac{V_oT_s}{8n^2L_r} + \frac{V_oT_s}{32L_s}\left(\frac{\varphi_a}{\pi} - \frac{\varphi_b}{\pi}\right) \end{cases} \tag{6.30}$$

$$\begin{cases} i_3(t_0) = \left(1 + \frac{2\varphi_a}{\pi} - \frac{2\varphi_b}{\pi}\right)\frac{V_oT_s}{8n^2L_r} - \left(1 + \frac{2\varphi_a}{\pi}\right)\frac{V_{\text{bat}}T_s}{4nL_r} + \frac{V_oT_s}{32L_s}\left(\frac{\varphi_a}{\pi} - \frac{\varphi_b}{\pi}\right) \\ i_3(t_1) = \left(D_1 - \frac{2\varphi_b}{\pi}\right)\frac{V_oT_s}{8n^2L_r} - \frac{V_{\text{bat}}D_1T_s}{4nL_r} + \frac{V_oT_s}{32L_s}\left(D_1 - 1 - \frac{\varphi_a}{\pi} - \frac{\varphi_b}{\pi}\right) \\ i_3(t_2) = \left(2 - \frac{2\varphi_b}{\pi} - D_1\right)\frac{V_oT_s}{8n^2L_r} - \frac{V_{\text{bat}}D_1T_s}{4nL_r} + \frac{V_oT_s}{32L_s}\left(1 - D_1 - \frac{\varphi_a}{\pi} - \frac{\varphi_b}{\pi}\right) \\ i_3(t_3) = \frac{V_oT_s}{8n^2L_r} - \left(1 - \frac{2\varphi_b}{\pi}\right)\frac{V_{\text{bat}}D_1T_s}{4nL_r} - \frac{V_oT_s}{32L_s}\left(\frac{\varphi_a}{\pi} - \frac{\varphi_b}{\pi}\right) \\ i_3(t_4) = \left(1 + \frac{2\varphi_a}{\pi}\right)\frac{V_{\text{bat}}T_s}{4nL_r} - \left(1 + \frac{2\varphi_a}{\pi} - \frac{2\varphi_b}{\pi}\right)\frac{V_oT_s}{8n^2L_r} - \frac{V_oT_s}{32L_s}\left(\frac{\varphi_a}{\pi} - \frac{\varphi_b}{\pi}\right) \end{cases} \tag{6.31}$$

References

1. Thomas S, Doncker RD, Lenke R (2015) Bidirectional DC–DC converter. United State Patent No. 9148065B2
2. Sha D, Lin Q, You F, Wang X, Xu G (2016) A ZVS bidirectional three-level dc–dc converter with direct current slew rate control of leakage inductance current. IEEE Trans Ind Appl 52 (3):2368–3377

3. Lu Y, Xing Y, Wu H (2016) A PWM plus phase-shift controlled interleaved isolated boost converter based on semiactive quadrupler rectifier for high step-up applications. IEEE Trans Ind Electron 63(7):4211–4221
4. Guo Z, Sun K, Sha D (2016) Improved ZVS three-level dc–dc converter with reduced circulating loss. IEEE Trans Power Electron 31(6):6394–6404
5. Kim J, Kwon J, Kwon B (2018) High-efficiency two-stage three-level grid-connected photovoltaic inverter. IEEE Trans Ind Electron 65(3):2368–2377
6. Pinheiro JR, Barbi I (1993) The three-level ZVS-PWM DC-to-DC converter. IEEE Trans Power Electron 8(4):486–492
7. Ruan X, Xu D, Zhou L, Li B, Chen Q (2002) Zero-voltage-switching PWM three-level converter with two clamping diodes. IEEE Trans Ind Electron 49(4):790–799

Chapter 7
A Current-Fed Dual Active Bridge DC–DC Converter Using Dual PWM Plus Double Phase Shifted Control

Abstract A double PWM plus double phase shifted (DPDPS) control is proposed for current-fed dual active bridge (DAB) bidirectional DC–DC converters. With which the circulating current during the non-power transfer stage can be minimized. The mode analysis of the current-fed bidirectional DC–DC converter using the proposed control strategy is given. The comparison concerning the peak current and RMS current for transformer windings and switches is made by using the conventional PWM plus phase shifted (PPS) control and the proposed DPDPS control. The mode analysis of the proposed control strategy and comparison between PPS control and the proposed control are given. With the proposed DPDPS control, the converter has lower conduction loss, lower peak current, and higher efficiency. A 1 kW prototype is built to verify the proposed topology employing the proposed DPDPS control.

Keywords Bidirectional DC–DC converter · PWM · Phase shifted
Conduction loss

7.1 Introduction to Current-Fed Dual Active Bridge

Energy storage systems using ultracapacitors and batteries usually feed a DC bus where the voltage has to be regulated. The voltage variation across the ultracapacitor or battery depends on its depth of discharge that can reach one half the rated voltage. This fact requires the existence of a power interface with a wide voltage conversion ratio capability. By using conventional voltage-fed DAB, if the conversion ratio varies too much, the inductor current slew rate during the power transfer stage cannot be maintained to be zero, thus the inductor RMS current cannot be optimized to be the minimum. For an energy storage system by using conventional voltage-fed DAB, the current ripple is significant. Thus, it reduces the lifetime of some energy storage devices [1].

D. Sha and G. Xu, *High-Frequency Isolated Bidirectional Dual Active Bridge DC–DC Converters with Wide Voltage Gain*, CPSS Power Electronics Series,
https://doi.org/10.1007/978-981-13-0259-6_7

In the previous chapters, possible solutions which by adding additional switches or transformers are introduced. From this chapter, another solution that using current-fed DAB converters is introduced.

A novel current-fed DAB characterized by reducing one-side current ripple dramatically was introduced in [2]. When the conversion ratio varies, the conduction loss during the power transfer stage may not be minimum. To deal with the voltage ratio variation issue, a PWM plus phase shifted (PPS) control was proposed, in which the low-voltage side (LVS) converter is PWM modulated while gate signals for the high-voltage side (HVS) converter switches are fixed as 50%. Thus, the active clamp voltage can be regulated according to the HVS voltage and the turns ratio. Therefore, the current slop during the power transferring stage can be maintained to be zero in facing the voltage conversion ratio variation [3–5]. Thus, minimum conduction loss during the power transfer stage can be obtained. However, for the non-power transfer stage, when LVS voltage is rather low, the circulation loss is significant. Besides, the very high current spike generated in the non-power transfer stage may damage power switches. This lowers the system reliability. An additional inductor can be added in series with the transformer to suppress the current spike, but the maximum transmitted power capability is limited and the non-power transfer stage length increases. A three-phase current-fed DAB can reduce the LVS current ripples significantly by interleaving technology [6]; it is suited for large power applications.

In this chapter, a dual PWM plus double phase shifted control strategy is proposed for current-fed DAB converter. With the proposed control strategy, the conduction loss and the circulation current can be reduced significantly. The current spike issue brought by the PPS control in facing the variation of the LVS voltage can be avoided in the full load range even when the transformer leakage inductance is designed to be pretty small. Besides, ZVS can be achieved well for LVS switches and ZVS or zero-current switching (ZCS) can be obtained for HVS switches.

7.2 Mode Analysis with the Proposed Control Strategy

The proposed topology is shown in Fig. 7.1. In the HVS, a full-bridge inverter is adopted. In the LVS, L_1 and L_2 are the DC inductors which can be treated as two constant current sources. Q_1 and Q_2 are the main switches with body diodes. Q_{1a}, C_{d1} and Q_{2a}, C_{d2} constitute the active clamp circuit. The AC inductor L_r represents the sum of the primary-referred transformer leakage inductance. In the HVS, S_1 – S_4 are the main switches with body diodes.

To reduce the conduction loss during the power transfer stage, PPS control can be used. As seen in Fig. 7.2, the clamp voltage is regulated to be the value V_c which is determined by the transformer turns ratios and the HVS voltage V_2. This makes the voltage imposed on the leakage inductance L_r equal to zero. Thereby, the slew rate of the leakage inductance current can be kept zero during the power transfer stage within the range $[(2d-1)\pi, \pi]$ and $[2d\pi, 2\pi]$. As shown in Fig. 7.2, when the

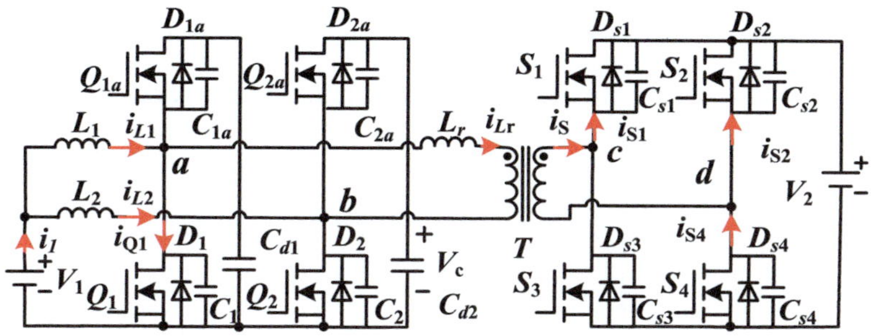

Fig. 7.1 Circuit of the current-fed bidirectional DC–DC converter

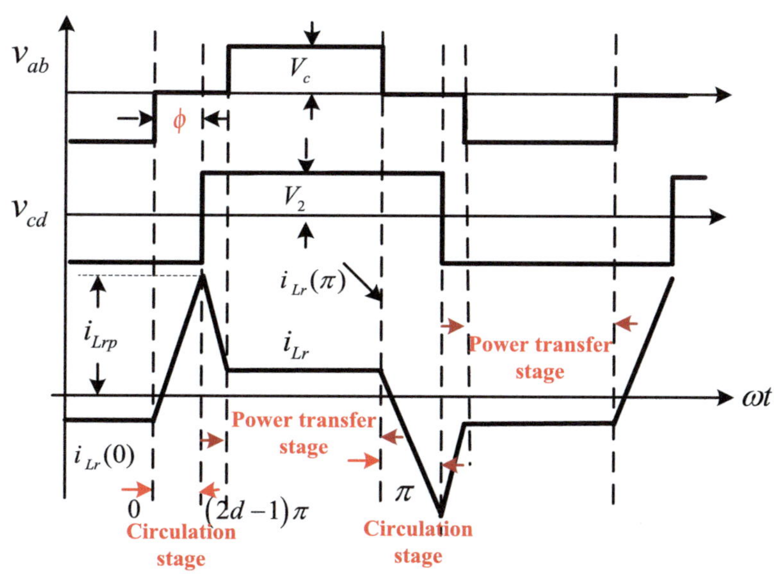

Fig. 7.2 PPS control for the converter as shown in Fig. 7.1 in boost mode

instantaneous value of the voltage v_{ab} is zero, the range can be defined as circulation stage in which no active power is transmitted between the LVS and the HVS. It is worth noting, during the circulation stage, the leakage current is not zero. Instead, it increases and decreases dramatically. When V_1 is low and the duty ratio for LVS bottom switch is very large, the circulation stage length becomes longer. During the circulation stage, the nonzero leakage current causes significant conduction loss for transformer windings and power switches. If V_1 varies widely, the conversion efficiency is reduced especially when V_1 is rather low, and the system reliability can even be affected due to the large current spike generated during the circulation stage.

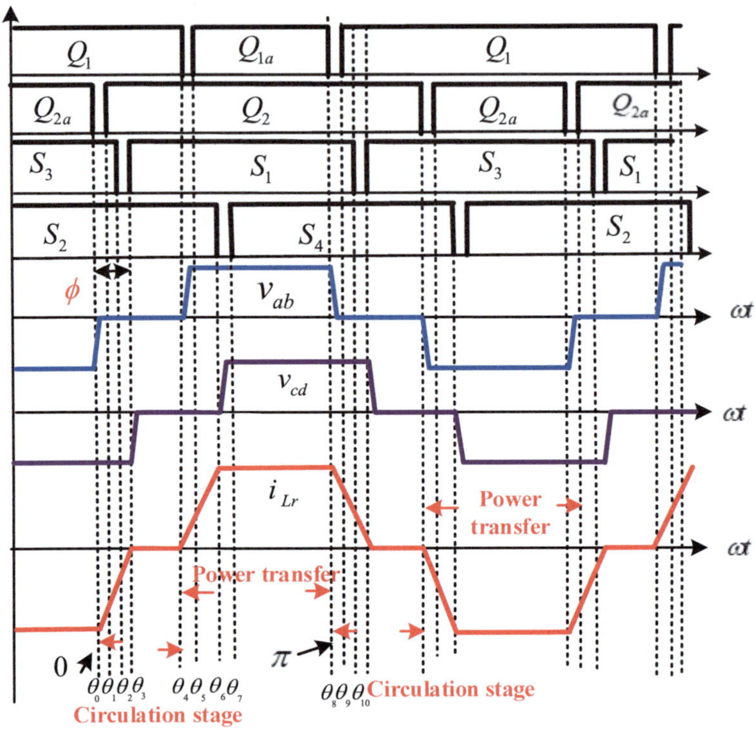

Fig. 7.3 Steady-state waveforms of the DPDPS control in boost mode

To minimize the conduction loss during the circulation stage, for the HVS, PWM control can be employed like the LVS with the same duty cycle. Thus, both v_{ab} and v_{cd} have the same shape but they are phase shifted according to the transmitted power magnitude and direction requirement. To achieve soft switching for the HVS switches, phase shifted gating sequence is used. The steady-state waveform in boost mode during one complete period can be seen in Fig. 7.3.

As seen, the voltage v_{ab} leads v_{cd}, during the circulation stage [θ_3, θ_4], and the leakage current drops to zero and stays at zero within the circulation stage range [θ_3, θ_4] until the power transfer stage begins. Thus, during the circulation stage, the conduction loss can be reduced. The detailed mode analysis can be described as follows.

The typical working modes are shown in Fig. 7.4, which will be described as follows

Stage 1 (Before θ_0): Q_1, Q_{2a}, S_2, and S_3 conduct. During this stage, the power flows from LVS (V_1) to HVS (V_2).

Stage 2 ($\theta_0 - \theta_1$): At θ_0, Q_{2a} is turned off. The difference between i_{L2} and i_{Lr} charge/discharge the junction capacitors C_{2a} and C_2.

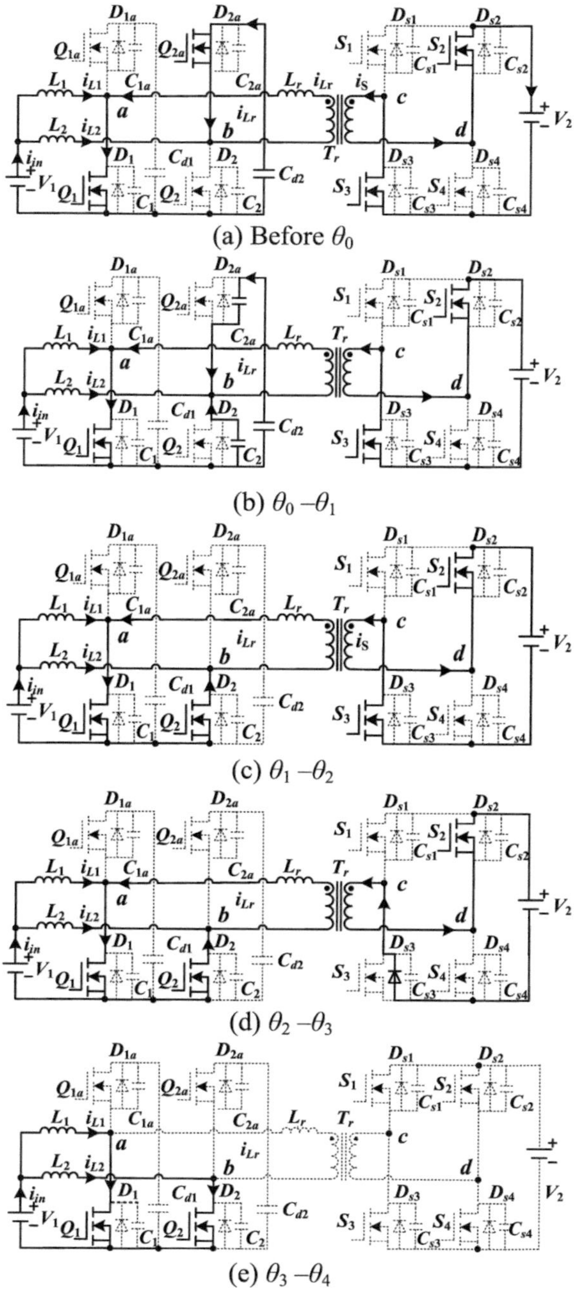

Fig. 7.4 Steady-state waveforms of the DPDPS control in boost mode

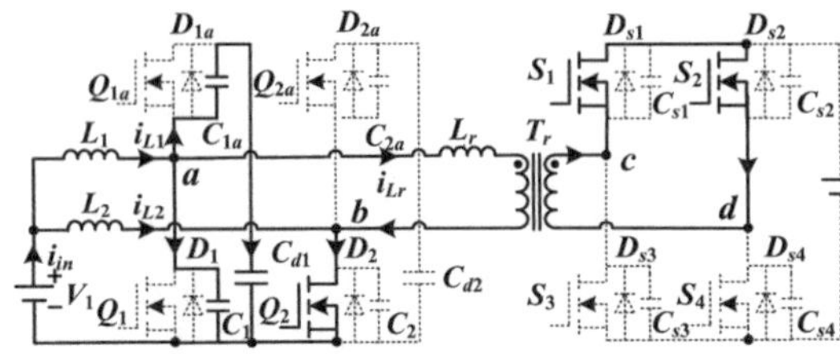

(f) θ_4 –θ_5

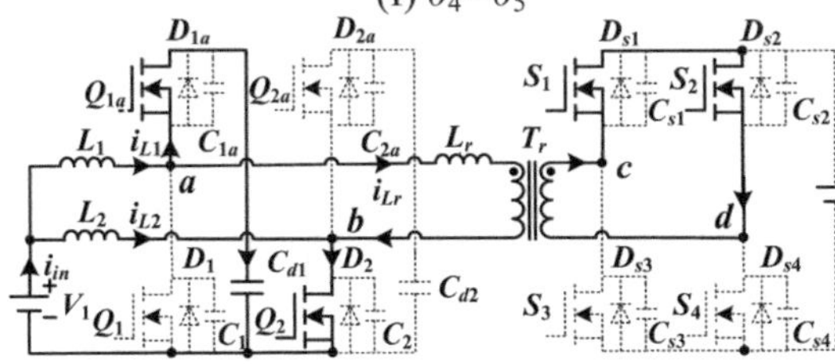

(g) θ_5 –θ_6

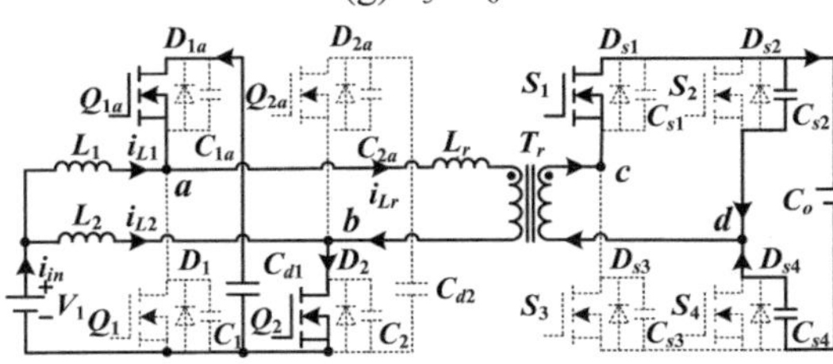

(h) θ_6 –θ_7

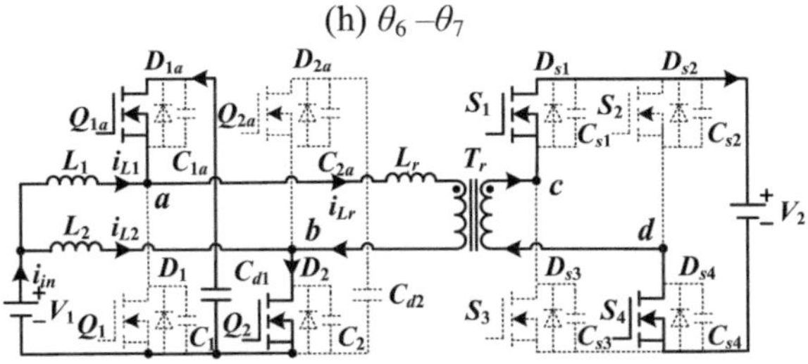

(i) θ_7 –θ_8

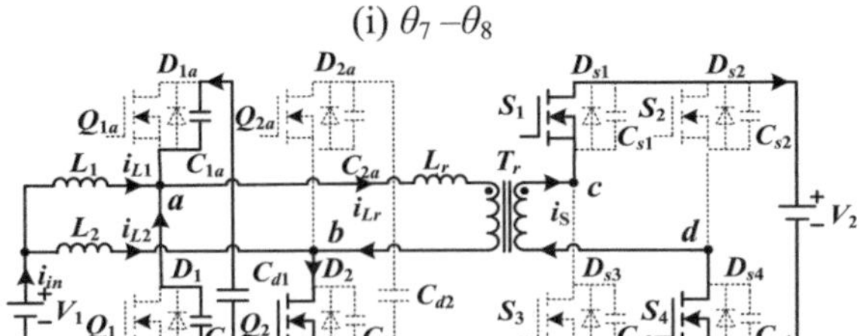

(j) θ_8 –θ_9

Fig. 7.4 (continued)

Stage 3 ($\theta_1 - \theta_2$): Before θ_1, when the voltage of C_2 reaches the clamp voltage V_c, then the body diode of Q_2 begins to conduct. At θ_1, Q_2 can be gated on at ZVS.

Stage 4 ($\theta_2 - \theta_3$): At θ_2, S_3 is turned off, and the body diode of S_3 becomes forward-biased.

Stage 5 ($\theta_3 - \theta_4$): At θ_3, S_1 is turned on at ZCS, and the voltage of C_{S3} rises to V_2 and the voltage of C_{S1} decays to zero. At the same time, the current i_{Lr} decays to zero and remains at zero.

Stage 6 ($\theta_4 - \theta_5$): At θ_4, Q_1 is turned off. The difference of DC current and the leakage inductance current charges C_1 and discharges C_{1a}.

Stage 7 ($\theta_5 - \theta_6$): Before θ_5, the voltage over the junction capacitor C_{1a} decreases to zero. The body diode of Q_{1a} therefore becomes forward-biased. At θ_5, Q_{1a} is gated on at ZVS.

Stage 8 ($\theta_6 - \theta_7$): At θ_6, S_2 is turned off,and the current i_s charges C_{s2} and discharges C_{s4}, causing the body diode of S_4 to conduct finally.

Stage 9 ($\theta_7 - \theta_8$): At θ_7, S_4 is turned on at ZVS.

Stage 10 ($\theta_8 - \theta_9$): At θ_8, Q_{1a} is turned off. The second half cycle similar to the first half cycle begins. The difference between the current i_{Lr} and i_{L1} charges and discharges junction capacitors C_{1a} and C_1 until the body diode of Q_1 begins to conduct.

The steady-state waveform in boost mode during one complete period can be seen in Fig. 7.5, and the corresponding working stages are shown in Fig. 7.6.

Stage 1 (Before θ_0): Q_1, Q_{2a}, S_2, and S_3 conduct. During this stage, the power flows from HVS (V_2) to LVS (V_1).

Stage 2 ($\theta_0 - \theta_1$): At θ_0, S_3 is turned off, and the current i_s charges and discharges the junction capacitors Cs_1 and Cs_3, respectively, until the body diode of S_1 begins to conduct current. Then S_1 can be gated on ZVS.

Stage 3 ($\theta_1 - \theta_2$): At θ_1, S_1 can be turned on with ZVS.

Stage 4 ($\theta_2 - \theta_3$): At θ_2, Q_{2a} is turned off, and the current of i_{Lr2} and i_{Lr} makes C_{2a} charged and C_2 discharged until the body diode of Q_2 begins to conduct. Then, the current i_{Lr} begins to decay to zero and remains at zero.

Stage 5 ($\theta_3 - \theta_4$): At θ_3, the switch Q_2 can be turned on with ZVS, and the current i_{Lr} remains at zero.

Stage 6 ($\theta_4 - \theta_5$): At θ_4, S_2 can be turned off at ZCS.

Stage 7 ($\theta_5 - \theta_6$): At θ_5, S_1 is turned on at ZCS. The current i_{Lr} increases in reverse direction.

Stage 8 ($\theta_6 - \theta_7$): At θ_6, Q_1 is turned off. The difference between of i_{L1} and i_{Lr} charges C_1 and discharges C_{1a} until the body diode of Q_{1a} begins to conduct current.

Stage 9 ($\theta_7 - \theta_8$): At θ_7, Q_{1a} can be gated on at ZVS. The current i_{Q1a} flows from source to drain at first, but flows from drain to source after $i_{L1} = i_{Lr}$. During this stage, the power flows from HVS (V_2) to LVS (V_1).

Stage 10 ($\theta_8 - \theta_9$): At θ_8, S_1 is turned off, and the current charges the junction capacitors C_{s1} and C_{s3}, respectively, until the body diode of S_3 begins to conduct.

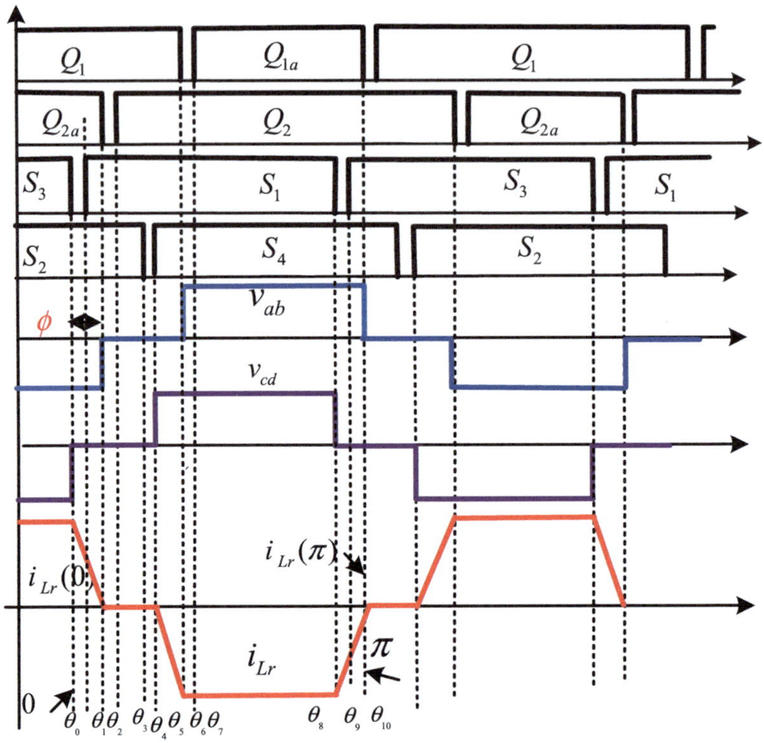

Fig. 7.5 Steady-state waveforms of the proposed DPDPS control in buck mode

7.3 Current Stress Comparison with PPS and DPDPS Control

To illustrate the conduction loss reduction by using the proposed PPS control and the proposed DPDPS control. Calculation comparison is made according to the specifications as shown in Table 7.1.

Take boost mode operation as an example, based on the leakage current waveform as shown in Fig. 7.2. The instantaneous value of the leakage current can be expressed as follows

$$i_{Lr}(\theta) = \begin{cases} \frac{V_c(\theta-\phi)}{\omega L_r}, & (0 \leq \theta \leq \phi) \\ 0, & (\phi \leq \theta \leq (2d-1)\pi) \\ \frac{V_c(\theta-2\pi D+\pi)}{\omega L_r}, & ((2d-1)\pi \leq \theta \leq (2d-1)\pi + \phi) \\ \frac{V_c\phi}{\omega L_r}, & ((2d-1)\pi + \phi \leq \theta \leq \pi) \\ -i_{Lr}(\theta-\pi), & (\pi \leq \theta \leq 2\pi) \end{cases} \tag{7.1}$$

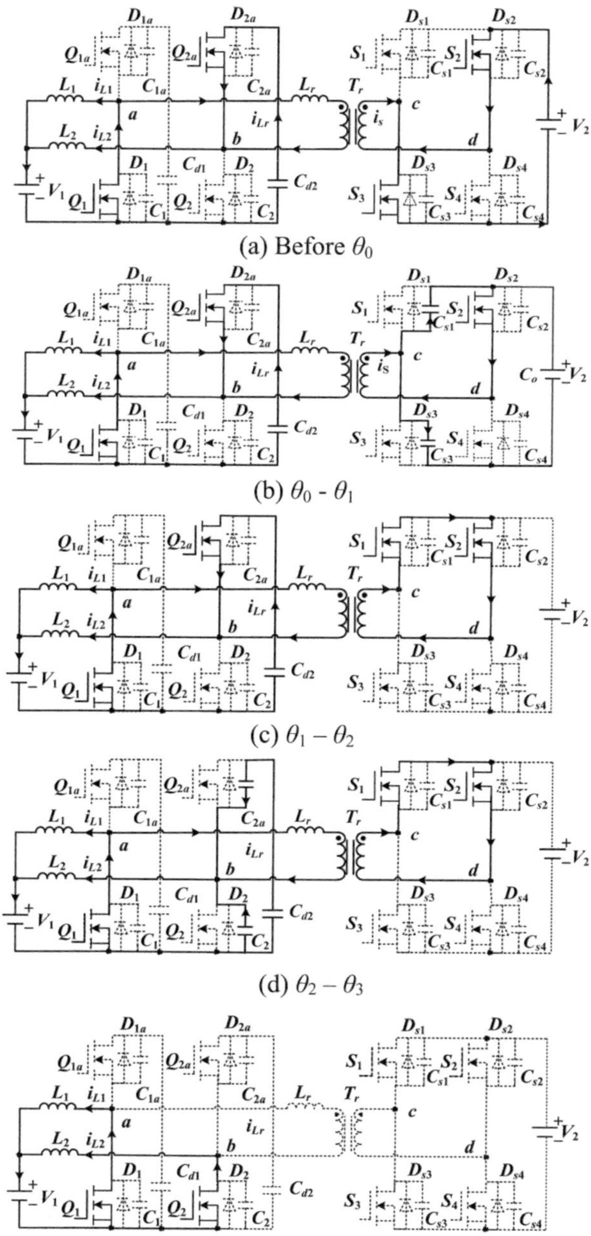

(a) Before θ_0

(b) θ_0 - θ_1

(c) $\theta_1 - \theta_2$

(d) $\theta_2 - \theta_3$

(e) $\theta_3 - \theta_4$

Fig. 7.6 Steady-state waveforms of the proposed DPDPS control in buck mode

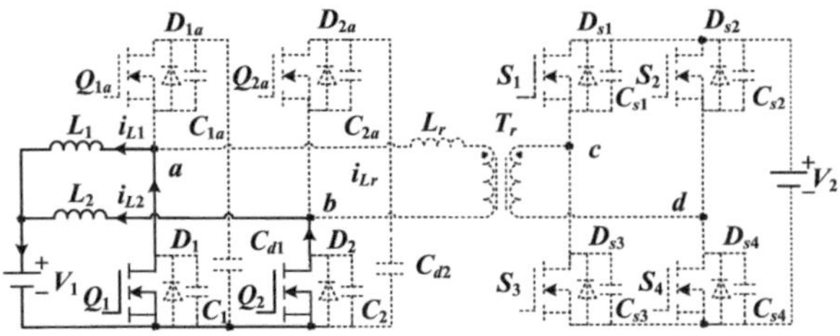

(f) $\theta_4 - \theta_5$

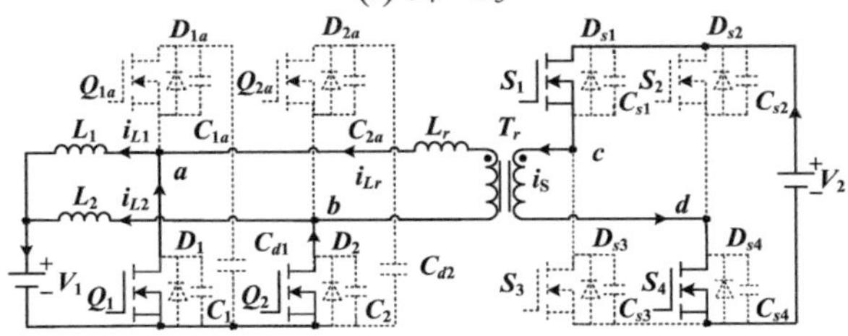

(g) $\theta_5 - \theta_6$

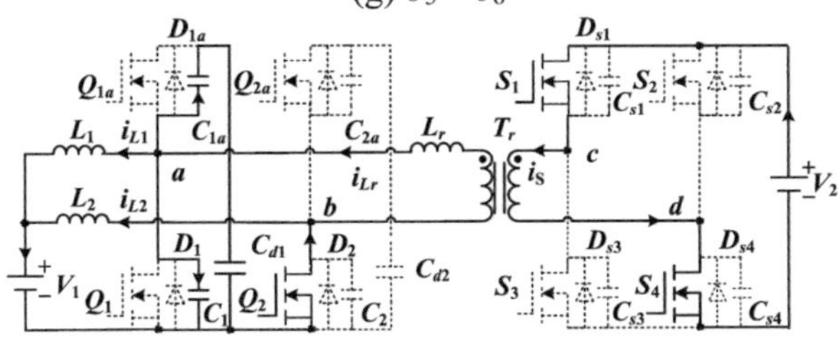

(h) $\theta_6 - \theta_7$

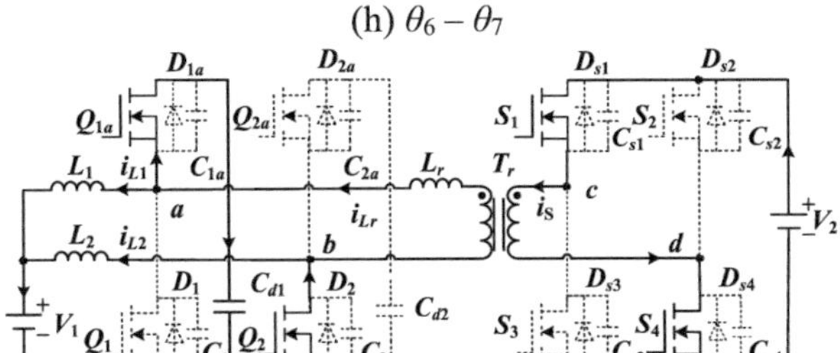

(i) $\theta_7 - \theta_8$

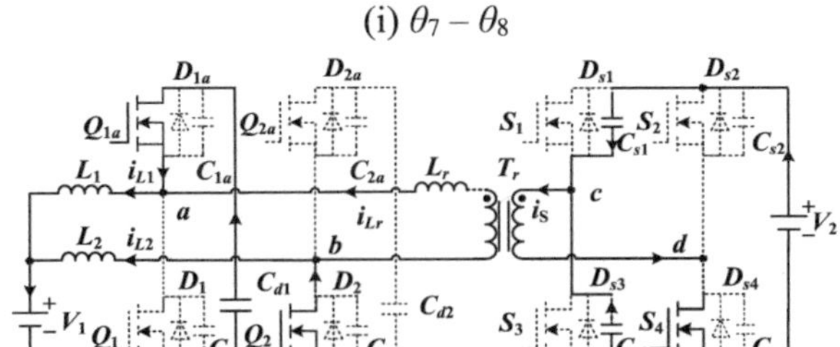

(j) $\theta_7 - \theta_8$

Fig. 7.6 (continued)

Table 7.1 Calculation specifications

Rated power P_o	1000 W
Voltage of HVS (V_2)	300 V
Voltage of LVS (V_1)	18–28 V
Switching frequency f	50 kHz
Inductance L_1, L_2	11 μH
Clamp capacitor C_{d1}, C_{d2}	30 μF
Turns ratio of T	2:10
Leakage inductance L_r	1.4 μH

where duty cycle D, clamp voltage V_c, angular frequency and phase shifted angle can be written by

$$D = 1 - \frac{N_2 V_1}{N_1 V_2}, \quad V_c = \frac{V_1}{1-D} = \frac{N_1}{N_2} V_2, \quad \omega = \frac{2\pi}{T} \tag{7.2}$$

For the output power P_o, the phase shifted angle can be obtained as

$$\phi = \frac{\left[\frac{2V_c^2(1-D)}{\omega L_r} - \sqrt{\left(\frac{2V_c^2(1-D)}{\omega L_r}\right)^2 - \frac{2V_c^2 P_o}{\pi\omega L_r}}\right] \times \pi\omega L_r}{V_c^2} \tag{7.3}$$

The leakage inductance current i_{Lr} RMS and peak values can be written by

$$\begin{cases} I_{Lr\text{RMS}} = \sqrt{\frac{\int_0^{2\pi} i_{Lr}(\theta)^2 d\theta}{2\pi}} \\ i_{Lr_{\text{peak}}} = \frac{V_c \phi}{\omega L_r} \end{cases} \tag{7.4}$$

The LVS inductor current i_{L2} is given by

$$i_{L2}(\theta) = \begin{cases} I_{L_{\min}} + K_1\theta, & \theta \in (0, 2\pi D) \\ I_{L_{\max}} + K_2(\theta - 2\pi D), & \theta \in (2\pi D, 2\pi) \end{cases} \tag{7.5}$$

where

$$\begin{cases} I_{L_{\min}} = \frac{P_o}{2V_c(1-D)} - \frac{V_c D(1-D)T}{2L} \\ I_{L_{\max}} = \frac{P_o}{2V_c(1-D)} + \frac{V_c D(1-D)T}{2L} \\ K_1 = \frac{V_c(1-D)T}{2\pi L}, \quad K_2 = -\frac{V_c D T}{2\pi L} \end{cases} \tag{7.6}$$

The instantaneous value through the switch Q_2 can be written by

$$i_{Q2}(\theta)=\begin{cases} i_{Lr}(\theta)+i_{L2}(\theta), & \theta\in(0,2\pi D) \\ 0, & \theta\in(2\pi D,2\pi) \end{cases} \tag{7.7}$$

The RMS and the peak values of the current i_{Q2} can be expressed as

$$I_{Q2a\text{RMS}}=\sqrt{\frac{\int_0^{2\pi} i_{Q2a}(\theta)^2 d\theta}{2\pi}},\quad i_{Q2a\text{peak}}=i_{Q2a}(2\pi D) \tag{7.8}$$

The instantaneous current in S_1 can be written by

$$i_{S1}(\theta)=\begin{cases} 0, & \theta\in(0,\phi) \\ -\frac{N_1}{N_2} i_{Lr}(\theta), & \theta\in(\phi,\pi+\phi) \\ 0, & \theta\in(\pi+\phi,2\pi) \end{cases} \tag{7.9}$$

The RMS and peak values of the current $i_{s1}(\theta)$ can be obtained as

$$I_{S1\text{RMS}}=\sqrt{\frac{\int_0^{2\pi} i_{S1}(\theta)^2 d\theta}{2\pi}}=\frac{\sqrt{2}}{2}\frac{N_1}{N_2} I_{Lr\text{RMS}},\quad i_{S1p}=\frac{N_1}{N_2} i_{Lrp} \tag{7.10}$$

7.3.1 Peak Current Analysis

Based on Eqs. (7.1)–(7.10), the RMS and peak values of the leakage inductance current i_{Lr} versus the LVS voltage V_1 in boost mode is illustrated in Fig. 7.7. As seen, with the proposed DPDPS control, compared with the PPS control, the RMS and peak values of the leakage inductance current can be reduced significantly with the same load and LVS voltage.

The peak current in a power switch can affect the reliability. Figure 7.8a, b shows the peak current of the LVS bottom switch Q_1 versus V_1 at light and heavy loads. As seen, with PPS control, the bottom switch has to undergo very large peak current especially when V_1 is very low. This leads to poor reliability. However, using the proposed DPDPS control, with the same V_1 input and the same output power, the peak current in the switch can be reduced dramatically. Since the bottom MOSFET and top MOSFET for one leg of the LVS have the same Rds(on), therefore the sum of RMS current square in one-leg switches is proportional to their conduction loss. As seen in Fig. 7.8c, d, the proposed DPDPS can reduce the conduction loss of the LVS switches especially when V_1 is very low.

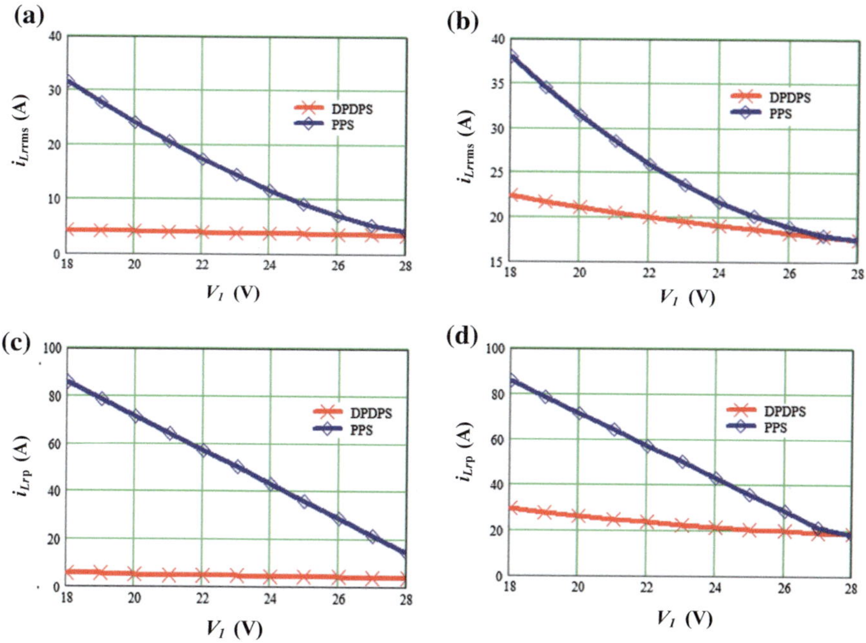

Fig. 7.7 Comparison of current i_{Lr} in boost mode: **a** i_{Lr} RMS value at 200 W output; **b** i_{Lr} RMS value at 1 kW output; **c** i_{Lr} peak value at 200 W output; **d** i_{Lr} peak value at 1 kW output

7.3.2 RMS Current Analysis

Figure 7.9 shows the RMS and peak currents in HVS switches with PPS control and the proposed control. As seen, in spite of the loads, the RMS current and peak current can be reduced by using the proposed DPDPS control, as compared with the PPS control. Obviously, the PPS control causes comparably high RMS and peak currents, in particular, at low LVS side voltage conditions, while the proposed DPDPS control can reduce the high RMS and peak current dramatically in this condition.

7.4 Implementation of the Control Strategy

The control block diagram is shown in Fig. 7.10. All the control part circled by the dashed line is implemented by a DSP. The PWM control loop is designed to regulate the clamp voltage V_c according to V_2 and the turns ratio. For the phase shift control loop, the outer loop is used to stabilize the HVS voltage V_2, and the output of the outer loop works as the reference for the inner current loop while the inner current loop output is used to generate the phase shift between the LVS and HVS. Since the duty ratio is determined by the PWM control loop, the HVS effective duty

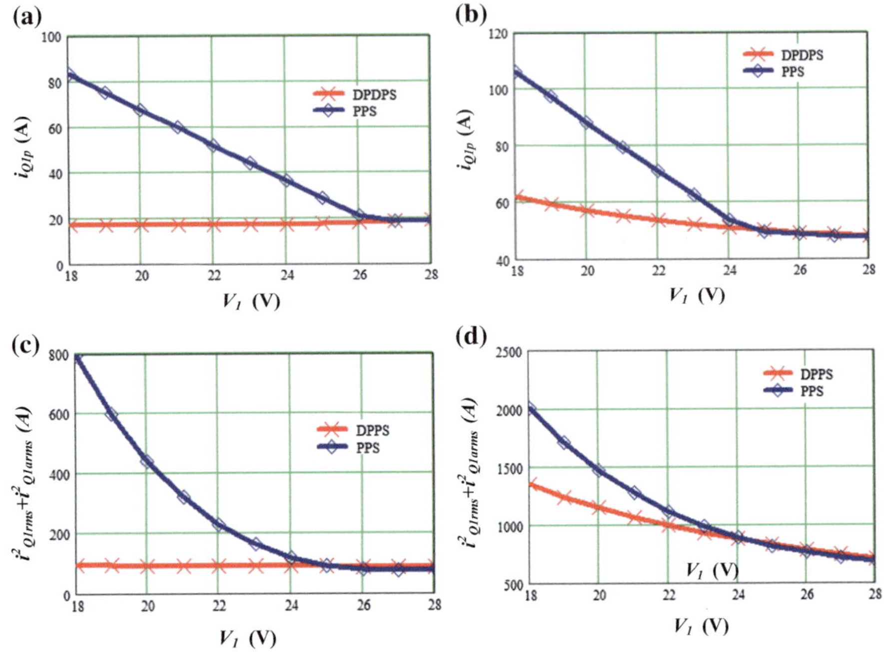

Fig. 7.8 Current comparison for LVS switches in boost mode: **a** peak current in bottom switch Q_1 at 200 W output; **b** peak current in Q_1 at 1 kW output; **c** sum of RMS current square for LVS one-leg switches at 200 W; **d** sum of RMS current square for LVS one-leg switches at 1 kW

cycle is identical with the LVS effective duty cycle, which can be seen in Figs. 7.3 and 7.5.

7.5 Experimental Results

7.5.1 Prototype

A 1 kW experimental prototype has been built in order to verify the effectiveness of the current-fed bidirectional DC–DC converter with the proposed DPDPS. Figure 7.11 gives the laboratory prototype picture.

The parameter specifications have been illustrated in Table 7.1. In addition, the switch type for LVS switches (Q_1–Q_2) is IXFN360N15T2, and the switch type for HVS switches (S_1–S_4) is IXFB94N50P. The high-frequency transformer core is EE70. The entire control of the system is implemented on a Texas Instruments TMS320F28335 DSP.

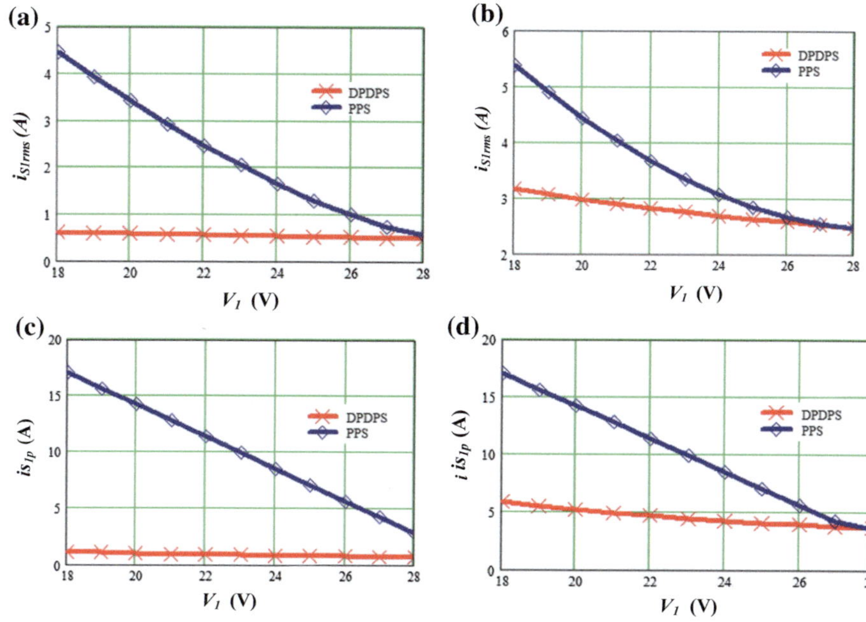

Fig. 7.9 Current comparison for HVS switches in boost mode: **a** S_1 RMS current at 200 W output; **b** S_1 RMS current at 1 kW output; **c** S_1 peak current at 200 W output; **d** S_1 peak current at 1 kW output

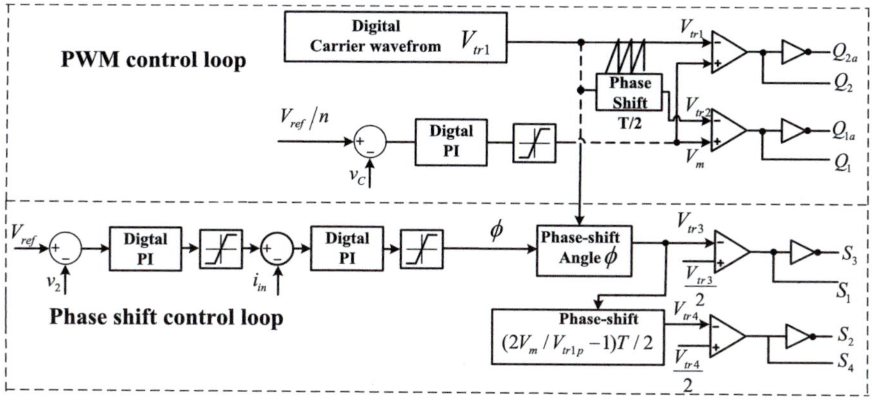

Fig. 7.10 Implementation of the proposed DPDPS control strategy

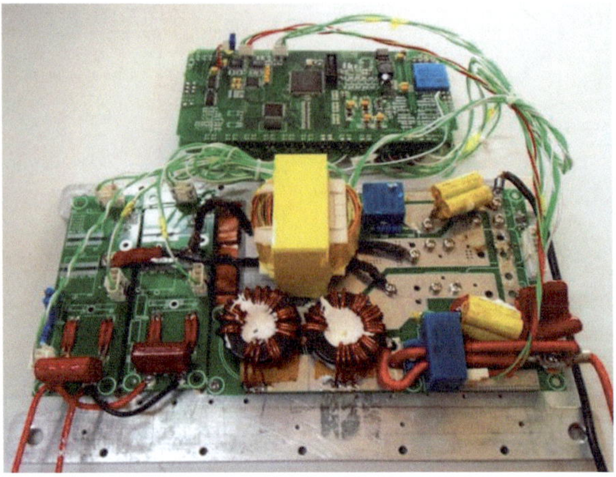

Fig. 7.11 Prototype picture

7.5.2 Boost Mode Operation

Figure 7.12 gives the steady-state waveform in boost mode under light-load (200 W) output with minimum and maximum input voltages. As seen, at the minimum input voltage, the leakage inductance peak current is as high as 80 A. But

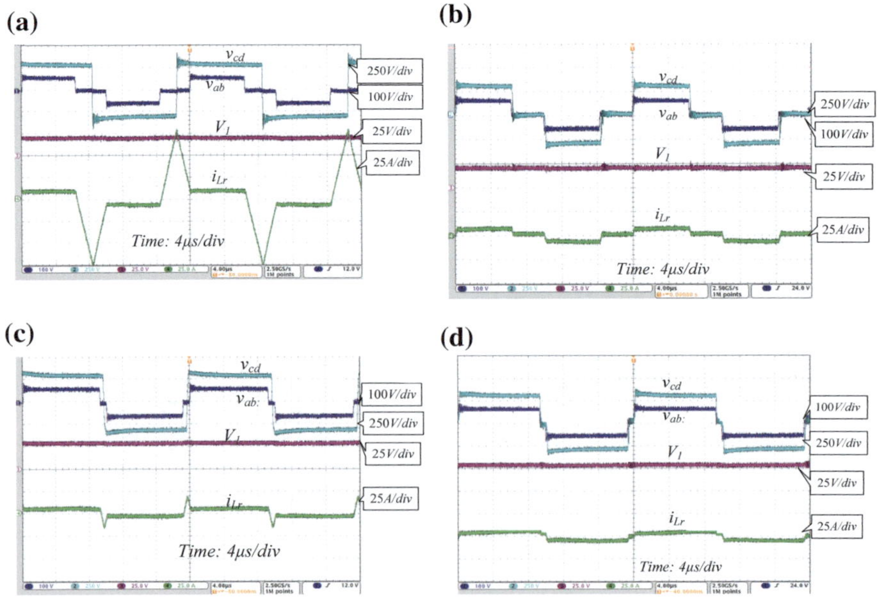

Fig. 7.12 Experimental results with PPS and DPDPS control in boost mode at 200 W output: **a** PPS control with $V_1 = 18$ V; **b** DPDPS control with $V_1 = 18$ V; **c** PPS control with $V_1 = 28$ V; **d** DPDPS control with $V_1 = 28$ V

with the proposed DPDPS control, the peak current can be reduced dramatically to no more than 10 A, which can be observed from Fig. 7.12a, b. Even though as the input voltage increases, the peak current value decreases, but the peak current of the leakage inductance is still higher than that with the DPDPS control. This can be seen from Fig. 7.12c, d.

Figure 7.13 gives the experimental result at full load (1 kW) in boost mode with the minimum and maximum input voltages. As seen from Fig. 7.13a, b, with the minimum input voltage (18 V), with the PPS control, the leakage inductance peak current is over 80 A. But with the proposed control, the leakage inductance peak current is just over 25 A. As seen in Fig. 7.13c, d, with the maximum input voltage (28 V), although the leakage inductance current differs not much, the circulation loss can be minimized by the proposed control because it remains as zero during the circulation stage.

Figure 7.14 gives the soft switching states for the switches at light load (200 W) with 24 V input in boost mode. Q_1 and Q_{1a} represent one leg of the LVS switches, and their switching waveforms are shown in Fig. 7.14a, b. As seen, both Q_1 and Q_{1a} can achieve with ZVS turn-on. Figure 7.14c, d illustrates the switching waveforms for HVS switches. As seen, the leading leg switch S_1 can achieve ZCS turn-on while the lagging switch S_2 can achieve ZVS turn-on.

Figure 7.15 gives the soft switching states for the switches at full load (1 kW) with 24 V input in boost mode. As shown in Fig. 7.15a, b, both Q_1 and Q_{1a} can be

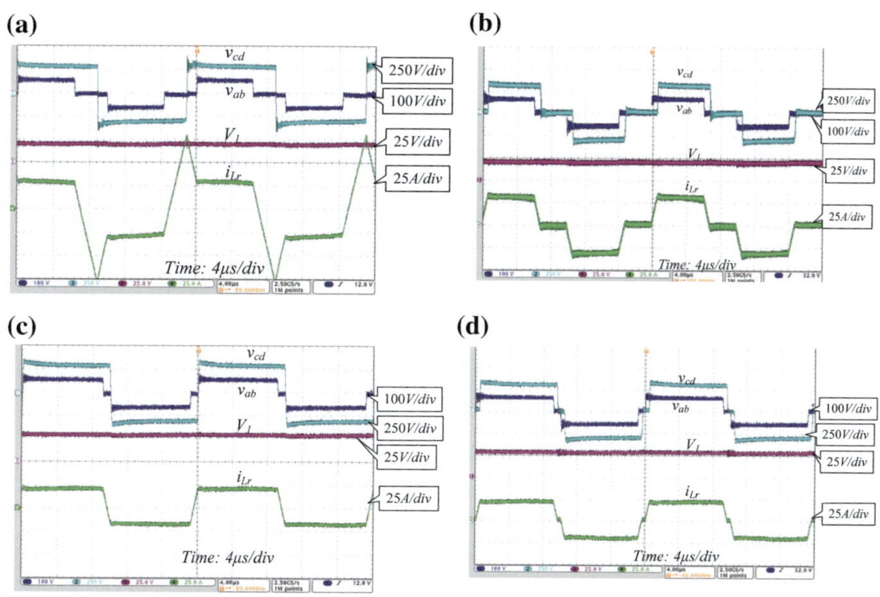

Fig. 7.13 Experimental results with PPS and DPDPS control in boost mode at 1000 W output: **a** PPS control with V_1 = 18 V; **b** DPDPS control with V_1 = 18 V; **c** PPS control with V_1 = 28 V; **d** DPDPS control with V_1 = 28 V

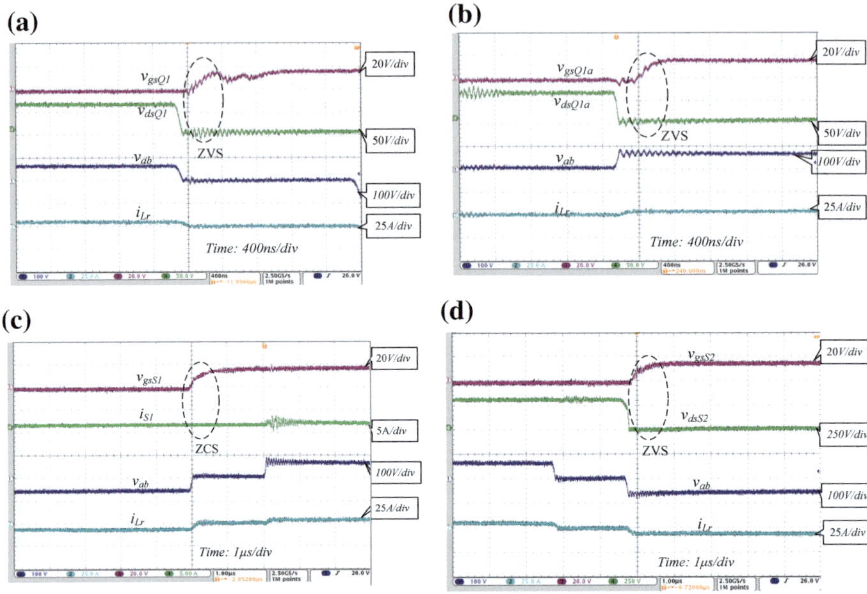

Fig. 7.14 Experimental results for soft switching with PPS and DPDPS control in boost mode at 24 V input and 200 W output condition: **a** ZVS of Q_1; **b** ZVS Q_{1a}; **c** ZCS of S_1; **d** ZVS of S_2

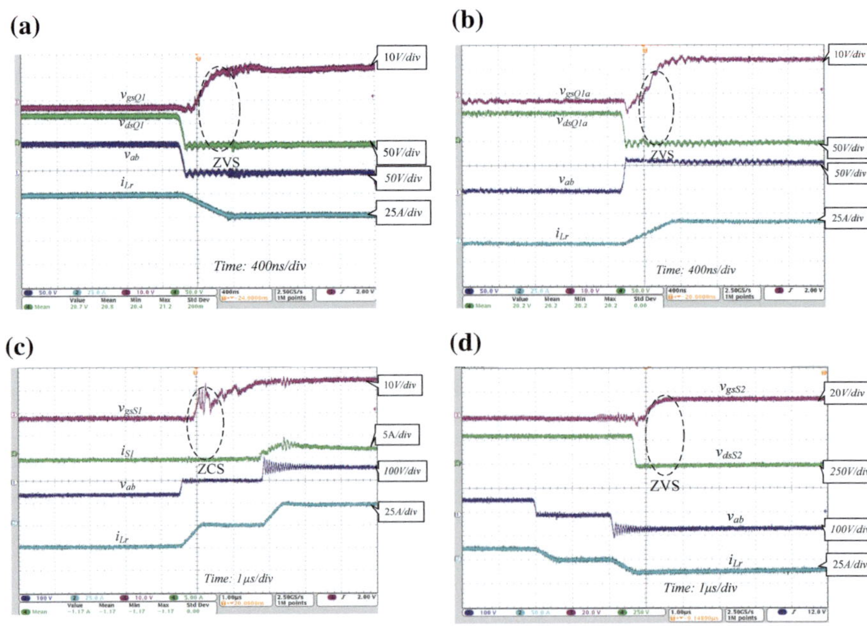

Fig. 7.15 Experimental results of the soft switching with PPS and DPDPS control in boost mode at 24 V input and 1 kW output condition: **a** ZVS of Q_1; **b** ZVS Q_{1a}; **c** ZCS of S_1; **d** ZVS of S_2

turned on with ZVS. As seen in Fig. 7.15c, d, for the HVS switches, the leading leg switch S_1 can achieve ZCS turn-on while the lagging switch S_2 can achieve ZVS turn-on.

7.5.3 Buck Mode Operation

Figure 7.16 gives the steady-state waveform in buck mode under light load (200 W) with minimum and maximum output voltages. As illustrated in Fig. 7.16a, b, at the minimum output voltage, the peak value of the leakage inductance current is higher than 80 A. But with the proposed control, the peak current can be reduced dramatically to no more than 10 A. Even though as the output voltage increases, the peak current value decreases with PPS control, but the peak current for the leakage inductance is still higher than that with the DPDPS control. This comparison can be seen from Fig. 7.16c, d.

Figure 7.17 gives the experimental result at full load (1 kW) in boost mode with the minimum and maximum output voltages. As seen from Fig. 7.17a, b with the minimum output voltage (18 V), using the PPS control, the peak current of the leakage inductance is over 80 A. Comparatively, with the proposed DPDPS control, the peak current is just over 25 A. At the maximum output voltage (28 V), although the leakage inductance current differs not much, the circulation loss can be minimized by the proposed DPDPS control.

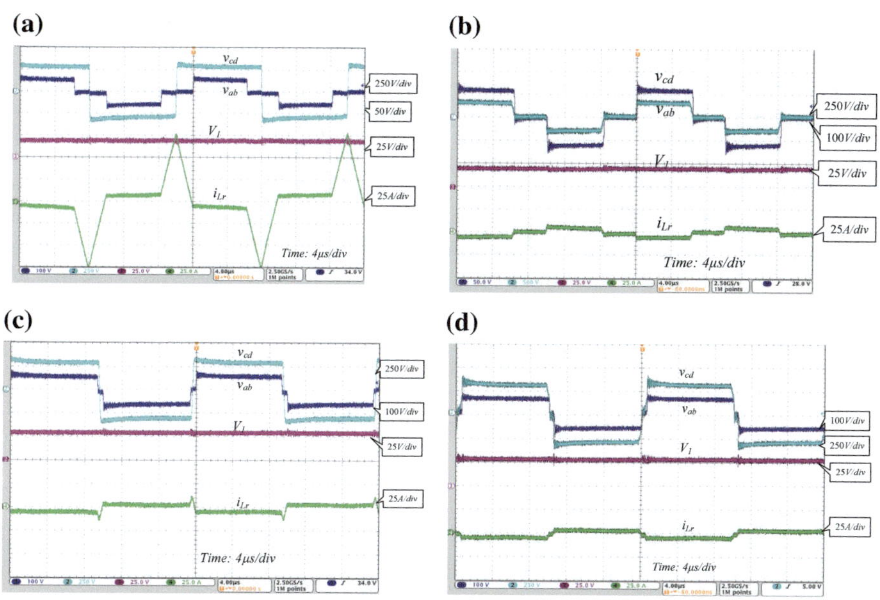

Fig. 7.16 Experimental results with PPS and DPDPS control in buck mode at 200 W output: **a** PPS control with V_1 = 18 V; **b** DPDPS control with V_1 = 18 V; **c** PPS control with V_1 = 28 V; **d** DPDPS control with V_1 = 28 V

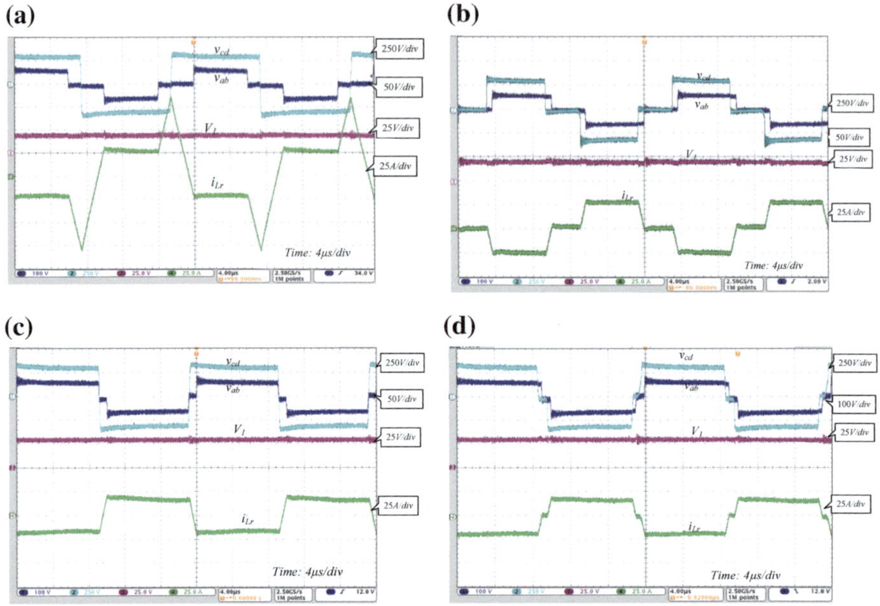

Fig. 7.17 Experimental results with PPS and DPDPS control in buck mode at 1000 W output: **a** PPS control with $V_1 = 18$ V; **b** DPDPS control with $V_1 = 18$ V; **c** PPS control with $V_1 = 28$ V; **d** DPDPS control with $V_1 = 28$ V

Figure 7.18 illustrates the buck mode soft switching states for the switches at light load (200 W) with 24 V output. As shown in Fig. 7.18a, b, the LVS switches Q_1 and Q_{1a} can achieve ZVS turn-on. Figure 7.18c, d illustrates the switching waveforms for HVS switches. As seen, the leading leg switch S_1 can achieve ZVS turn-on while lagging switch S_2 can be turned on with ZCS.

Figure 7.19 gives the soft switching states for the switches at full load (1 kW) with 24 V output in buck mode. As seen in Fig. 7.19a, b, both Q_1 and $Q_{1\text{a}}$ can be gated on with ZVS. For the HVS switches, as shown in Fig. 7.19c, d, the leading leg switch S_1 can be turned with ZVS turn-on while lagging switch S_2 can achieve ZCS turn-on.

7.5.4 *Operation Mode Transition and Efficiency Comparison*

Figure 7.20 illustrates the transition between the boost mode and buck mode with the proposed control. As seen in Fig. 7.20a, if V_2 is higher than 300 V, the working mode changes from boost mode to buck mode automatically. While, as shown in Fig. 7.20b, if V_2 decreases to be lower than 300 V, the converter working mode will

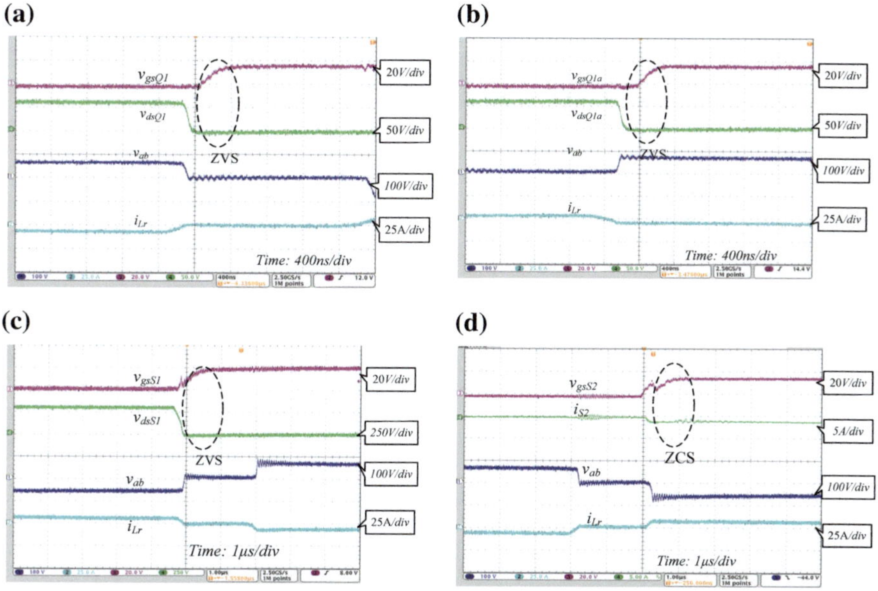

Fig. 7.18 Experimental results of the soft switching with PPS and DPDPS control in buck mode at 24 V output and 200 W output condition: **a** ZVS of Q_1; **b** ZVS of Q_{1a}; **c** ZVS of S_1; **d** ZCS of S_2

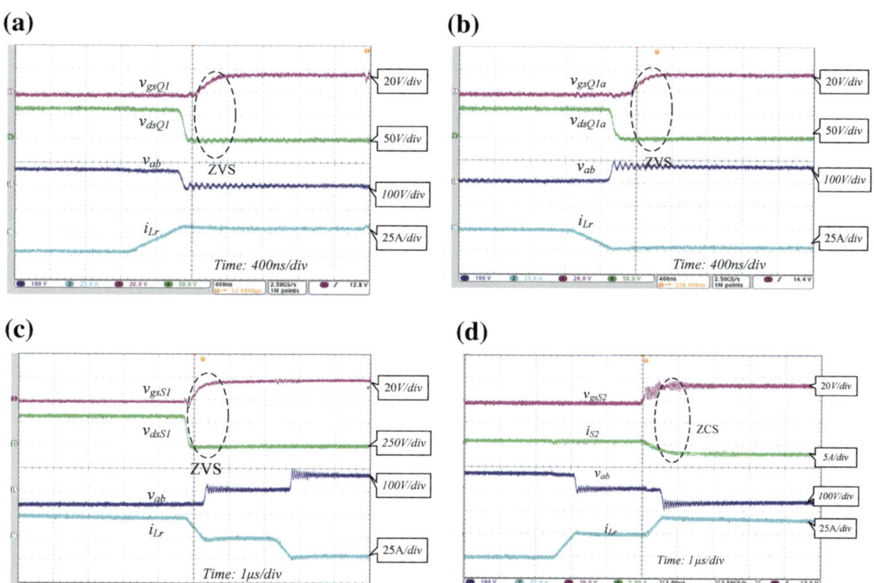

Fig. 7.19 Experimental results of soft switching with PPS and DPDPS control in buck mode at 24 V output and 1 kW output condition: **a** ZVS of Q_1; **b** ZVS Q_{1a}; **c** ZVS of S_1; **d** ZCS of S_2

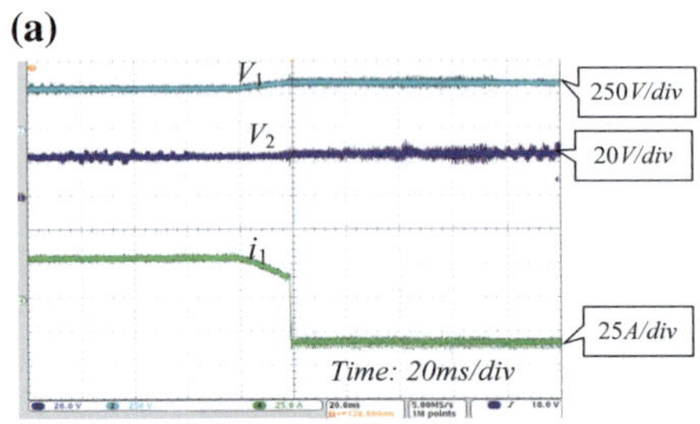

Fig. 7.20 Experimental results of operation mode transition with the proposed DPDPS control: **a** boost to buck; **b** buck to boost

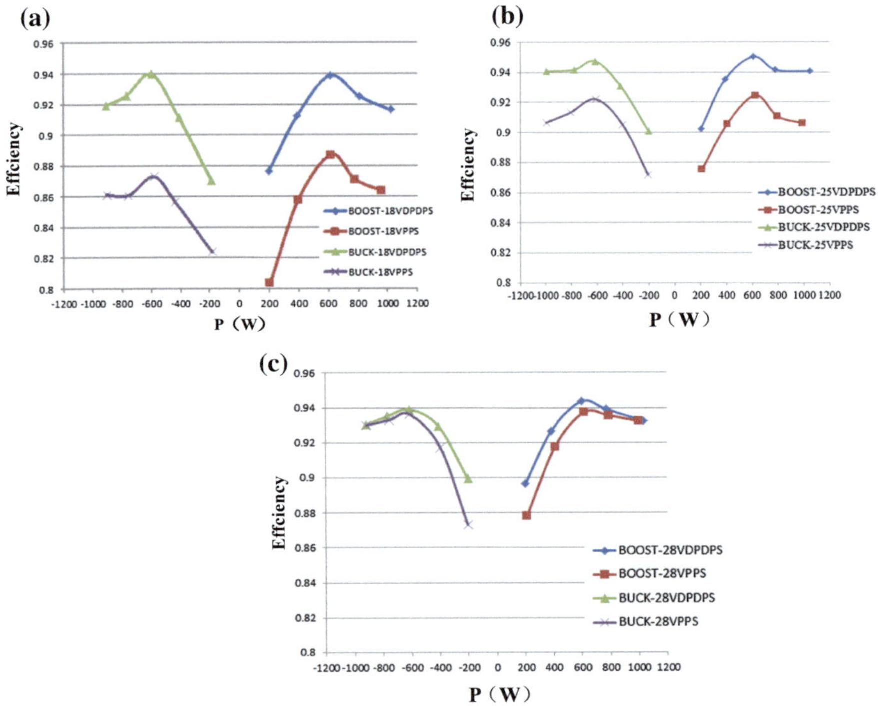

Fig. 7.21 Efficiency comparison at different V_1: **a** $V_1 = 18$ V; **b** $V_1 = 24$ V; **c** $V_1 = 28$ V

quickly change from buck mode to boost mode. As shown, the transition from boost to buck or vice versa is very smooth by using the proposed DPDPS control.

Figure 7.21 shows the conversion efficiency comparison at different loads, operation modes, and different LVS voltage V_1 using the PPS control and the proposed DPDPS control. As illustrated in the same working condition, the conversion efficiency obtained by the proposed DPDPS control is higher than the PPS control, particularly at lower LVS voltages.

7.6 Conclusion

The chapter proposes a dual PWM plus double phase shift control strategy for current-fed dual active bridge bidirectional DC–DC converters. With the proposed control strategy, the current stress in transformer windings and power switches especially during the circulation stage can be suppressed significantly. Thereby, the conduction loss can be reduced and the system reliability can be improved. Besides, for LVS power switches, ZVS can be achieved for all switches within a wide load range. For the HVS switches, in boost mode, ZCS can be achieved for leading leg switches and ZVS can be achieved for lagging switches. However in buck mode, leading leg switches can achieve ZVS and lagging leg switches can achieve ZCS. Thereby, the conversion efficiency is higher than the PPS control especially when V_1 is low. Besides, with the proposed DPDPS control, seamless transition between buck and boost mode can be obtained. The validity of the proposed control strategy and converter has been verified by experimental results of a 1 kW prototype.

References

1. Yoo H, Sul S, Park Y, Jeong J (2008) System integration and power flow management for a series hybrid electric vehicle using super capacitors and batteries. IEEE Trans Ind Appl 44(1): 108–114
2. Peng FZ, Li H, Su G, Lawler JS (2004) A new ZVS bidirectional DC-DC converter for fuel cell and battery application. IEEE Trans Power Electron 19(1):54–65
3. Xu DH, Zhao CH, Fan HF (2004) A PWM plus phase-shift control bidirectional DC-DC converter. IEEE Trans Power Electron 19(3):666–675
4. Xiao H, Xie S (2008) A ZVS bidirectional DC-DC converter with phase shift plus PWM control scheme. IEEE Trans Power Electron 23(2):813–823
5. Li W, Wu H, Yu H, He X (2011) Isolated winding-coupled bidirectional ZVS converter with PWM plus phase-shift (PPS) control strategy. IEEE Trans Power Electron 26(12):3560–3570
6. Wang Z, Li H (2012) A soft switching three-phase current-fed bidirectional DC-DC converter with high efficiency over a wide input voltage range. IEEE Trans Power Electron 27(2): 669–684

Chapter 8
High Efficiency Current-Fed Dual Active Bridge DC–DC Converter with ZVS Achievement Throughout Full Range of Load Using Optimized Switching Patterns

Abstract For current-fed dual active bridge bidirectional DC–DC converters, all the possible switching patterns are summarized in view of the combinations of both side PWM duty cycles and phase shift angle. A control strategy is proposed for the current-fed dual active bridge converter to operate with the optimized patterns. The equivalent duty cycle of the secondary side is smaller than that of the primary side by a fixed value, which is optimized based on the soft switching achievement and the circulating current minimization. The closed-loop control is easy to be implemented since there are only two independent variables, one-side duty cycle and the phase shift angle. With the proposed control, zero-voltage switching (ZVS) can be achieved for all power switches throughout full range of load even at no load condition. The typical working modes with the proposed control are given. The optimal design of the system parameters including the fixed time delta and dead time is illustrated. The effectiveness of the proposed control is verified by the experimental results of a 1 kW laboratory prototype.

Keywords Bidirectional DC–DC converter · PWM control · ZVS
Battery charging/discharging · Circulation loss

8.1 Operation Principle of the Control

As discussed in the previous section, to adapt to the wide voltage conversion ratio situations, a PWM plus phase shift control was proposed for current-fed dual active bridge converter, in which the battery side is PWM modulated while duty cycles for the high-voltage side (HVS) switches are fixed as 50%. Since the active clamp voltage can be regulated based on the HVS voltage value and the turns ratio, the current stress during the power transfer stage can be suppressed as the voltage conversion ratio varies widely [1–3]. Although minimum conduction loss during the power transfer stage can be obtained, for the non-power transfer stage, especially when the battery voltage is rather low, the current spike and circulation loss

D. Sha and G. Xu, *High-Frequency Isolated Bidirectional Dual Active Bridge DC–DC Converters with Wide Voltage Gain*, CPSS Power Electronics Series,
https://doi.org/10.1007/978-981-13-0259-6_8

may be high, causing the reduction of the conversion efficiency and the system reliability issues. Of course, the current stress can be suppressed by adding an additional inductor in series connected with the transformer to increase the equivalent leakage inductance value, but the maximum transferred power capability is limited due to the large inductance value. Besides, the circulation stage length increases and results in low conversion efficiency. PWM control can be also employed for HVS switches to reduce the circulation loss [4–6]. To ensure ZVS for HVS switches, unequal duty cycles for both side switches is a very good option with unmatched voltage control [4]. But the current stress is relatively high due to the unmatched voltage control. For voltage-matched control, identical duty cycle control can achieve ZVS for all the power switches for current-fed semi-DAB converter employing diodes with voltage match control; for current-fed DAB composed of only fully control power devices, its effectiveness is unknown [5]. For current-fed dual active bridge (DAB), the duty cycle relationship for both sides has to vary according to the specific working conditions. This complicates the control system design [6].

In this chapter, for current-fed DAB, all the possible switching patterns have been illustrated and analyzed with voltage matching control. Based on the analysis, optimized switching patterns are selected and employed in control.

8.1.1 *Topology of the Current-Fed DAB and the Operating Modes with Voltage Matching Control*

Figure 8.1 illustrates the power circuit of the current-fed DAB. In the low-voltage side (LVS), there are two interleaved buck/boost circuits in parallel connection. L_1 and L_2 are DC inductors with the same inductance, and C_c is the clamp capacitor. The duty cycle D_p of bottom switches makes the clamp voltage V_c to be matched with the secondary side voltage V_2, which means the slew rate of leakage inductance current during the main power transfer stage maintains zero in spite of the V_1 variation. The primary side voltage of the transformer is a three-level waveform

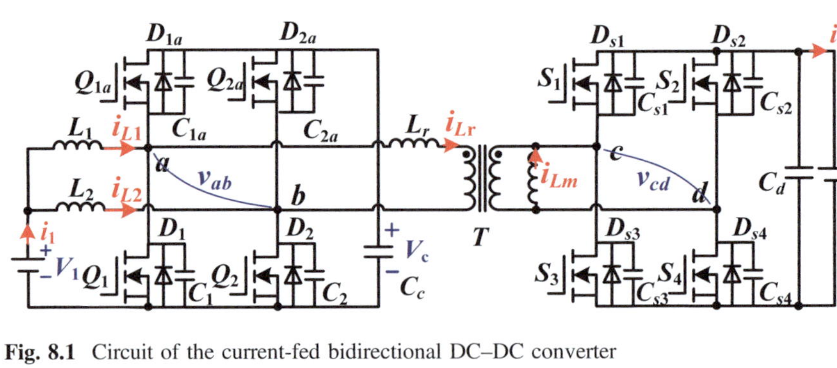

Fig. 8.1 Circuit of the current-fed bidirectional DC–DC converter

employing the PWM control. The AC inductor L_r represents the sum of the external inductance and the high-frequency transformer leakage inductance. In the HVS, L_m is the secondary side magnetizing inductance. An inner phase shift is employed in the full bridge, and the secondary side voltage of the transformer also becomes a three-level waveform. So an equivalent duty cycle D_s can be defined similarly as the primary duty cycle D_p. The output power P varies as phase shift angle φ_E between primary side and secondary side varies. The power flow is bidirectional. Boost mode is defined as the power flows from LVS to HVS, while the reversed power flow is referred to as buck mode.

For the current-fed DAB, there are totally twelve patterns when the primary and secondary side voltage waveforms are both three-level waveforms with voltage matching control. However, not all of them are practical operation modes. The six practical patterns shown in Fig. 8.2 are divided into two groups according to the relationship between two-side duty cycles. Mode IA–VIA are the switching patterns when $D_p > D_s$, while Mode IB–VIB occur when $D_p < D_s$. In Fig. 8.2, v_{ab} is the voltage of primary side, v_{cd} is the voltage of secondary side, and i_{Lr} is the leakage inductance current. Mode I–III happen in boost mode at light load, medium load, and heavy load, respectively, while Mode IV–VI happen in buck mode at the aforementioned three kinds of loads, respectively.

However, ZVS for secondary side switches cannot be achieved in half of the above modes. To achieve ZVS, there should be a reversed current to discharge the parasitic junction capacitor before the switch turns on. The ZVS condition is described as $i_{Lr}(\theta_a) > 0$ and $i_{Lr}(\theta_b) > 0$ when the magnetizing current and parasitic junction output capacitance are neglected. As seen in Fig. 8.2, the conditions of $i_{Lr}(\theta_a) > 0$ and $i_{Lr}(\theta_b) > 0$ can be ensured for all the A modes. That means for $D_p > D_s$, ZVS can be achieved even at light load theoretically. However, when $D_p < D_s$, ZVS cannot be achieved at light load or medium load because the value of I_b is negative. Even at heavy load, ZVS cannot be ensured since I_y might be negative in some cases. As a result, it is better to let $D_p > D_s$ for the ZVS achievement.

The six unpractical patterns Mode VII–XII are listed in Fig. 8.3, and the leakage inductance current i_{Lr} during circulation stages is shown in shadow. As can be seen, the circulating current is high in these modes, but the transferred power is limited. In addition, for most of the DAB converters, the range of φ_E is limited within $[-\pi/2, \pi/2]$ to ensure that the output power P increases as φ_E is increased. However, the output power P is constant in Mode VII and VIII regardless of the phase shift angle variation and thus the power is out of control. In Mode IX–XII, the output power P decreases as φ_E is increased, making the transferred power curve non-monotonic. Hence, these patterns should be avoided in practice. Normally, φ_E is selected within the range of $[-\pi/2, \pi/2]$ so that power is a monotonously increasing function of the phase shift angle.

To sum up, for Mode VII–XII, the transferred power does not increase as φ_E is increased, and the circulating current is high. For Mode IB–VIB, the ZVS is lost for HVS switches. Mode IA–VIA are the optimized patterns in which ZVS can be

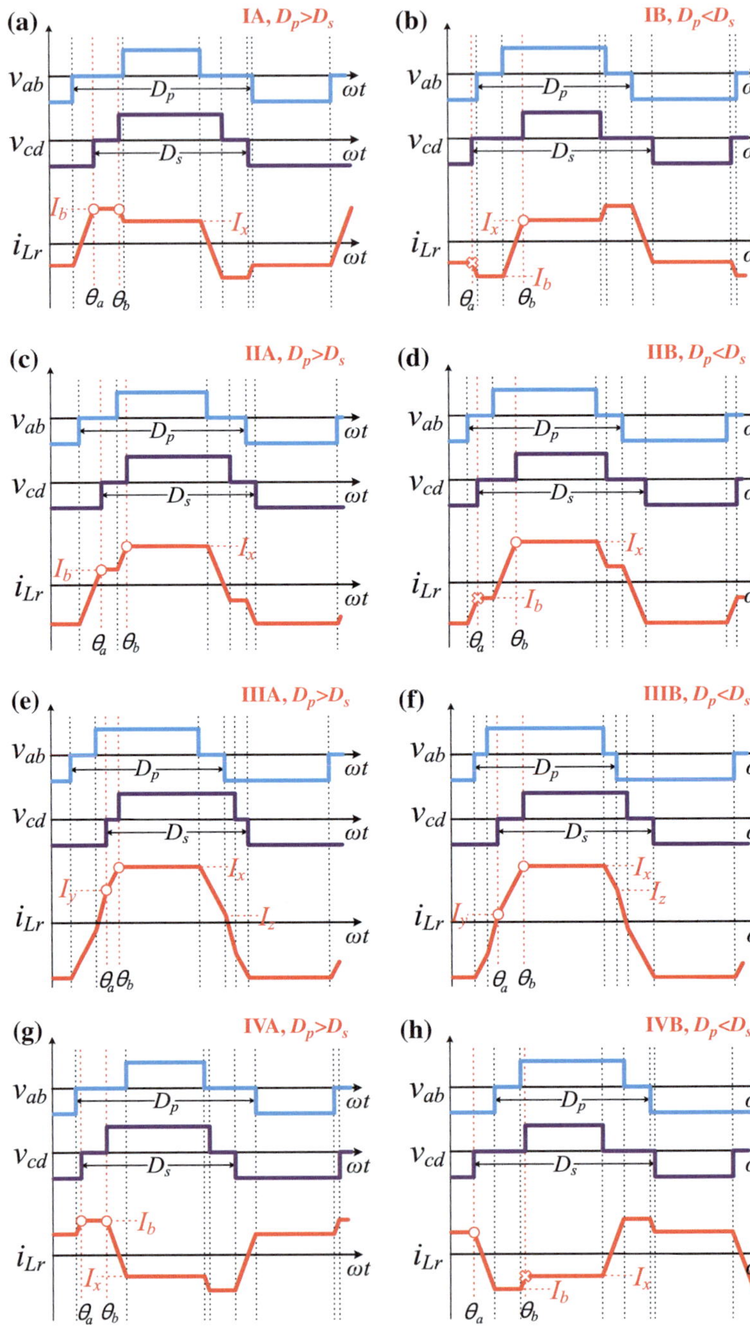

Fig. 8.2 All working patterns for the current-fed DAB with voltage matching control: **a** Mode IA; **b** Mode IB; **c** Mode IIA; **d** Mode IIB; **e** Mode IIIA; **f** Mode IIIB; **g** Mode IVA; **h** Mode IVB; **i** Mode VA; **j** Mode VB; **k** Mode VIA; **l** Mode VIB

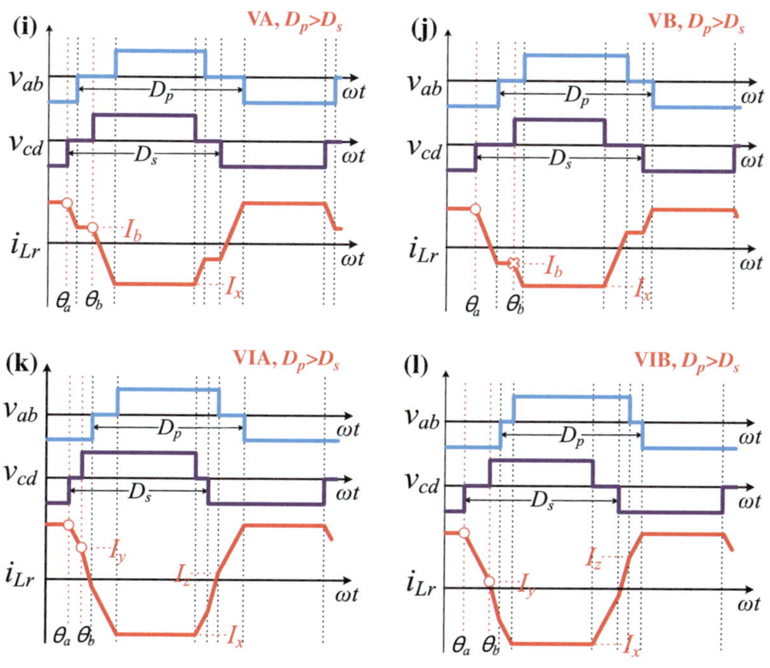

Fig. 8.2 (continued)

achieved for all switches with low circulation loss. In this chapter, a novel control strategy is proposed to make the converter works in these optimized patterns.

8.1.2 Power Expressions of the Proposed Control

In this chapter, a PWM plus dual phase shift with fixed duty cycle delta control is developed to make the converter operating in optimized patterns. The difference between duty cycles of the two ports is fixed at any conditions, which is defined as the fixed delta in this chapter. The closed-loop design is simple to be implemented because the HVS equivalent duty cycle is determined by that of the LVS. The fixed delta $\Delta D = D_\mathrm{p} - D_\mathrm{s}$ is chosen to be a positive constant, so that ZVS of all the switches in the HVS can be obtained in wide load range. With the proposed control, the converter works in Mode IA–VIA. The unpractical modes and modes without wide ZVS capability are avoided. All the possible working patterns are shown in Fig. 8.2. Among them, the six unpractical working patterns are summarized in Fig. 8.3.

The expressions of switching currents marked in Fig. 8.2 are listed in Table 8.1. Applying these switching currents, the power expression of each mode can be obtained as shown in Table 8.2, where φ_E is the phase shift angle, $\Delta\varphi = 2\pi\Delta D$ is

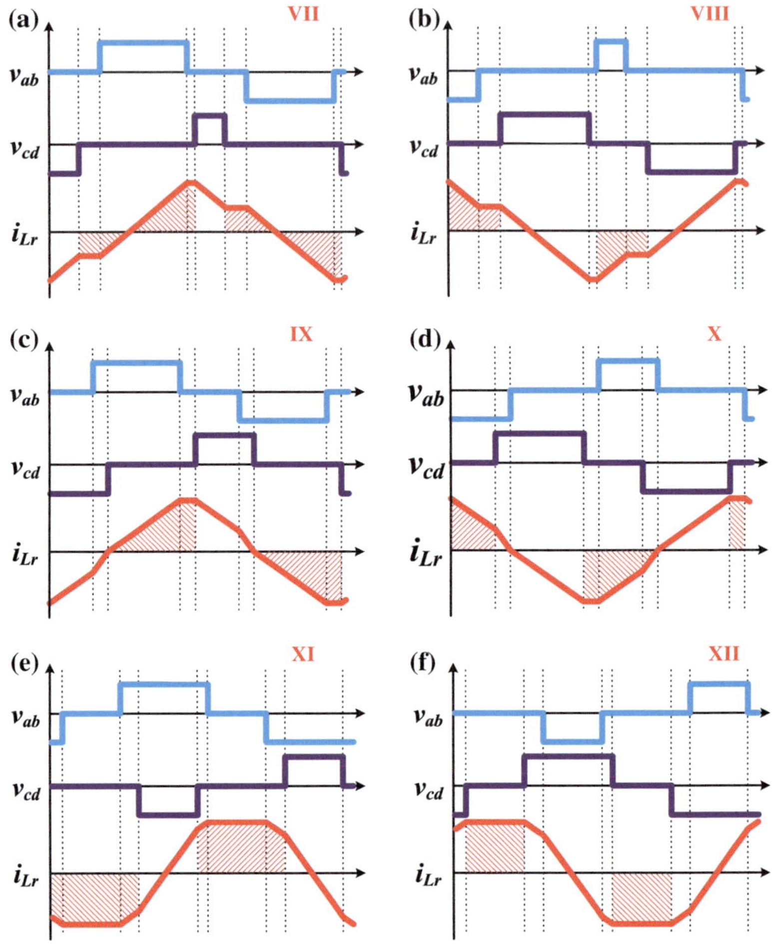

Fig. 8.3 Six unpractical patterns for the current-fed DAB with voltage matching control: **a** Mode VII; **b** Mode VIII; **c** Mode IX; **d** Mode X; **e** Mode XI; **f** Mode XII

Table 8.1 Switching currents in each mode

$I_b = \frac{V_c T \Delta\varphi}{4\pi L_r}$	$I_x = \frac{V_c T \varphi_E}{2\pi L_r}$
$I_y = \frac{V_c T}{2\pi L_r}\left(\lvert\varphi_E\rvert - \xi_s\right)$	$I_z = \frac{V_c T}{2\pi L_r}\left(\lvert\varphi_E\rvert - \xi_p\right)$

the fixed delta, ξ_p and ξ_s are the freewheeling intervals of v_{ab} and v_{cd}, which are denoted in Fig. 8.5. The three-dimensional (3D) power curve has been plotted according to the power expressions in Table 8.2. Figure 8.4 illustrates output power curve versus primary side duty cycle D_p and phase shift angle φ_E with the proposed control.

Table 8.2 Power expressions in each mode

Description	Power expression
Mode VIA Buck Heavy load	$\frac{V_c^2 T}{4\pi^2 L_r}\left[2\pi\varphi_E + 2\varphi_E^2 + 2\pi^2\left(D_p^2 + D_s^2 - D_p - D_s\right) + \pi^2\right]$
Mode VA Buck Medium load	$\frac{V_c^2 T}{4\pi^2 L_r}\left[\varphi_E^2 + 2\pi(2 - D_p - D_s)\varphi_E + \pi^2(D_p - D_s)^2\right]$
Mode IA, IVA Buck/Boost Light load	$\frac{V_c^2 T}{4\pi^2 L_r}\left[4\pi\varphi_E(1 - D_p)\right]$
Mode IIA Boost Medium load	$\frac{V_c^2 T}{4\pi^2 L_r}\left[-\varphi_E^2 + 2\pi(2 - D_p - D_s)\varphi_E - \pi^2(D_p - D_s)^2\right]$
Mode IIIA Boost Heavy load	$\frac{V_c^2 T}{4\pi^2 L_r}\left[2\pi\varphi_E - 2\varphi_E^2 - 2\pi^2\left(D_p^2 + D_s^2 - D_p - D_s\right) - \pi^2\right]$

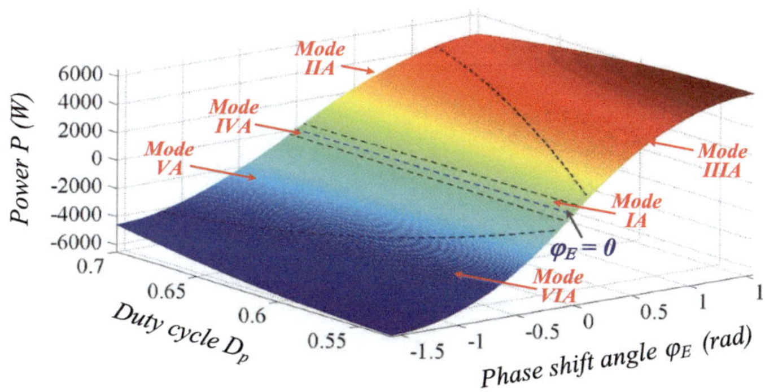

Fig. 8.4 *P* curve versus D_p and φ_E when $\Delta\varphi = 0.1256$

As seen, the converter with the proposed control can work in the expected Mode IA–VIA in a wide battery voltage range (18–28 V, corresponding duty cycle range is 0.7–0.53). The output power increases monotonously as the phase shift angle increases in the range of $-\pi/2$ to $\pi/2$. As shown in Fig. 8.4, Mode IIIA may happen more often at lighter load conditions at higher battery voltage than lower ones. Besides, the maximum output power is higher as well because of shorter freewheeling intervals.

8.1.3 Working Principle of the Proposed Switching Pattern

The proposed control shows better performance especially at light-load and medium-load conditions. Mode IIA can be chosen as an example to illustrate the

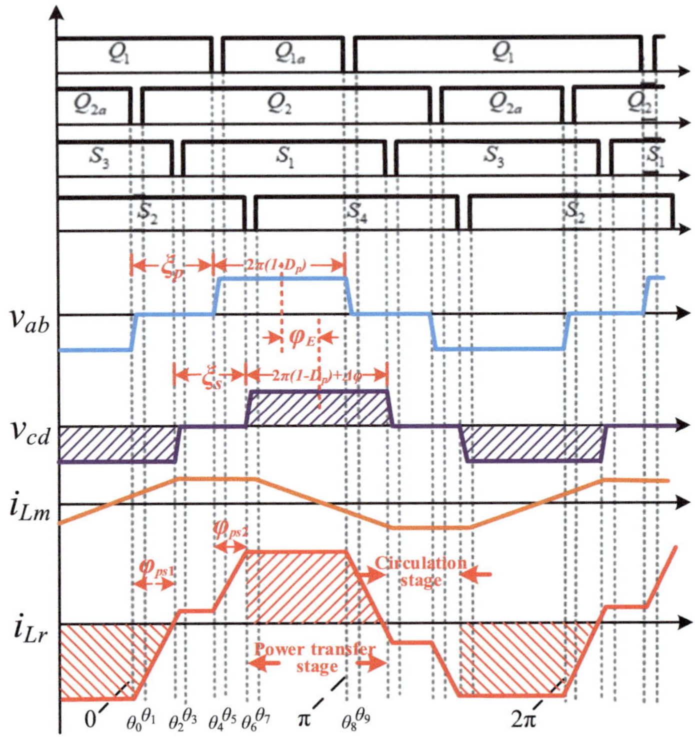

Fig. 8.5 Steady-state waveforms of the proposed control in Mode IIA

working principle of the converter with the proposed control at medium load in boost mode. Detailed waveforms and mode analysis of Mode IIA are given to show the operation principle. Figure 8.5 gives detailed steady-state waveforms of the proposed control working at Mode IIA. There are nine working stages in a half switching period according to the state of switches. In addition to the waveforms of v_{ab}, v_{cd}, and i_{Lr}, the waveform of magnetizing current i_{Lm} is also shown. Besides, φ_E is the phase shift angle, ξ_p and ξ_s are the freewheeling intervals, φ_{ps1} and φ_{ps2} are the phase shift between θ_0–θ_2 and θ_4–θ_6, respectively. The fixed delta $\Delta\varphi$ is the difference between φ_{ps1} and φ_{ps2}.

Among all the working stages, the stages that power is transferred to the high-voltage DC bus are defined as power transfer stages. They have been shown in shadow in Fig. 8.5. Other stages can be referred to as circulation stages.

The detailed mode analysis is shown in Fig. 8.6, and they are described as follows.

Stage 1 (Before θ_0): Q_1, Q_{2a}, S_2, and S_3 conduct. During this stage, the power flows from LVS to HVS.

Stage 2 (θ_0–θ_1): At θ_0, Q_{2a} turns off. The sum of i_{L2} and i_{Lr} charges C_{2a} and discharges C_2 until the body diode of Q_2 begins to conduct. Then, Q_2 turns on with ZVS.

Stage 3 (θ_1–θ_2): At θ_1, Q_2 turns on with ZVS.

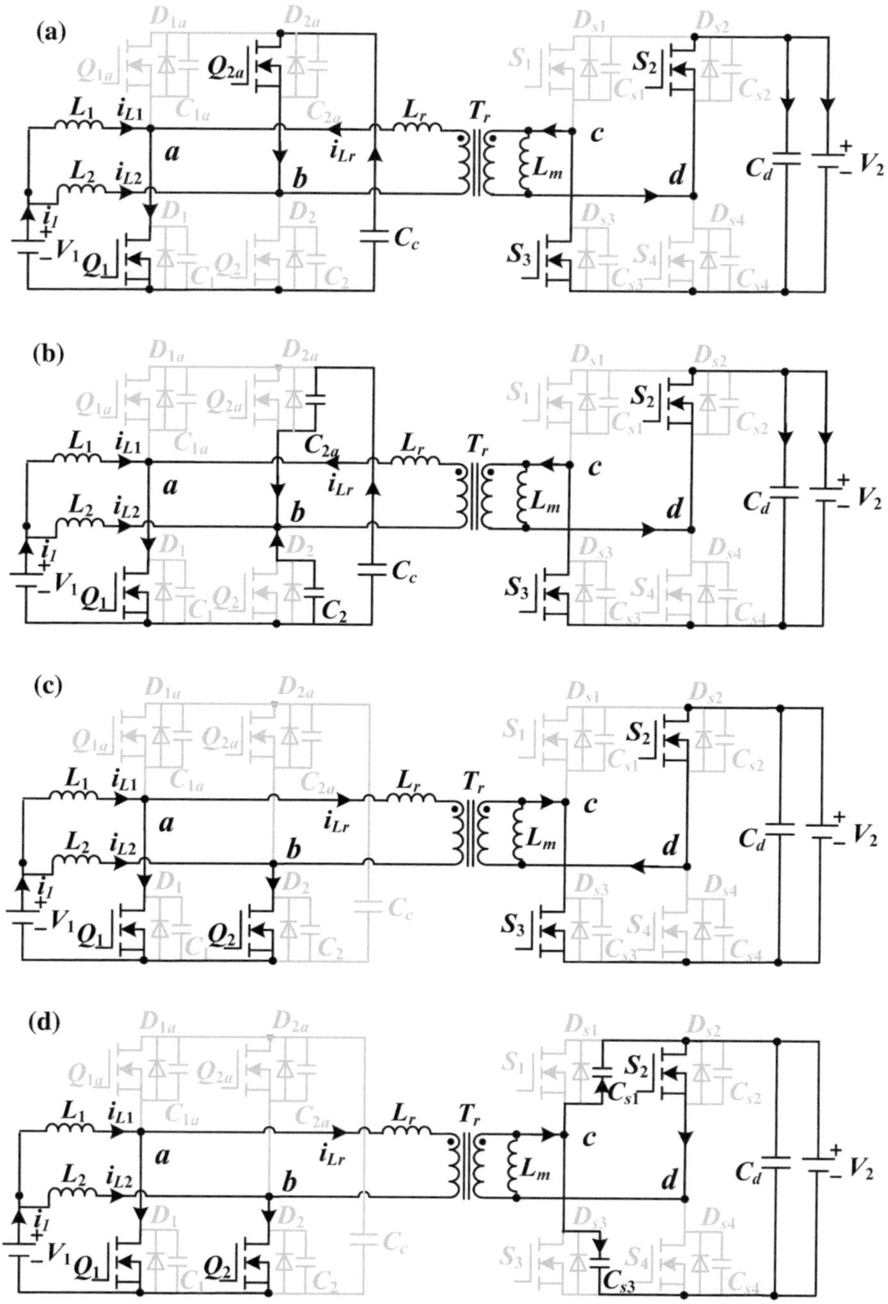

Fig. 8.6 Working modes for half period in Mode IIA. **a** Stage 1; **b** Stage 2; **c** Stage 3; **d** Stage 4; **e** Stage 5; **f** Stage 6; **g** Stage 7; **h** Stage 8; **i** Stage 9; **j** Stage 10

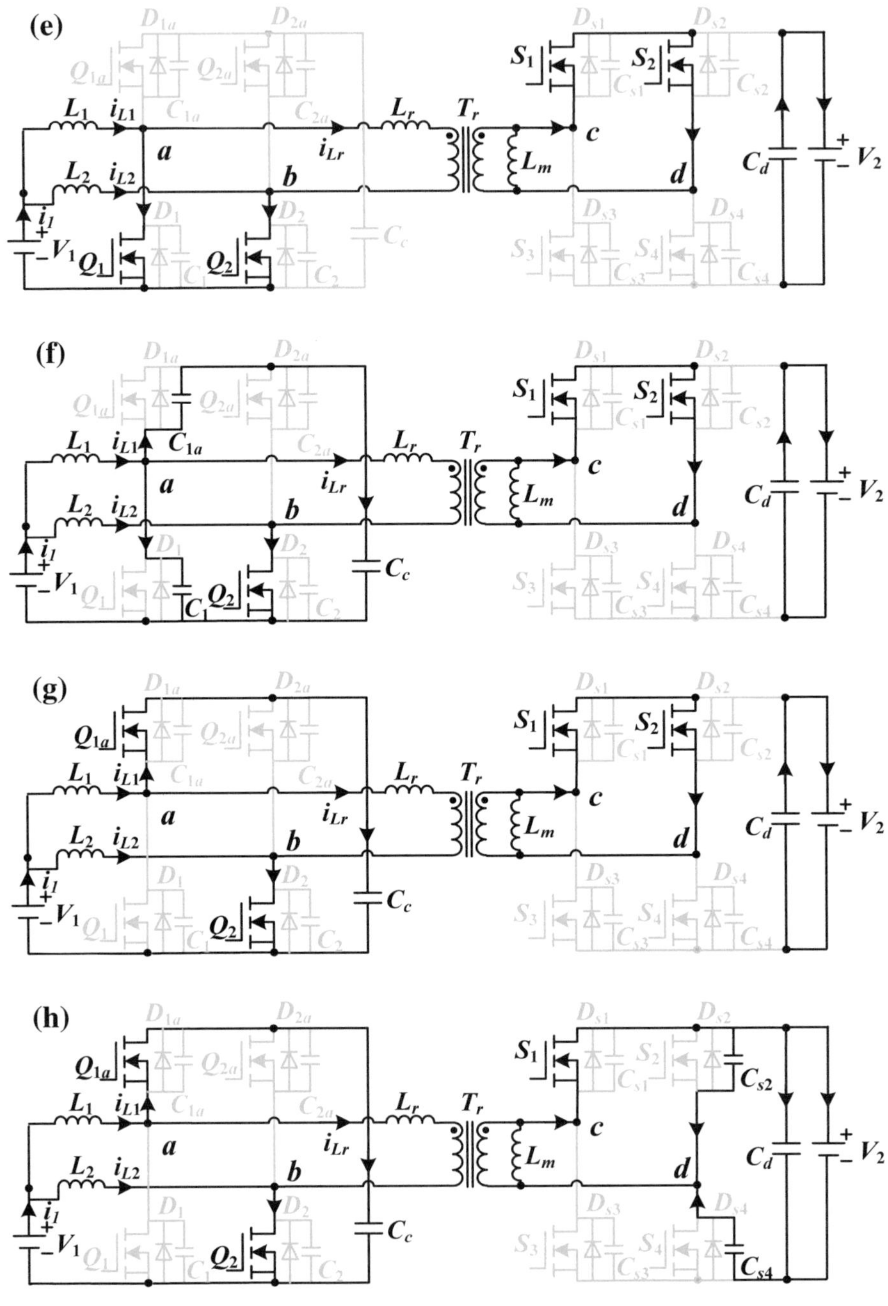

Fig. 8.6 (continued)

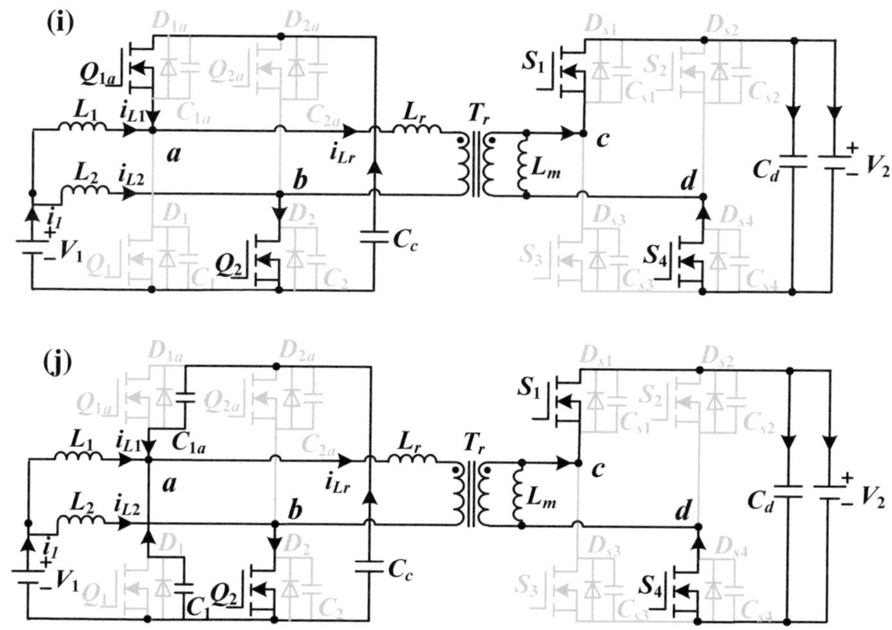

Fig. 8.6 (continued)

Stage 4 (θ_2–θ_3): At θ_2, S_3 turns off after the current i_{Lr} flows in the opposite direction. C_{s3} is charged and C_{s1} is discharged until the body diode of S_1 begins to conduct.

Stage 5 (θ_3–θ_4): At θ_3, S_1 turns on under ZVS.

Stage 6 (θ_4–θ_5): At θ_4, Q_1 turns off. The difference of i_{L1} and i_{Lr} charges C_{1a} and discharges C_1 until the body diode of Q_{1a} begins to conduct.

Stage 7 (θ_5–θ_6): At θ_5, Q_{1a} turns on with ZVS.

Stage 8 (θ_6–θ_7): At θ_6, S_2 turns off. Then, the current i_s charges C_{s2} and discharges C_{s4}, causing the body diode of S_4 conduct finally.

Stage 9 (θ_7–θ_8): At θ_7, S_4 turns on under ZVS. The current i_{Q1a} flows from source to drain at first, but flows from drain to source after $i_{L1} = i_{Lr}$. During this stage, the power flows from LVS to HVS.

Stage 10 (θ_8–θ_9): At t_8, Q_{1a} turns off. The second half cycle similar to the first half cycle. The difference between the current i_{Lr} and i_{L1} charges and discharges junction capacitors of Q_{1a} and Q_1 until the body diode of Q_1 begins to conduct.

8.1.4 Discussion of the Circulating Current

The fixed delta value needs to be designed with a trade-off in view of the ZVS achievement and circulation loss reduction. On one hand, the fixed delta should be

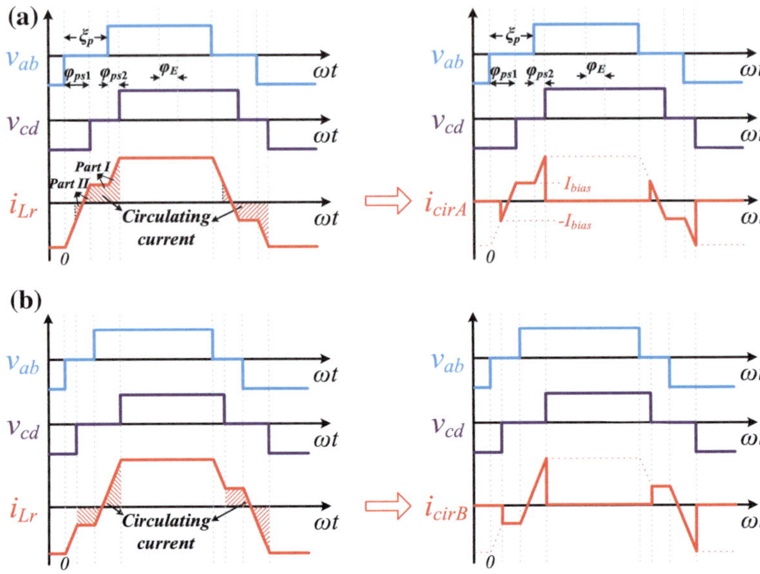

Fig. 8.7 Discussion of the circulating current: **a** non-power transfer stages and circulating current in Mode IIA, **b** non-power transfer stages and circulating current in Mode IIB

designed with large value to discharge the output capacitor of the MOSFETs in the HVS. On the other hand, large fixed delta causes higher circulation loss. The circulating current is defined as the i_{Lr} in the non-power transfer stages as shown in Fig. 8.7. The circulating current includes two parts in Mode IIA. One is the leakage inductance current during the freewheeling intervals, and another is the non-active power stage when power transfers. There is no active power transfer in both these two stages.

The resonant process during dead time interval is neglected to simplify the analysis. According to the definition, the circulating current in Mode IIA can be described as

$$i_{\text{cirA}} = \begin{cases} i_{Lr}, & \varphi_E - \frac{1}{2}\Delta\varphi + k\pi < \theta < \varphi_E - \frac{1}{2}\Delta\varphi + \xi_p + k\pi \\ 0, & \text{others} \end{cases} \tag{8.1}$$

where k can be any arbitrary. The circulating current i_{cirA} in a half switching cycle can be expressed as follows

$$i_{\text{cirA}}(\theta) \begin{cases} 0, & 0 < \theta < \varphi_E - \frac{1}{2}\Delta\varphi \\ \frac{V_cT}{2\pi L_r}(\theta - \varphi_E), & \varphi_E - \frac{1}{2}\Delta\varphi < \theta < \varphi_E + \frac{1}{2}\Delta\varphi \\ \frac{V_cT}{2\pi L_r}\frac{1}{2}\Delta\varphi, & \varphi_E + \frac{1}{2}\Delta\varphi < \theta < \xi_p \\ \frac{V_cT}{2\pi L_r}\left(\frac{1}{2}\Delta\varphi + \theta - \xi_p\right), & \xi_p < \theta < \varphi_E - \frac{1}{2}\Delta\varphi + \xi_p \\ 0, & \varphi_E - \frac{1}{2}\Delta\varphi + \xi_p < \theta < \pi \end{cases} \tag{8.2}$$

Then, its RMS value is written by

$$I_{\mathrm{cirA\_RMS}} = \sqrt{\frac{1}{2\pi}\int_{0}^{2\pi} i_{\mathrm{cirA}}^{2}\mathrm{d}\theta} = \frac{V_c T}{4\pi L_r}\sqrt{\frac{-3\varphi_E\Delta\varphi^2 - \Delta\varphi^3 + 4\varphi_E^3 + 3\Delta\varphi^2\xi_p}{3\pi}} \tag{8.3}$$

Likewise, the circulating current and RMS value in Mode IIB can be written by, respectively,

$$i_{\mathrm{cirB}} = \begin{cases} i_{Lr}, & \varphi_E + \frac{1}{2}\Delta\varphi + k\pi < \theta < \varphi_E - \frac{1}{2}\Delta\varphi + \xi_p + k\pi \\ 0, & \text{others} \end{cases} \tag{8.4}$$

$$I_{\mathrm{cirB\_RMS}} = \sqrt{\frac{1}{2\pi}\int_{0}^{2\pi} i_{\mathrm{cirB}}^{2}\mathrm{d}\theta} = \frac{V_c T}{4\pi L_r}\sqrt{\frac{-3\varphi_E\Delta\varphi^2 - 2\Delta\varphi^3 + 4\varphi_E^3 + 3\Delta\varphi^2\xi_p}{3\pi}} \tag{8.5}$$

Figure 8.8 plots the RMS value of circulating current curve versus the LVS voltage V_1 and delta $\Delta\varphi$. The circulation loss is high at low LVS voltage because of the large freewheeling intervals. As seen, when $\Delta\varphi$ is too large or too small, the circulating current is high. The circulation loss becomes minimum when $\Delta\varphi = 0$ in spite of the LVS voltage. However, there should be a positive $\Delta\varphi$ to ensure ZVS for HVS switches. So the fixed delta should be designed as small as possible as long as ZVS can be obtained. Besides, a large fixed delta results in a large difference between the primary side duty cycle D_{p} and the secondary side equivalent duty cycle D_{s}. Since D_{s} must be larger than 50%, the range of LVS voltage will be limited. A small fixed delta also leads to a wide voltage range for the converter.

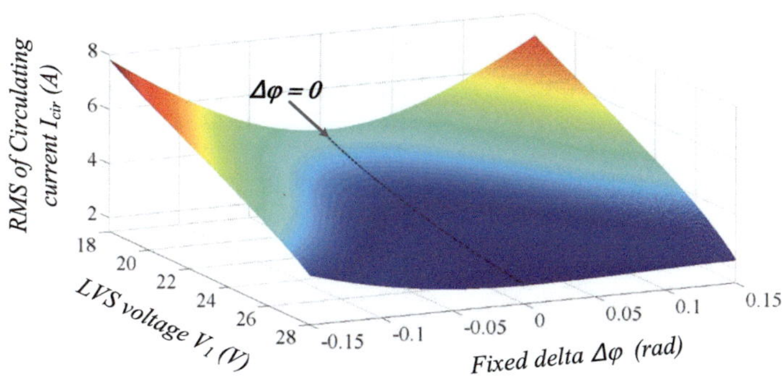

Fig. 8.8 Circulating current with respect to V_1 and $\Delta\varphi$

8.2 Soft Switching Condition

ZVS can be achieved for all switches within whole load range with the proposed control. For the battery side, the output capacitances of the switches are charged/discharged by the difference between the DC inductance current i_L and the leakage inductance current i_{Lr}. ZVS can be achieved easily as long as the DC inductance value is selected properly [6].

The key is to achieve ZVS for HVS switches. Just like most of the DAB converters, soft switching can be achieved easily at heavy load, but it is difficult to ensure ZVS at light load. With the proposed control, a bias current generated by the given fixed delta ΔT helps to charge/discharge the junction capacitors of HVS switches, and ZVS can be obtained even at no load. The resonant process and how to design the fixed delta to ensure ZVS are discussed in detail in this section.

8.2.1 Resonant Process Analysis

At medium-load or light-load conditions, the bias current helps to charge/discharge the output capacitances of the HVS switches. This resonant process is analyzed in detail.

Take boost mode operation at medium load as an example, the resonant process curve is zoomed in and plotted in Fig. 8.9. The working modes have already been shown in θ_1–θ_3 in Fig. 8.6. It is assumed that the leading leg switch S_3 turns off at t_2 when the leakage inductance current $i_{Lr} = I_{LrA}$. The resonance begins at t_2, and the magnetizing current i_{Lm} and i_{Lr} reflected to HVS charges C_{s3} and discharges C_{s1} until the leakage inductance current slew rate decays to zero. If the dead time is chosen large enough, S_1 turns on under ZVS. In fact, i_{Lr} itself can be designed large enough to discharge the parasitic output capacitance, and the magnetizing current i_{Lm} is not necessary for the ZVS implementation. So the magnetizing inductance

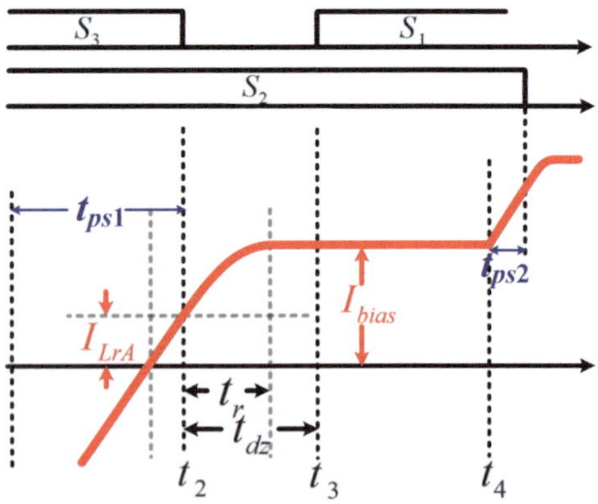

Fig. 8.9 Expanded waveforms of the resonant process in boost mode

can be designed freely to derive an optimal transformer design with the lowest total losses, making the transformer design easier.

This resonant process can be described by differential equations. The moment when S_3 turns off can be defined as $t = 0$. The magnetizing current i_{Lm} can be viewed constant and equal to maximum value i_{Lmmax} in this resonant process. The value of leakage inductance current when the resonant process is completed is denoted as the bias current I_{bias}, and the resonance ends at time t_r. The relationship between i_{Lr} and the drain-to-source voltage v_{ds1} of the switch S_1 can be written by

$$i_{Lmmax} + \frac{N_1}{N_2} i_{Lr} = -2C_{oss.eff} \frac{dv_{ds1}}{dt} \tag{8.6}$$

$$\frac{N_1 v_{ds1}}{N_2} = L_r \frac{di_{Lr}}{dt} \tag{8.7}$$

where $C_{oss.eff}$ is the charge equivalent parasitic output capacitance of the MOSFETs under the condition that v_{ds} = 0–300 V. i_{Lmmax} is the peak value of the magnetizing current which can be written by $i_{Lmmax} = V_2(1 - D_p)T/2L_m$. The initial conditions for the differential equations are $v_{ds1}(0) = V_2$ and $i_{Lr}(0) = I_{LrA}$.

The angle frequency of the resonant process can be expressed by

$$\omega_1 = \frac{N_1}{N_2} \sqrt{\frac{1}{2C_{oss.eff} L_r}} \tag{8.8}$$

Then, expressions of the resonant current and voltage can be written by, respectively,

$$\begin{cases} i_{L_r}(t) = \sqrt{\left(\frac{N_2}{N_1} i_{L_m\max} + i_{L_rA}\right)^2 + \frac{2C_{oss.eff} V_2^2}{L_r}} \\ \quad \times \sin\left(\omega_1 t + \arctan\left(\left(\frac{N_2}{N_1} i_{L_m\max} + i_{LrA}\right)\sqrt{\frac{L_r}{2C_{oss.eff} V_2^2}}\right)\right) - \frac{N_2}{N_1} i_{L_m\max} \\ v_{ds1}(t) = \sqrt{V_2^2 + \left(\frac{N_2}{N_1} i_{L_m\max} + i_{L_rA}\right)^2 \frac{L_r}{2C_{oss.eff}}} \\ \quad \times \sin\left(\omega_1 t - \arctan\left(\frac{N_1 V_2}{N_2 i_{Lmmax} + N_1 i_{LrA}} \sqrt{\frac{2C_{oss.eff}}{L_r}}\right)\right) \end{cases} \tag{8.9}$$

When $v_{ds1}(t) = 0$, the resonant process is completed. The resonant time duration length t_r and the bias current I_{bias} can be obtained from (8.9), and they are expressed as, respectively,

$$\begin{cases} t_r = \dfrac{\arctan\left(\frac{N_1 V_2}{N_2 i_{L_m\max} + N_1 I_{L_rA}} \sqrt{\frac{2C_{oss.eff}}{L_r}}\right)}{\omega_1} \\ I_{bias} = \sqrt{\left(\frac{N_2}{N_1} i_{L_m\max} + I_{LrA}\right)^2 + \frac{2C_{oss.eff} V_2^2}{L_r}} - \frac{N_2}{N_1} i_{L_m\max} \end{cases} \tag{8.10}$$

As seen, the both t_r and the I_{bias} are affected with the initial current I_{LrA} which is determined by the fixed delta and operation mode. The bias current I_{bias} increases as I_{LrA} is increased while t_r decreases with the increase of I_{LrA}.

The resonant processes in other cases are similar to this since the equivalent resonant circuits are identical. In buck mode, the resonant process begins at $i_{Lr} = I_{\mathrm{bias}}$ and ends at $i_{Lr} = I_{\mathrm{LrA}}$; thus, the results shown in (8.10) are also suitable for buck mode.

8.2.2 *Soft Switching Condition*

To achieve soft switching, the given fixed delta ΔT should be large enough to ensure $I_{\mathrm{LrA}} \geq 0$. Then if the dead time is selected properly, the HVS switches can be turned on with ZVS. However, the relationships between ΔT and I_{LrA} are not same in different operation modes. In this section, the resonant processes of primary side are neglected because of the low drain-to-source voltages and the large switching currents.

As aforementioned, the fixed delta ΔT is the difference between t_{ps1} and t_{ps2}. According to Fig. 8.9, if the resonance in t_{ps1} is neglected, the fixed time delta ΔT in boost mode can be expressed as,

$$\Delta T = t_{\mathrm{ps1}} - t_{\mathrm{ps2}} = L_r \frac{I_{\mathrm{bias}} + I_{\mathrm{LrA}}}{V_c} \tag{8.11}$$

The detailed resonant procedure in buck mode is illustrated (Fig. 8.10). It is worth noting that t_{ps1} and t_{ps2} are negative here. As can be seen, both t_{ps1} and t_{ps2} include resonant process, which are different from those of the boost mode. When the resonance in t_{ps1} is neglected, the fixed time delta ΔT can be written by

$$\Delta T = t_{\mathrm{ps1}} - t_{\mathrm{ps2}} = L_r \frac{I_{\mathrm{bias}} + I_{\mathrm{LrA}}}{V_c} + t_r \tag{8.12}$$

As seen in (8.11) and (8.12), when the fixed time delta ΔT is identical, the bias current in buck mode is less than that in boost mode. In other words, once the ZVS condition is ensured for buck mode, the ZVS condition for boost mode can be achieved absolutely. So the fixed delta ΔT should be designed large enough to ensure $I_{\mathrm{LrA}} \geq 0$ in (8.12).

The dead time also should be selected properly to ensure ZVS. As shown in Fig. 8.10, the dead time t_{dz} should be large enough to make the switch turn on after the resonant process is completed. On the other hand, t_{dz} should be less than a value to avoid the charging of HVS switch junction capacitors in reverse direction. In summary, to achieve ZVS in spite of the working modes, the fixed time delta ΔT and the dead time t_{dz} for HVS can be designed as follows

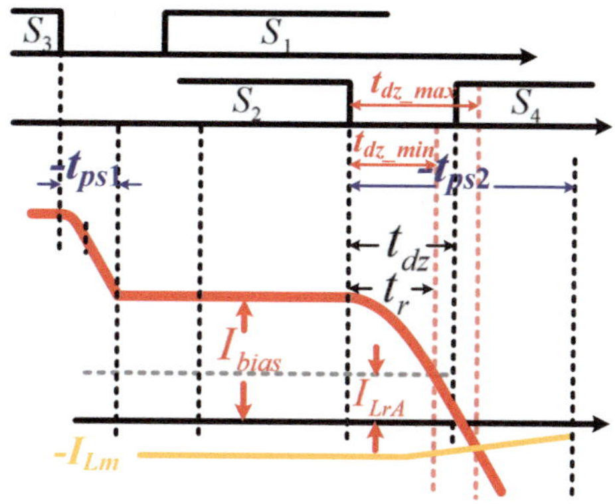

Fig. 8.10 Expanded waveform of the resonant process in buck mode

$$
\begin{cases} \Delta T \geq L_r \frac{I_{\text{bias}}}{V_c} + t_r \\ t_r \leq t_{\text{dz}} \leq t_r + \frac{N_2 L_r i_{L m\max}}{N_1 V_c} \end{cases} \tag{8.13}
$$

where I_{bias} and t_r are the bias current and the resonant time duration length when $I_{\text{LrA}} = 0$ as given in (8.10). The relationship among the boundary of ΔT and t_{dz}, the magnetizing inductance L_m, and the leakage inductance L_r is plotted in Fig. 8.11 in

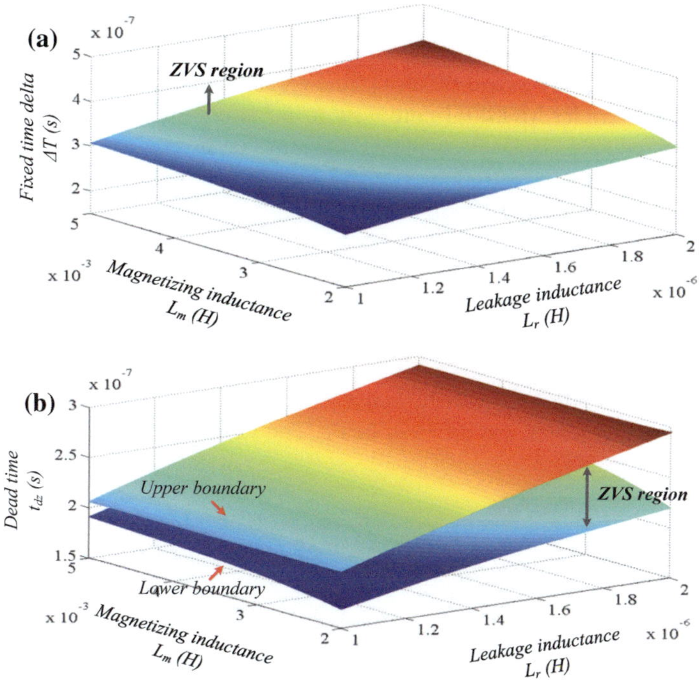

Fig. 8.11 ZVS condition for HVS. **a** Fixed time delta ΔT; **b** dead time t_{dz}

3D form when $I_{LrA} = 0$ and $C_{oss.eff} = 342$ pF. Figure 8.11a indicates that ZVS can be achieved as the fixed time delta is larger than the boundary value. Figure 8.11b shows that the dead time of HVS switches must be selected between the two boundaries in order to achieve ZVS.

If ΔT is large enough, ZVS of HVS MOSFETs can be obtained. However, as ΔT increases, I_{bias} increases and conduction loss increases as well. Therefore, the fixed delta should be designed as small as possible as long as ZVS can be achieved.

8.3 Experimental Results

8.3.1 Prototype and Specifications

A 1 kW experimental prototype has been built in order to verify the effectiveness of the current-fed bidirectional DC–DC converter with the proposed control. Figure 8.12 gives the laboratory prototype picture.

The parameter specifications are illustrated in Table 8.3. It must be pointed out the DC bus is 300 V to supply the voltage for aviation static inverter which is used to achieve integrated starter and generator function [6]. Besides, battery side switches: IPT015N10N5; HVS switches: FDA50N50, whose charge equivalent output capacitance $C_{oss.eff}$ is about 342 pF according to the datasheet. The secondary side magnetizing inductance $L_m = 5$ mH, and then the minimum i_{Lmmax} can be calculated to be 0.18 A. According to Fig. 8.11, the fixed time delta should be selected as $\Delta T \geq 368$ ns, and the dead time t_{dz} should be selected around 250 ns. For LVS, two parallel MOSFETs are used as a bottom switch, so the conduction loss can be reduced. Besides, a planar transformer is used and the conversion efficiency can be improved further. The entire control of the system is implemented with a Texas Instruments TM320F28335 DSP.

8.3.2 Steady-State Operation

Figure 8.13 gives the steady-state waveforms with the proposed control under different battery voltages and different loads. The waveforms at no load are shown

Table 8.3 System specifications

V_1	18–28 V	V_2	300 V
P	1 kW	f	50 kHz
L_1, L_2	11 μH	C_c	30 μF
L_r	1.5 μH	L_m	1 mH
N_1:N_2	2:10	$C_{oss.eff}$	342 pF
ΔT	400 ns	t_{dz}	250 ns

Fig. 8.12 Experimental prototype

in Fig. 8.13a, b. As seen, although the leakage inductance current in the power transfer stage is zero, there is a bias current to help achieve ZVS for HVS switches. The duty cycles are different with different battery voltages, but the delta between primary and secondary side duty cycles is identical. This causes almost identical bias currents. Figure 8.13c–f shows the waveforms at rated power 1 kW in boost mode and buck mode, respectively. The primary side bias current is about 7A under any conditions. ZVS can be achieved, and the circulation loss is low. Thus, the proposed control strategy can adapt to the power flow direction and battery voltage variation automatically.

8.3.3 *Soft Switching Waveforms*

All the switches can achieve ZVS in full-load range employing the proposed control. Because the magnetizing current is minimum with the minimum LVS voltage (18 V), ZVS can be ensured in the full-load range once ZVS can be obtained with minimum LVS voltage. ZVS achievement for the HVS switches with V_1 = 18 V at no load is illustrated in Fig. 8.14.

For all the power switches, under no load condition, they can achieve ZVS turn on.

Figure 8.15 shows the ZVS achievement with the proposed control at 1 kW output. Only the waveforms of upper switches are illustrated in Fig. 8.15 since the upper and lower switches in each switching leg operate symmetrically. As shown, ZVS can be achieved in both boost mode and buck mode at 1 kW output.

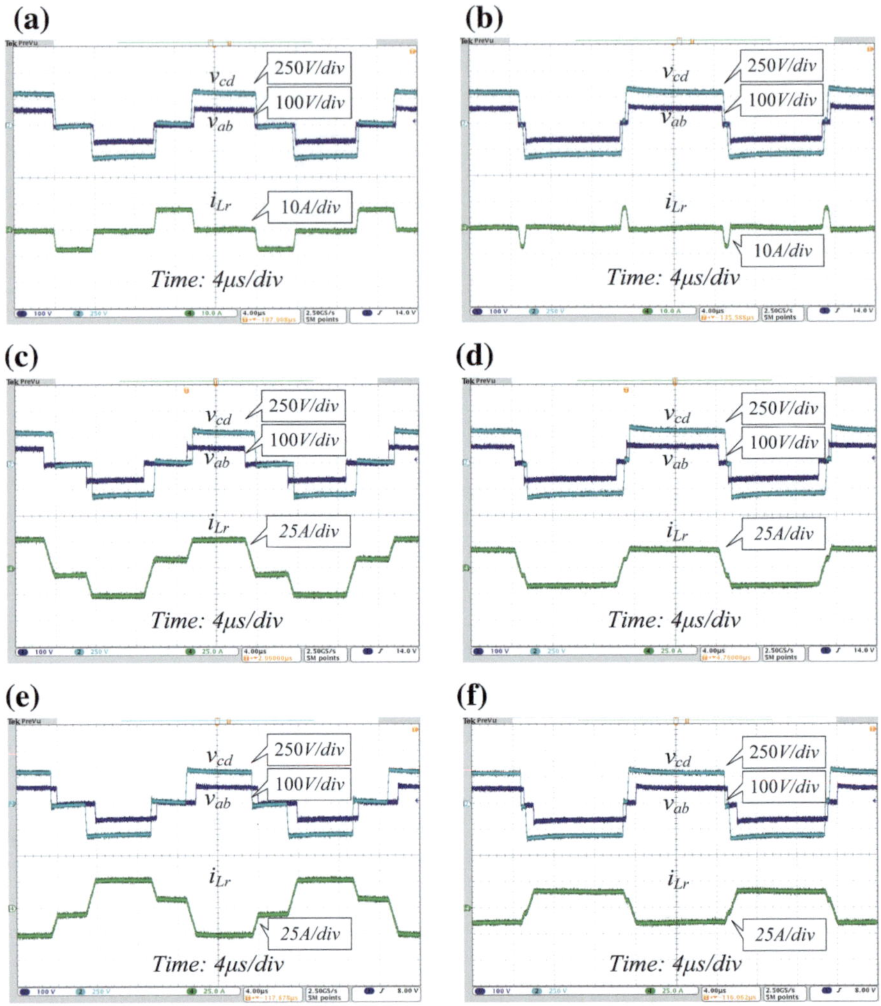

Fig. 8.13 Steady-state operating waveforms with the proposed control: **a** at no load with V_1 = 18 V; **b** at no load with V_1 = 28 V; **c** at 1 kW output with V_1 = 18 V in boost mode; **d** at 1 kW output with V_1 = 28 V in boost mode; **e** at 1 kW output with V_1 = 18 V in buck mode; **f** at 1 kW output with V_1 = 28 V in buck mode

According to the aforementioned analysis, the fixed delta ΔT should be large enough to ensure ZVS of HVS switches, and ΔT is selected to be 400 ns in the experiment. Figure 8.16a illustrates the switching waveforms of secondary side switch S_4 when ΔT = 300 ns. As shown, the bias current is about 4A in this case, and the leakage inductance current reflected to secondary side is not large enough to discharge the junction capacitor completely within the dead time, and the MOSFETs turn on with hard switching. In contrast, ZVS can be obtained when

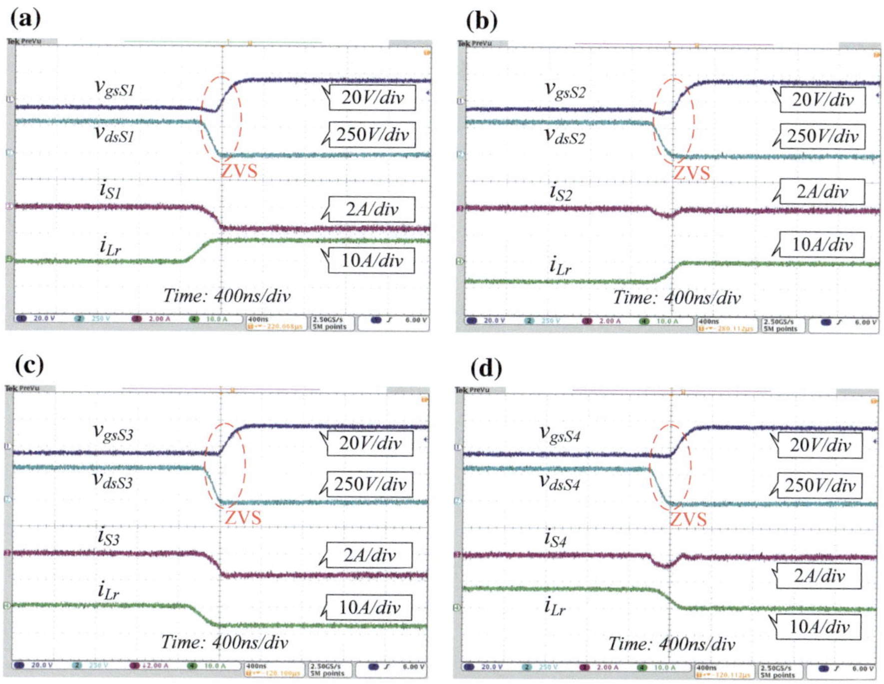

Fig. 8.14 Experimental results of soft switching waveforms with the proposed control at no load with V_1 = 18 V: **a** ZVS of S_1; **b** ZVS of S_2; **c** ZVS of S_3; **d** ZVS of S_4

ΔT = 400 ns as shown in Fig. 8.16b. The experimental results agree well with the theoretical analysis.

8.3.4 Dynamical Operation

Figure 8.17 illustrates the load-step experimental result from 250 to 750 W with V_1 = 23 V in boost mode and the zoomed waveforms are shown on the right side. In Fig. 8.17, V_2 and i_2 are the voltage and current of HVS, respectively. The working mode switches from Mode IA to Mode IIA as the out power increases, indicating that the working mode can be switched between different switching patterns automatically. The output current i_2 is increased from 0.83 to 2.5 A for the load-step change. V_2 drops a little bit and returns to the steady-state value within 20 ms, meaning that the system has better dynamic performance employing the proposed control.

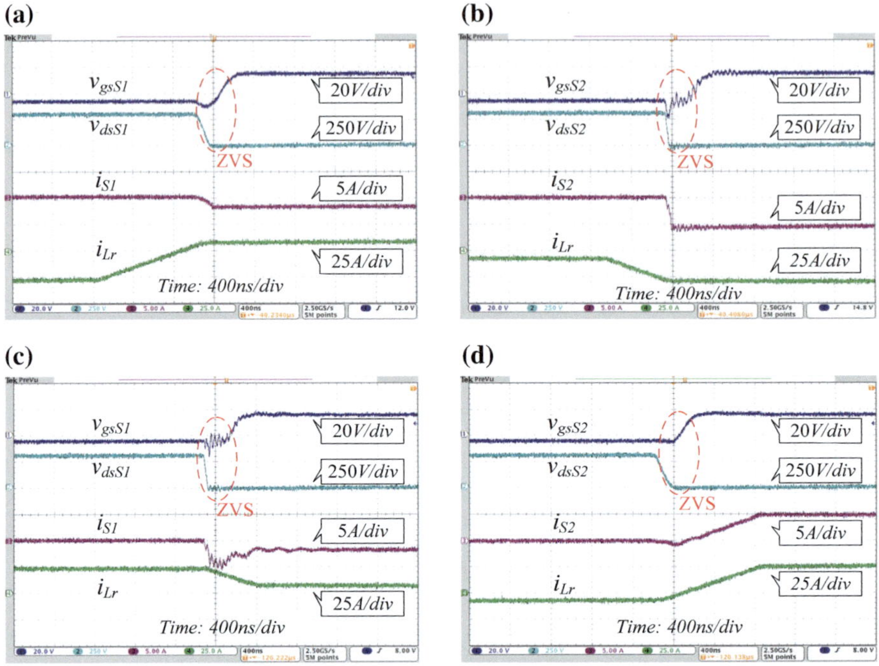

Fig. 8.15 Soft switching waveforms with the proposed control at 1 kW output with V_1 = 18 V: **a** ZVS of S_1 in boost mode; **b** ZVS of S_2 in boost mode; **c** ZVS of S_1 in buck mode; **d** ZVS of S_2 in buck mode

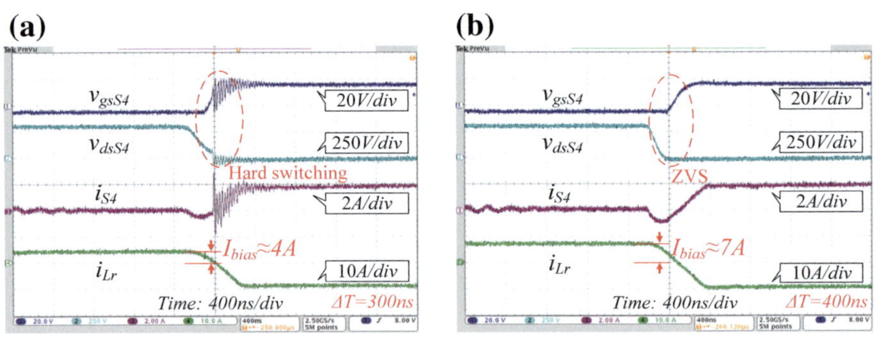

Fig. 8.16 Switching waveforms with different ΔT at 400 W output with V_1 = 23 V in buck mode: **a** ΔT = 300 ns; **b** ΔT = 400 ns

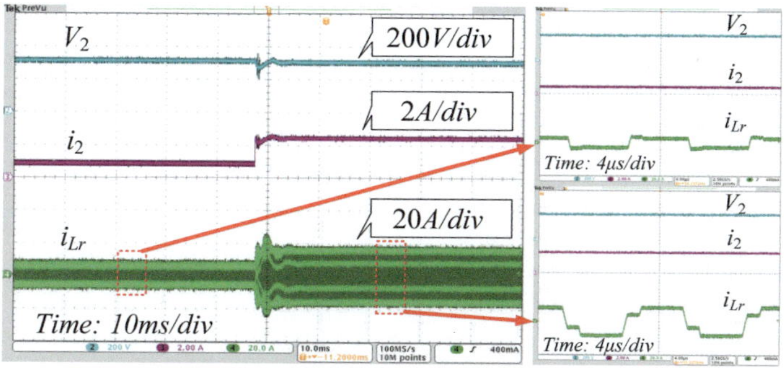

Fig. 8.17 Load-step experimental result from 250 to 750 W with V_1 = 23 V in boost mode

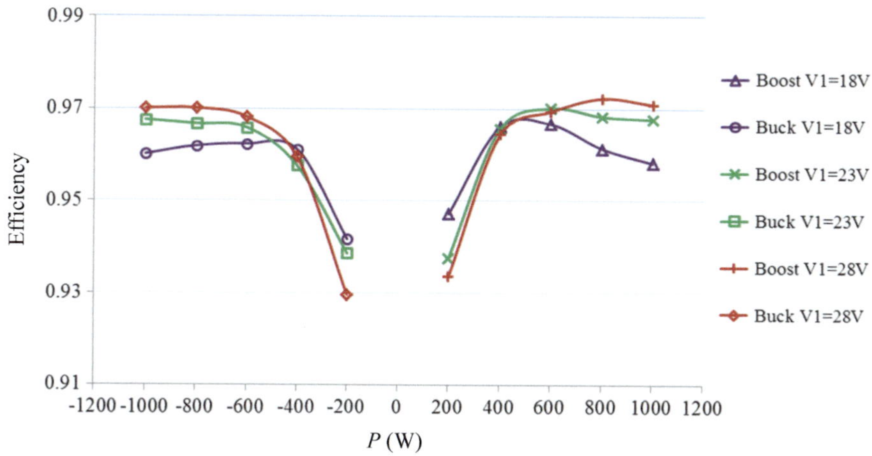

Fig. 8.18 Power stage conversion efficiency

8.3.5 *Conversion Efficiency and Loss Breakdown Analysis*

Figure 8.18 shows the power stage conversion efficiency at different loads and different battery voltages using the proposed optimized control. The overall conversion efficiency is high. At heavy loads, the conduction loss makes up the majority of the total loss, so the efficiency at 28 V battery voltage is higher than that at 18 V battery voltage. However, core losses of the DC inductors and the transformer dominate at light loads, so the efficiency at 28 V battery voltage is higher than that at 18 V battery voltage. Figure 8.19 illustrates the loss breakdown analysis with different LVS voltages at 1 kW load. It is worth noting that control circuit and driving loss are not included. The predicted loss is a little bit lower than measured loss because of stray loss in the circuit like resistance of PCB layout lines and connectors, ESR of capacitors. As seen, the switching loss is low due to ZVS

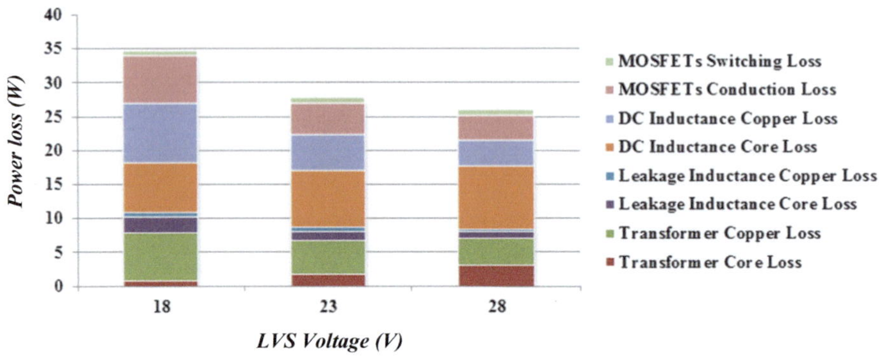

Fig. 8.19 Loss breakdown with different LVS voltages at 1 kW rated power

achievement, and the conduction loss is also reduced significantly with the proposed control. Copper losses are higher at low LVS voltage, while the core losses are higher at high LVS voltage.

8.4 Conclusion

All the possible combinations of switching patterns employing voltage matching control including practical and unpractical ones for current-fed DAB have been studied extensively in this chapter. Among all the possibilities, optimized switching patterns are proposed with which the equivalent duty cycle for the HVS bridge is smaller than that of the battery side bridge with a constant delta regardless of the power flow direction, load, and battery voltage variations. Thus, closed-loop control is simplified and very easy to be implemented on line. The duty cycle delta is optimized to obtain the minimum circulating current under conditions where ZVS can be achieved for all switches. The key parameter design to achieve ZVS was addressed in detail in this chapter. With the proposed modulation, ZVS can be achieved for all switches throughout the full range of load in spite of the variation of operation mode and the battery voltage with reduced circulating current. Thus, the overall conversion efficiency is high. Besides, the converter has very good dynamic performance employing the proposed control. Experimental results from a 1 kW prototype have verified the theoretical analysis and the effectiveness of the proposed switching pattern.

References

1. Xu D, Zhao C, Fan H (2004) A PWM plus phase-shift control bidirectional DC-DC converter. IEEE Trans Power Electron 19(3):666–675

2. Xiao H, Xie S (2008) A ZVS bidirectional DC-DC converter with phase shift plus PWM control scheme. IEEE Trans Power Electron 23(2):813–823
3. Li W, Wu H, Yu H, He X (2011) Isolated winding-coupled bidirectional ZVS converter with PWM plus phase-shift (PPS) control strategy. IEEE Trans Power Electron 26(12):3560–3570
4. Shi Y, Li R, Xue Y, Li H (2015) Optimized operation of current-fed dual active bridge DC-DC converter for PV applications. IEEE Trans Ind Electron 62(11):6986–6995
5. Sha D, You F, Wang X (2016) A high efficiency current fed semi dual active bridge DC-DC converter for low input voltage applications. IEEE Trans Ind Electron 63(4):2155–2164
6. Ding Z, Yang C, Zhang Z, Wang C, Xie S (2014) A novel soft-switching multiport bidirectional DC–DC converter for hybrid energy storage system. IEEE Trans Power Electron 29(4):1595–1609

Chapter 9
A ZVS Bidirectional Three-Level DC–DC Converter with Direct Current Slew Rate Control of Leakage Inductance Current

Abstract A high-frequency isolated bidirectional three-level DC–DC converter is proposed for high-voltage applications. Direct current slew rate (DCSR) control of leakage inductance is proposed to minimize conduction loss and current stress in facing the load variation, the mismatch of turns ratio and circuit parasitic parameters. The mode analysis and the disadvantages of conventional PWM plus phase shift (PPS) control are addressed while these disadvantages can be dealt with the proposed control. Comprehensive comparison between conventional PPS control and the proposed DCSR control are made within the designed low-voltage side (LVS) voltage range. Besides, the implementation of the proposed DCSR control is also given. With the proposed DCSR control, lower conduction loss, lower peak current, lower voltage spike over switches can be obtained in spite of the turns ratio mismatch, load variation, and system parasitic parameters. Zero-voltage switching (ZVS) can be achieved for all power switches in spite of the power flow direction. The effectiveness of the proposed DCSR control on the proposed topology is verified by simulation and experimental results.

Keywords DC–DC converter · Bidirectional · Three-level

9.1 Introduction to Current-Fed Three-Level DAB Converter

In some applications, one side of the bidirectional DC–DC converter is high voltage. IGBTs can be used to undergo the voltage stress of the high-voltage side (HVS) [1].

However, the switching frequency of IGBTs is recommended to be used for situations where the switching frequency is no more than 20 kHz. This reduces the power density, and the cost may be increased. Comparatively, MOSFETs can be switched at higher frequency, but their voltage rating is limited. Besides, with the increase of the voltage rating of a MOSFET, its on-state resistor Rds(on) increases dramatically. For example, a high-voltage rating MOSFET such as 900 V level, its

D. Sha and G. Xu, *High-Frequency Isolated Bidirectional Dual Active Bridge DC–DC Converters with Wide Voltage Gain*, CPSS Power Electronics Series,
https://doi.org/10.1007/978-981-13-0259-6_9

on-state resistance Rds(on) can exceed 1Ω, which reduces the system conversion efficiency and reliability. While, with the decrease of voltage rating, Rds(on) will drop significantly which could improve the efficiency and save the cost. To use low-voltage rating MOSFETs in high-input-voltage situations, two half-bridge converters can be series connected in the HVS [2], but sensing the HVS individual voltages has to be done in order to achieve active voltage balance. The three-level DC–DC converter has been widely used for unidirectional power flow applications [3, 4]. Besides, it can also be used for non-isolated bidirectional applications [5, 6]. In many applications, the galvanic isolation is needed due to safety considerations, but the galvanic isolation is seldom used for three-level bidirectional power flow situations.

PWM plus phase shift (PPS) control [7, 8] is used for current-fed DAB to make the magnitudes of active clamp voltage and the voltage-fed voltage match. Thus, conduction loss can be reduced when input-/output-side voltages vary because the active clamp voltage reference is determined by the HVS voltage and the transformer turns ratio.

Actually, during power transfer period of the dual active bridge (DAB), the slew rate of the leakage inductance current is controlled indirectly in the literature. With the development of the processing speed of DSP, it is feasible to implement the direct control of the leakage inductance current during power transfer period. In this chapter, a three-level current-fed DAB is presented for high-voltage applications.

9.2 Proposed Bidirectional DC–DC Converter

The proposed topology is shown in Fig. 9.1. In the HVS, a three-level half-bridge topology is used to reduce the voltage stress of power switches. L_1 and L_2 are the LVS inductors. Q_1 and Q_2 are the main switches. Q_{1a}, C_{d1} and Q_{2a}, C_{d2} constitute the clamp circuit to achieve ZVS of Q_1 and Q_2, respectively. L_r is the transformer leakage inductance. In the HVS, S_1–S_4 are the main switches. D_1, D_2 and C_c are clamping diodes and the clamping capacitor to achieve voltage balance for main switches, respectively. C_3 and C_4 are the HVS capacitors to divide HVS voltage.

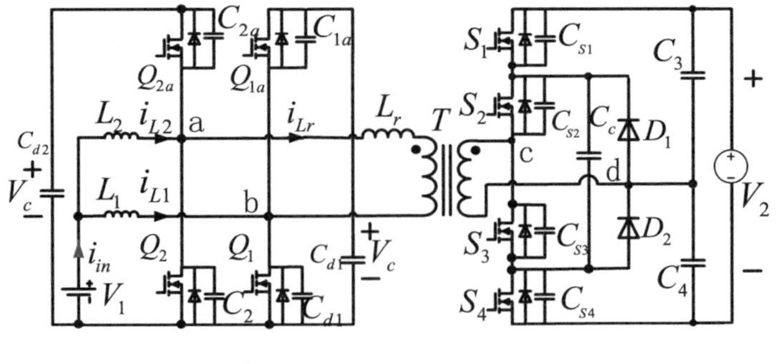

Fig. 9.1 Circuit of proposed topology

There are two operation modes including buck mode (power flows from HVS to LVS) and boost mode (power flows from LVS to HVS). Figure 9.2 gives the various stages (θ_0–θ_{16}) of the circuit operation during one switching period in the boost mode and buck mode. For simplicity of analysis, all the devices are regarded as ideal.

In a switching period of the boost mode, there are sixteen working stages. Among all the working stages, some stages can be defined as current circulating stages in which power is not transmitted between the LVS and HVS. For example, during the range of θ_1–θ_6 or θ_9–θ_{14}, the average power transmitted between LVS and HVS is zero. Other stages can be defined as power transfer stages in which both the AC voltage v_{ab} and AC voltage v_{cd} have the same polarity. During one switching cycle, the power transfer stages are θ_6–θ_9 and θ_{14}–θ_{17}.

The equivalent circuits of all working stages are shown in Fig. 9.3, which is described as follows:

Stage 1 (Before θ_0): Q_2, Q_{1a}, S_3, and S_4 conduct. During this stage, the power flows from LVS (V_1) to HVS (V_2).

Stage 2 (θ_0–θ_1): At θ_0, Q_{1a} is turned off. C_1, C_{1a}, and L_r begin to resonate, making C_{1a} charged and C_1 discharged.

Stage 3 (θ_1–θ_2): At θ_1, the voltage of C_{1a} attempts to overshoot to clamp voltage C_{d1}. The body diode of Q_1 is therefore forward-biased, and the drain-source voltage of Q_{1a} is clamped effectively by the clamp capacitance C_{d1}. During this stage, Q_1 can be gated on at ZVS.

Stage 4 (θ_2–θ_3): At θ_2, Q_2 is turned off. C_2, C_{2a}, and L_r begin to resonate, making C_2 charged and C_{2a} discharged.

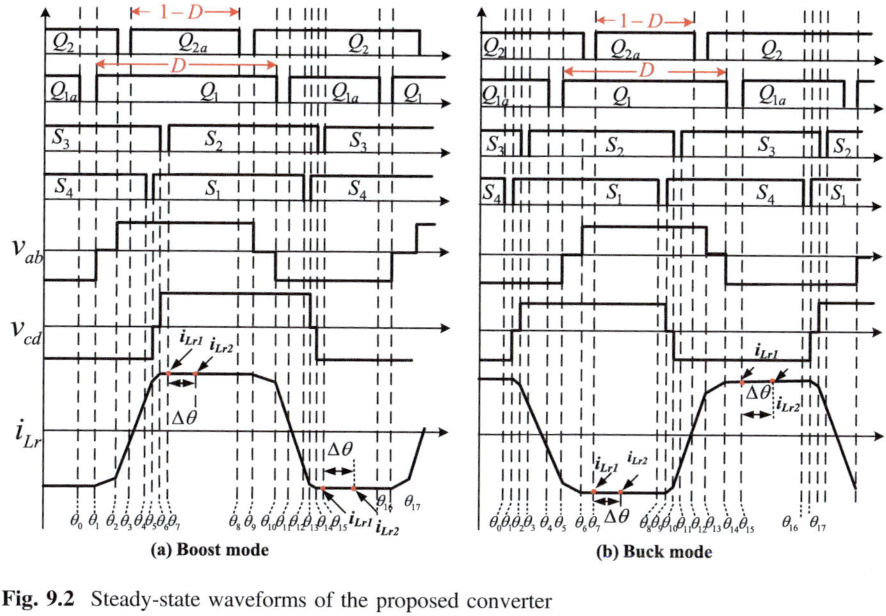

Fig. 9.2 Steady-state waveforms of the proposed converter

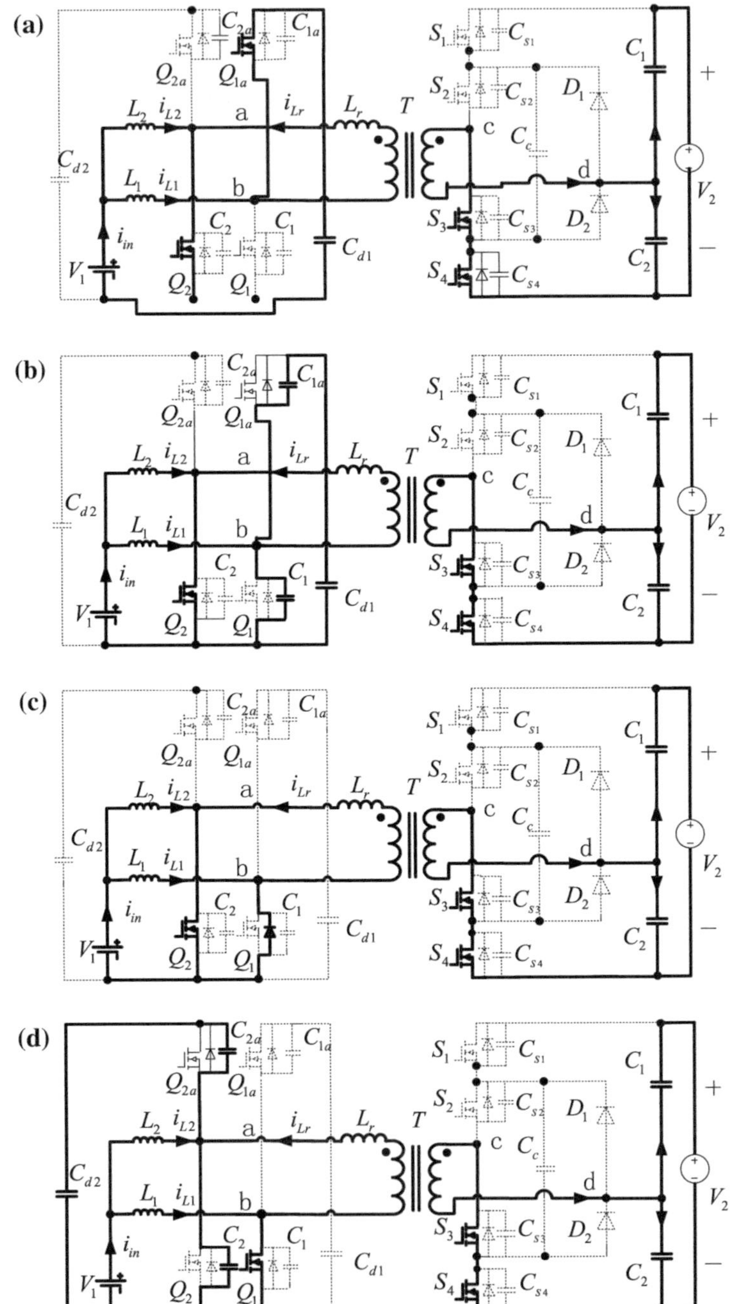

Fig. 9.3 Equivalent circuits of the switching modes. **a** Stage 1; **b** stage 2; **c** stage 3; **d** stage 4; **e** stage 5; **f** stage 6; **g** stage 7; **h** stage 8; **i** stage 9

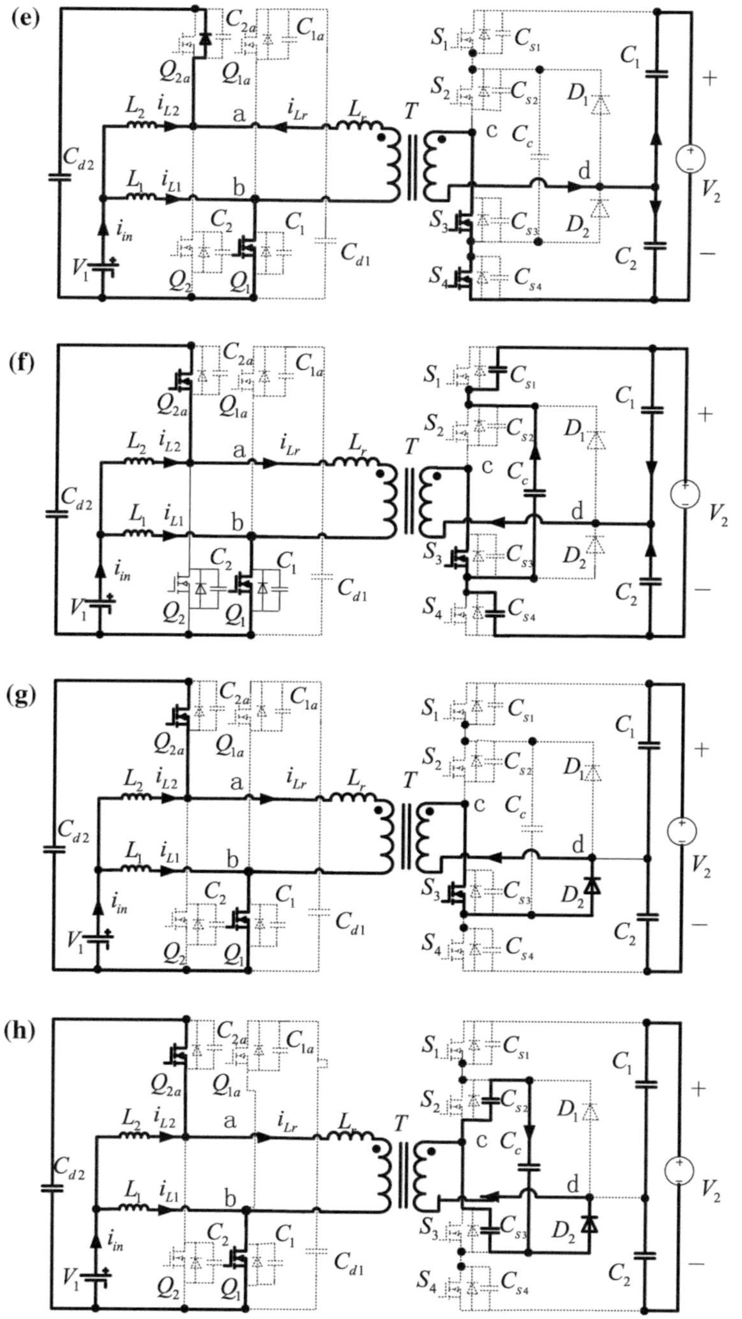

Fig. 9.3 (continued)

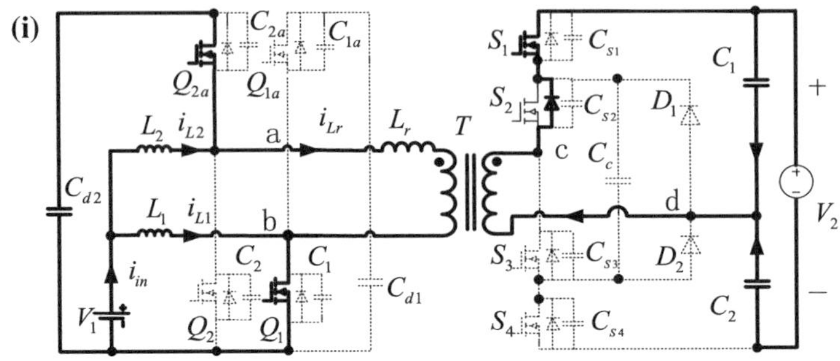

Fig. 9.3 (continued)

Stage 5 (θ_3–θ_4): At θ_3, the voltage of C_{2a} attempts to overshoot to active clamp voltage V_c. The body diode of Q_{2a} is therefore forward-biased, and the drain-source voltage of Q_2 is clamped effectively by the clamping capacitance C_{d2}. In this stage, Q_{2a} can be gated on at ZVS.

Stage 6 (θ_4–θ_5): At θ_4, S_4 is turned off. C_{S4}, C_{S1}, and L_r begin to resonate via C_c, making C_{S4} charged and C_{S1} discharged.

Stage 7 (θ_5–θ_6): At θ_5, the voltage of C_{S4} rises to $V_2/2$ and the voltage of C_{S1} decays to zero. At the same time, D_2 conducts naturally and $V_{cd} = 0$. During this period, the voltage of C_{S1} is clamped to zero, so S_1 can be gated on at ZVS.

Stage 8 (θ_6–θ_7): At θ_6, S_3 is turned off. C_{S3}, C_{S2}, and L_r begin to resonate via C_c, making C_{S3} charged and C_{S2} discharged.

Stage 9 (θ_7–θ_8): At θ_7, the voltage of C_{S3} rises to $V_2/2$ and the voltage of C_{S2} decays to zero. At the same time, the body diode of S_2 is therefore forward-biased and $V_{cd} = V_2/2$. During this period, the voltage of C_{S2} is clamped to zero, so S_2 can be gated on at ZVS. At θ_8, Q_{2a} is turned off, and the second half cycle (θ_8–θ_{16}) is similar to the first half cycle.

9.3 Comparison of PPS and DCSR Controls

9.3.1 Physical Turns Ratio Mismatch Considerations

The control block diagram of the PPS control is shown in Fig. 9.3a. For the active clamp voltage control loop, the control target is to stabilize the active clamp voltage according to the HVS voltage and the turns ratio. The equivalent circuit with conventional PPS control for ideal circuit is shown in Fig. 9.4b. Variable duty cycle obtained by the active clamp voltage PWM control loop acts as an electronic transformer between V_1 and V_c. According to KVL, during the power transfer stage, one can obtain

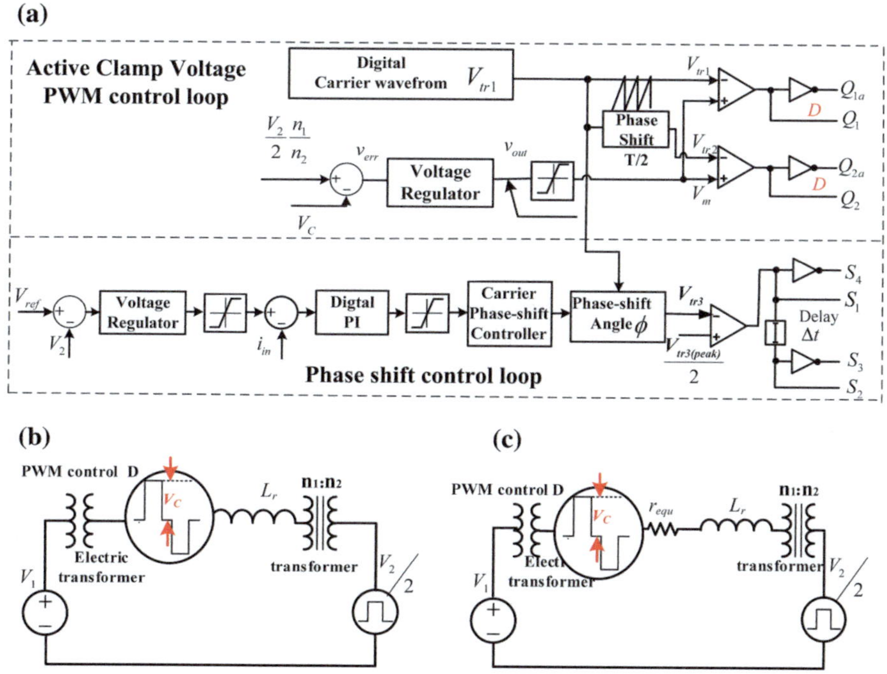

Fig. 9.4 **a** Conventional PPS control; **b** ideal circuit for PPS control; **c** actual circuit for PPS control

$$L_r \frac{\mathrm{d}i_{Lr}}{\mathrm{d}t} = V_C - \frac{V_2}{2} \times \frac{n_1}{n_2} \tag{9.1}$$

With PPS control, to minimize conduction loss, the reference of active clamp voltage using PPS control is equal to the product of the $V_2/2$ and the turns ratio n_1: n_2. This implies that (9.1) is equal to zero. It should be noted that transformer turns ratio value has to be obtained precisely. In fact, the transformer cannot be manufactured with a precise value for mass production. Once the reference of PWM control loop has been set, the turns ratio mismatch can result in variation of current slew rate of leakage inductance.

For example, if the actual turns ratio n_1:n_2 is larger than its predefined value, then the following can be obtained as,

$$L_r \frac{\mathrm{d}i_{Lr}}{\mathrm{d}t} = V_C - \frac{V_2}{2} \times \frac{n_1}{n_2} \tag{9.2}$$

Thus, the current slew rate of the leakage inductance is not zero during the power transfer stage. Otherwise, if the actual turns ratio is less than its predefined value, one can obtain

$$L_r \frac{di_{Lr}}{dt} = V_C - \frac{V_2}{2} \times \frac{n_1}{n_2} > 0 \quad (9.3)$$

Thus, the current slew rate of the leakage inductance is also not zero during the power transfer stage.

To verify this, a simulation is made. The simulation specifications are specified in Table 9.1.

The simulation results with different turns ratios for PPS control are shown in Fig. 9.5. Supposing the active clamp voltage reference is determined by the turns ratio 2:10 shown in Fig. 9.5b, if there is a slight mismatch for the turns ratio, the current slew rate of the leakage inductance will not be zero, which is shown in Fig. 9.5a, c respectively. This will increase the conduction loss.

The simulation results with different turns ratios with DCSR control are shown in Fig. 9.6. If the turns ratio n_1:n_2 = 2:9, the active clamp voltage is regulated to be higher as 66.67 V. And if the turns ratio n_1:n_2 = 2:11, the active clamp voltage is regulated to be lower as 54.54 V.

Table 9.1 Simulation specifications

Rated power	1200 W	L_1, L_2	15 μH
HVS (V_2)	600 V	C_{d1}, C_{d2}	30 μF
LVS (V_1)	19–24.5 V	C_c	10 μF
Turns ratio	2:9/2:10/2:11	C_3, C_4	100 μF
Switching frequency	50 kHz	L_r	2.2 μH

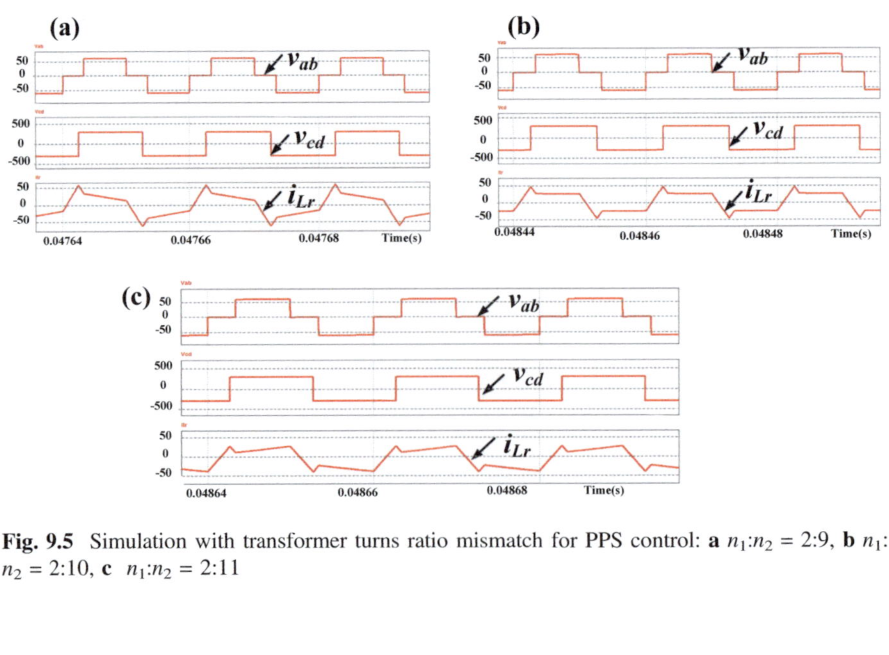

Fig. 9.5 Simulation with transformer turns ratio mismatch for PPS control: **a** n_1:n_2 = 2:9, **b** n_1:n_2 = 2:10, **c** n_1:n_2 = 2:11

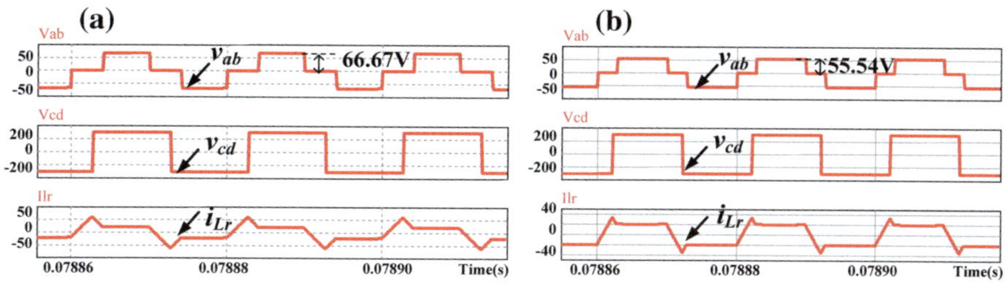

Fig. 9.6 Simulation with transformer turns ratio mismatch for DCSR control: **a** n_1:n_2 = 2:9, **b** n_1:n_2 = 2:11

9.3.2 Actual Equivalent Circuit

Even if the turns ratio of transformer can be designed precisely and the reference of PWM control loop is given accurately, the slew rate of leakage inductance current cannot be kept zero in facing the load variation due to the circuit equivalent resistance including on-state resistance of power MOSFETs of both sides, equivalent series resistance (ESR) of transformer windings, and line resistance. Figure 9.4b shows the ideal circuit for PPS control, i.e., all the components including power devices and transformers are ideal. Practically, the equivalent circuit is shown in Fig. 9.4c in view of the circuit equivalent resistance r_{equ} which is the sum of the on-state resistance of MOSFETs and equivalent resistance for inductor, transformer, and electrical line. According to KVL, the following can be obtained

$$L_r \frac{\mathrm{d}i_{Lr}}{\mathrm{d}t} = V_C - i_{Lr} \times r_{\text{equ}} - \frac{V_2}{2} \times \frac{n_1}{n_2} \tag{9.4}$$

Assuming the turns ratio of transformer can be designed precisely, and with PPS control, the clamp voltage V_{C} is regulated to be fixed. Under a certain output power, $L_r \mathrm{d}i/\mathrm{d}t = 0$, and therefore the slew rate of the leakage inductance current can be kept zero. However, once the load varies, the leakage inductance current i_{Lr} during the power transfer stage varies as well, but the clamp voltage V_{C} and turns ratio are kept constant. This makes $L_r \mathrm{d}i/\mathrm{d}t \neq 0$, and the slew rate of the leakage inductance current is not equal to zero any more. For example, if the output power increases, i_{Lr} increases as well, leading to

$$L_r \frac{\mathrm{d}i_{Lr}}{\mathrm{d}t} = V_C - i_{Lr} \times r_{\text{equ}} - \frac{V_2}{2} \times \frac{n_1}{n_2} < 0 \tag{9.5}$$

Thus, the slew rate of leakage inductance current cannot be kept zero during the power transfer stage.

In view of the voltage drop over the equivalent resistance, even if the clamp voltage V_C can be regulated to be fixed, the slew rate of the leakage inductance current may not be equal to zero as the load varies during the power transfer stage.

To verify this, simulation in view of the circuit parasitic parameters was made. The devices are listed as follows: Q_1, Q_{1a}, Q_2, Q_{2a}: IXFN230N10 whose Rds(on) is 6 mΩ; S_1–S_4: IXFX100N50P whose Rds(on) is 49 mΩ. Type of diodes for the three-level side is MUR3060T. The turns ratio is fixed as 2:10. The ESR of the transformer winding and line impedance is 6 mΩ.

Figure 9.7 shows the simulation results with different power outputs at 24 V input with the PPS control and the proposed DCSR control. As seen in Fig. 9.7a, with the conventional PPS control, when the output power is 200 W, the slew rate of the leakage inductance can be controlled almost to be zero during the power transfer stage. However, as seen in Fig. 9.7b, when the output power increases up to 1.2 kW, the slew rate of the leakage inductance current cannot be kept zero anymore due to the larger voltage drop of power devices, ESR of equivalent circuits. As seen, with conventional PPS control, the active clamp voltage V_C is fixed in spite of the load variation. Thus, at 200 W and 1.2 kW outputs, the active clamp voltage V_C is regulated to be 60 V according to the relationship of the output voltage and the turns ratio. However, with the proposed DCSR control, at 200 W output, the active clamp voltage is regulated to be a little bit higher to be 60.3 V. At 1.2 kW output, the active clamp voltage is regulated to be even higher to be 61.4 V. With the increase of the output power, the voltage drop over power MOSFETs, line impedance, and transformer windings increases. Therefore, the active clamp voltage should vary as well. But with the conventional PPS control, the clamp voltage has been set fixed despite the load variation. Thus, the leakage inductance current slew rate cannot be maintained zero as the load varies.

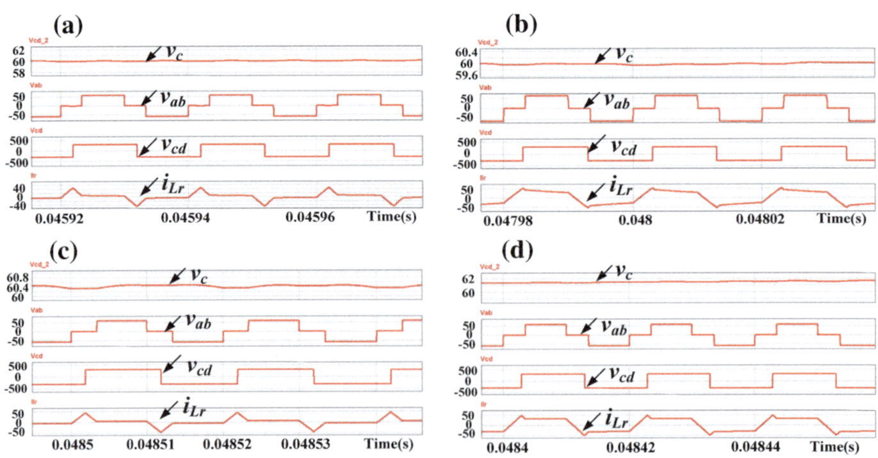

Fig. 9.7 Simulation in boost mode (turns ratio is 2:10) at 24 V input: **a** PPS at 200 W output; **b** PPS at 1.2 kW output; **c** DCSR at 200 W output; **d** DCSR at 1.2 kW output

9.3.3 RMS Current Comparison

Due to the above limitations, it is difficult to control the slew rate of leakage inductance current to be zero with the PPS control in view of turns ratio mismatch and load variation. With the proposed DCSR control, the slew rate of the leakage inductance current can be controlled to be zero during power transfer stage. To compare the RMS current in view of the turns ratio mismatch, the expression of the instantaneous current can be calculated as follows. Because the zero state of v_{cd} is very short, the state is ignored. Taking the boost mode as an example.

When $(2d-1)\ \pi \leq \phi \leq \pi$, according to Fig. 9.8a, the expression of the instantaneous leakage current can be given by

$$i_{Lr}(\theta) = \begin{cases} i_{Lr}(0) + \frac{n_1 V_2 \theta}{2n_2 \omega L_r}, & (0 \leq \theta \leq (2d-1)\,\pi) \\ i_{Lr}(0) + \frac{n_1 V_2 (2d-1)\pi}{2n_2 \omega L_r} + \left(\frac{V_1}{1-d} + \frac{n_1 V_2}{2n_2}\right)\left[\frac{\theta-(2d-1)\pi}{\omega L_r}\right], & ((2d-1)\,\pi \leq \theta \leq \phi) \\ i_{Lr}(0) + \frac{n_1 V_2 (2d-1)\pi}{2n_2 \omega L_r} + \left(\frac{V_1}{1-d} + \frac{n_1 V_2}{2n_2}\right)\left[\frac{\phi-(2d-1)\pi}{\omega L_r}\right] + \\ \left(\frac{V_1}{1-d} - \frac{n_1 V_2}{2n_2}\right)\left[\frac{\theta-\phi}{\omega L_r}\right], & (\phi \leq \theta \leq \pi) \\ -i_{Lr}(\theta-\pi), & (\pi \leq \theta \leq 2\pi) \end{cases} \tag{9.6}$$

where

$$i_{Lr}(\pi) = -i_{Lr}(0) + \frac{n_1 V_2 (2d-1)\pi}{2n_2 \omega L_r} + \left(\frac{V_1}{1-d} + \frac{n_1 V_2}{2n_2}\right)\left[\frac{\phi-(2d-1)\pi}{\omega L_r}\right] + \left(\frac{V_1}{1-d} - \frac{n_1 V_2}{2n_2}\right)\left[\frac{\pi-\phi}{\omega L_r}\right]$$

when $\pi \leq \phi \leq (2d-1)\pi$, according to Fig. 9.8b,

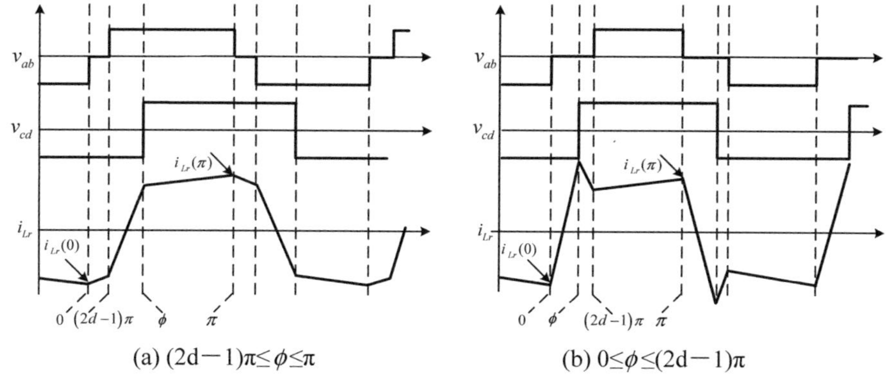

Fig. 9.8 Steady-state waveforms of the proposed converter in boost mode

$$i_{Lr}(\theta) = \begin{cases} -i_{Lr}(0) + \frac{n_1 V_2 \theta}{2n_2 \omega L_r} & (0 \leq \theta \leq \phi) \\ -i_{Lr}(0) + \frac{n_1 V_2 \phi}{2n_2 \omega L_r} - \frac{n_1 V_2}{2n_2}\left(\frac{\theta - \phi}{\omega L_r}\right) & (\phi \leq \theta \leq (2d-1)\,\pi) \\ -i_{Lr}(0) + \frac{n_1 V_2 \phi}{2n_2 \omega L_r} - \frac{n_1 V_2}{2n_2}\left[\frac{(2d-1)\pi - \phi}{\omega L_r}\right] & \\ + \left(\frac{V_1}{1-d} - \frac{n_1 V_2}{2n_2}\right)\left[\frac{\theta - (2d-1)\pi}{\omega L_r}\right] & ((2d-1)\,\pi \leq \theta \leq \pi) \\ -i_{Lr}(\theta - \pi) \quad (\pi \leq \theta \leq 2\pi) & \end{cases} \tag{9.7}$$

where

$$\begin{aligned} i_{Lr}(\pi) = -i_{Lr}(0) + \frac{n_1 V_2 \phi}{2n_2 \omega L_r} - \frac{n_1 V_2}{2n_2}\left[\frac{(2d-1)\pi - \phi}{\omega L_r}\right] \\ + \left(\frac{V_1}{1-d} - \frac{n_1 V_2}{2n_2}\right)\left[\frac{\pi - (2d-1)\pi}{\omega L_r}\right] \end{aligned} \tag{9.8}$$

For the current of LVS bottom switches, taking Q_1 as an example, its instantaneous current can be written by

$$i_{Q_1}(\theta) = \begin{cases} \frac{P}{V_1} - \frac{V_1 d\pi}{L_1 \omega} + \frac{V_1 \theta}{L_1 \omega} + i_{L_r}(\theta) & (0 \leq \theta \leq 2d\pi) \\ 0 \quad (2d\pi \leq \theta \leq 2\pi) & \end{cases} \tag{9.9}$$

The RMS current through the bottom switch Q_1 is given by

$$I_{\text{RMS\_}Q_1} = \sqrt{\int_0^{2\pi} \frac{i_{Q_1}^2(\theta)}{2\pi} d\theta} \tag{9.10}$$

Based on (9.6)–(9.10), the RMS current of LVS bottom switch Q_1 can be derived. In the same way, the RMS current of LVS top switch Q_{1a} can be calculated.

Figure 9.9 shows the RMS current of LVS bottom switch Q_1 in boost mode with different control strategies (DCSR and PPS). With PPS control, the turns ratio used to calculate the voltage reference of the PWM control loop is 2:10. Two kinds of situations with different turns ratios are considered. One is 2:9 and the other is 2:11. With the proposed DCSR control, the current slew rate of the transformer leakage inductance is kept zero in spite of the turns ratio mismatch. With the PPS control, the slew rate cannot be kept zero if the turns ratio is not equal to the predefined turns ratio.

To compare fairly, the RMS value of weighted current for both top and bottom switches can be calculated because the weighted current can reveal the total conduction of switches of one leg. The weighting factors for both the top switch and the bottom switch are same with the value of 0.5. As seen in Fig. 9.10, in facing the turns ratio mismatch, the proposed DCSR control can minimize the conduction loss of switches compared with using the PPS control.

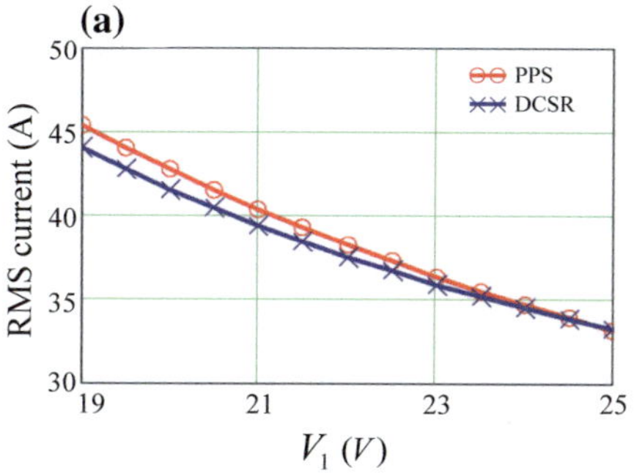

Fig. 9.9 RMS value of LVS bottom switch Q_1 with different turns ratios. **a** n_1:n_2 = 2:9; **b** n_1: n_2 = 2:11

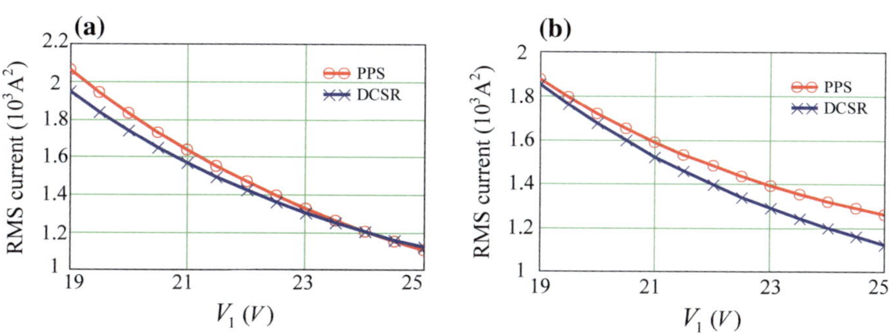

Fig. 9.10 RMS value of weighted current for one-leg switches (Q_1 & Q_{1a}) with different turns ratios. **a** n_1:n_2 = 2:9; **b** n_1:n_2 = 2:11

9.3.4 The Peak Current of Main Switches

Since the LVS bottom switch handles much more current than the top one, only the peak current distribution of the bottom switch is analyzed with different control strategies. Figure 9.11 shows the peak current of LVS bottom switches with the proposed DCSR and PPS controls. As seen, using the proposed DCSR control can minimize the peak current compared with the PPS control in facing the turns ratio mismatch.

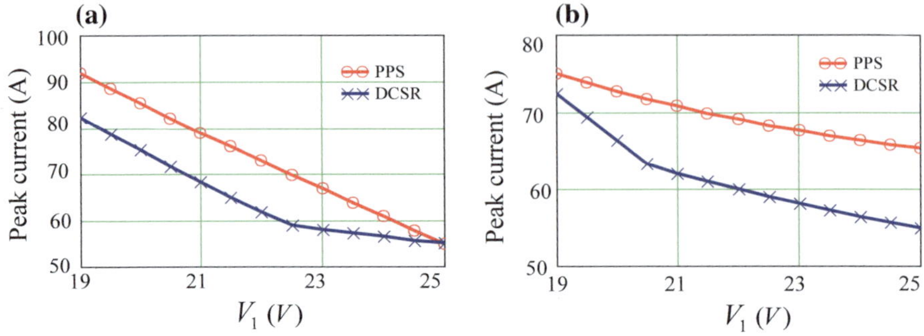

Fig. 9.11 Peak value of LVS bottom switch current with different turns ratios. **a** n_1:n_2 = 2:9; **b** n_1: n_2 = 2:11

9.4 Implementation of the DCSR Control

9.4.1 Voltage Balance Issue for the Three-Level HVS

The three-level half-bridge topology is adopted in the HVS. For the HVS, S_1, S_2, $S_{3,}$ and S_4 are in series connection in order to share the HVS voltage. S_1, S_4 are defined as leading leg switches, and S_2, S_3 are lagging leg switches. The voltages of HVS power devices cannot be balanced well due to the inconsistency and stray parameters of the drive circuit. To ensure voltage balance, clamping capacitance (C_c) and diodes (D_1 and D_2) are introduced. The control strategy based on the proposed topology is easy to be implemented.

The PWM signals of S_1, S_2, $S_{3,}$ and S_4 are shown in Fig. 9.2a. S_1, S_4 are turned on and off before S_2 and S_3. When S_1 is turned off, C_{S1} is charged and C_{S4} is discharged via C_c. When the voltage of C_{S1} rises to $V_2/2$, the clamp diode D_1 will conduct and the voltage of C_{S1} is clamped. By means of the control strategy based on the proposed topology, the voltages of HVS power devices are balanced perfectly.

9.4.2 Implementation of the Proposed Control Strategy

The control block diagram of the proposed control strategy is shown in Fig. 9.12. The analog circuit (circled by the green dashed line) is composed of a current transformer and a level shift circuit to eliminate the negative value for the input of inner A/D of the DSP. The turns ratio of current transformer is 1:250, and the sampling resistance is 10 Ω. By this means, the leakage inductance current i_{Lr} is sampled as a voltage value from −1.5 to 1.5 V. Through the level shift circuit, a 1.5 V bias voltage is added and the value input to the inner A/D varies from 0 to 3 V. The CPLD is responsible for logical operation (circled by the red dashed line)

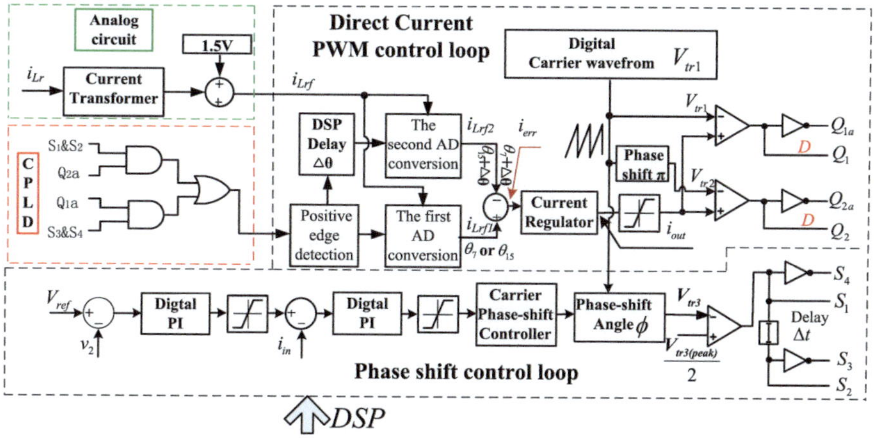

Fig. 9.12 Control block diagram of the proposed DCSR control

to generate a positive edge for DSP as the start of the first A/D conversion of i_{Lrf}. As shown in Fig. 9.2a, the slew rate of the leakage inductance current i_{Lr} is expected to be regulated to be zero during the range (θ_6–θ_9 and θ_{14}–θ_{17}). Hence, in this range, i_{Lr} should be sampled twice to get the slew rate.

All the control parts circled by the black dashed line are implemented by floating-point DSP TMS320F28335. The PWM control loop is designed to control the slew rate of i_{Lr} and the phase shift control loop is used to stabilize the HVS voltage and the direction of power flow. As shown in Fig. 9.2a, the slew rate of the leakage inductance current i_{Lr} can be regulated to be zero during the range (θ_6–θ_9 and θ_{14}–θ_{17}). For example, during the time interval (θ_6–θ_9) of a switching period, i_{Lr} is sampled at points θ_7 and $\theta_7 + \Delta\theta$ to obtain the currents i_{Lrf1} and i_{Lrf2}, respectively. The control target is to let i_{Lrf1} equals to i_{Lrf2} and its regulation output is compared with carrier waveform V_{tr1} and V_{tr2}, respectively, to generate gate sequence for LVS switches. To achieve interleaving to minimize the LVS total current ripple, the phase of V_{tr2} is opposite to that of V_{tr1}. In this way, the slew rate of i_{Lr} can be controlled to be zero by regulating the duty cycle in spite of the mismatch of converter parameters.

The regulation process to smooth the current slew rate can be seen in Fig. 9.13. It is assumed that $di_{Lr}/dt < 0$ during the power transfer stage, the second sampling value i_{Lrf2} is less than the first sampling value i_{Lrf1}, thus the current comparator

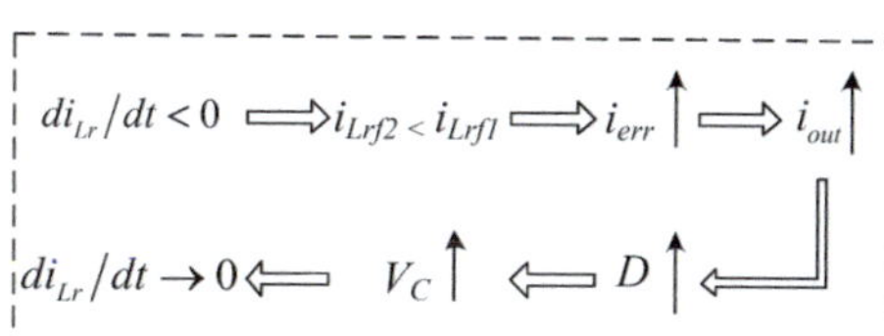

Fig. 9.13 Regulation process when $di_{Lr}/dt < 0$ during power transfer stage

output increases and the duty cycle D increases as well. The clamp capacitor voltage V_C increases. This makes the di_{Lr}/dt increases until it approaches zero. Therefore, the proposed DCSR control can smooth the leakage inductance current slew rate during the power transfer stage.

On the contrary, if $di_{Lr}/dt > 0$ during the power transfer stage, the proposed control can also smooth the leakage inductance current.

The current of LVS i_{in} and HVS voltage V_2 are sampled for the phase shift control loop. V_{ref} is the HVS voltage reference. Through voltage and current loop control, the phase shift Φ is obtained. The carrier waveform V_{tr3} is generated by phase shifting V_{tr1} with the angle of Φ. $V_{tr3(peak)}$ is the peak value of the carrier waveform V_{tr3}. As seen, duty cycles for HVS switches are all 0.5. As shown in Fig. 9.12, if the converter operates in boost mode, the voltage regulator output is a positive value. If V_2 is higher than its reference V_{ref}, the voltage control loop output becomes negative and the converter will operate in buck mode automatically.

It should be noted that sensing the LVS active clamp DC voltage is avoided with the proposed DCSR control. Thus, voltage sensors used to sense active clamp voltages are not needed any more, reducing the cost. Comparatively, with the DCSR control, the leakage inductance current is sensed through a low-cost current transformer which does not need extra auxiliary power. The sensed current is then processed by the DSP inner A/D whose processing speed is faster than external one and costs no extra money.

9.5 Experimental Results

A 1.2 kW experimental prototype switched at 50 kHz has been built for the ZVS bidirectional three-level DC–DC converter with the proposed DCSR control. An IGBT is not suitable to be used due to that high switching frequency. The system parameters have been shown in Table 9.1, and the power devices are the same for the aforementioned simulation in Sect. 9.3. It should be noted that even though the HVS voltage is 600 V, the voltage rating for the HVS MOSFET IXFX100N50P is only 500 V, besides its Rds(on) is only 49 mΩ. The system efficiency can be improved by using that low-voltage rating power MOSFET in that high-voltage application, which indicates the advantage of the three-level topology. Two transformers are purposely manufactured. One transformer whose turns ratio is 2:10 is used as the predefined turn ratio for the conventional PPS control. The turns ratio of another one is 2:11. Actually, the measured values of them are 2:10.10 and 2:11.05, respectively.

The laboratory prototype is shown in Fig. 9.14. The digital signal processor is TMS320F28335. Leakage inductance current is sampled with a current transformer whose turns ratio is 1:250, and the sampling resistance is 10 Ω. The sampling signal is added to 1.5 V DC voltage. Then this analog signal is converted to digital signal by the inner A/D of digital controller twice in one switching cycle. For the bidirectional operation, LVS DC source is a battery pack composed of two 12 V

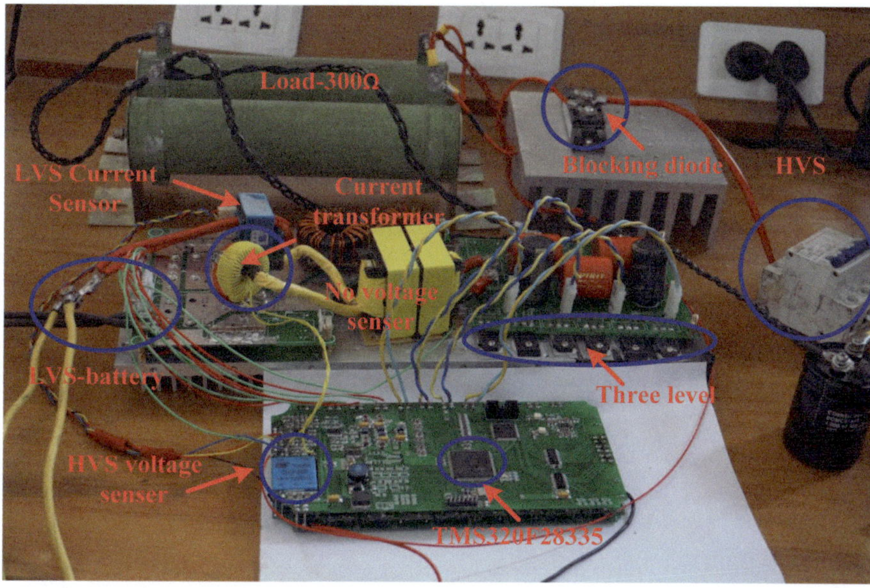

Fig. 9.14 Picture of the laboratory prototype

batteries in series connection. The power devices are the same with those for the simulation as shown in Fig. 9.6.

For the implementation of the proposed DCSR control, the key is the fast and accurate sampling of the leakage inductance current. As previously stated, a low-cost current transformer has been used to detect the leakage inductance current since it is high-frequency AC. Actually, although the detected leakage inductance current varies within one switching cycle, the value needs to be sensed during the power transfer stage in which the instantaneous value of v_{ab} is not zero. Once the gating sequence for the LVS is known, it is easy for the DSP to know the power transfer stage. Theoretically, leakage inductance currents can be sensed at arbitrary two points within the power transfer stage can be used to achieve the proposed flat slew rate control.

For ease of implementation, the first inner A/D conversion is done when DSP detects the positive edge based on the logical operation of switching sequence using CPLD as shown in Fig. 9.12, and after a short period of time delay, the second inner A/D conversion is accomplished.

Figure 9.15 illustrates the experimental results of the leakage inductance current sampling. From Fig. 9.15a–c, V_{sa} is an indicator whose positive edge is the first sample time and V_{sb} positive edge is the second sample time. Figure 9.15a, b, c shows the sampling times together with gating sequence, both-side AC voltages, and leakage inductance current. As seen, both the two sampling times occur during the power transfer stage in which the leakage inductance current is flat. V_{gsQ2a} is the gate drive signal of Q_{2a}, and V_{gsS2} is the gate drive signal of S_2. v_{ab} is the primary

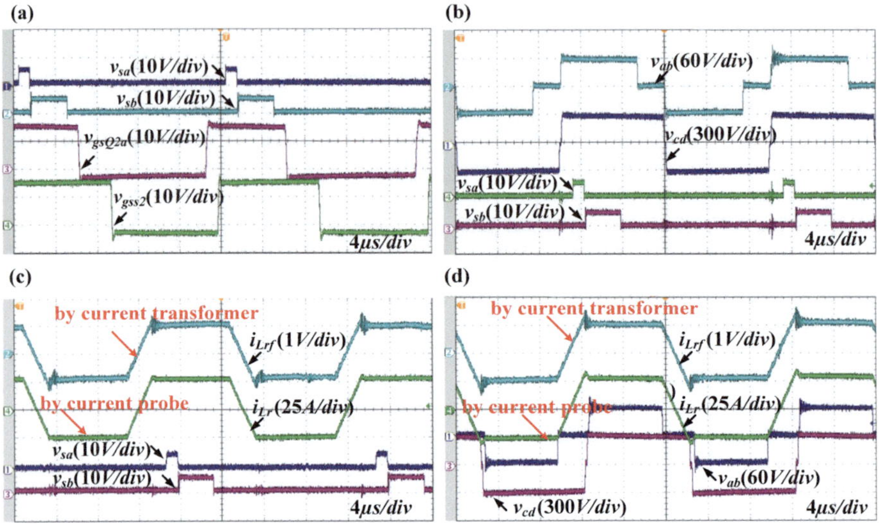

Fig. 9.15 Experimental results of leakage inductance current detection: **a** sampling times and switching sequence; **b** detection time and voltages; **c** detection time and leakage inductance current; **d** leakage inductance current and voltages

voltage of the transformer, and v_{cd} is the secondary voltage. i_{Lr} is the leakage inductance current, and i_{Lrf} is its sampling signal of leakage inductance current. i_{Lr} can be sampled well through the current transformer as shown in Fig. 9.14. For the boost mode, the first sampling point is obtained in one switching cycle when V_{gsQ2a} and V_{gsS2} are both positive. The second sampling point happens after 1 μs. The value of i_{Lrf} is converted to a digital value twice in one switching cycle by the inner A/D of DSP (TMS320F28335) at the instant of positive edges for V_{sa} and V_{sb}. It should be noted that in Fig. 9.15c, d, the leakage inductance current obtained by the current transformer is almost the same as the value got by the scope current probe pretty well. The low-cost transformer is effective in sensing the high-frequency leakage inductance current.

The experimental results of the proposed converter with DCSR control are shown in Fig. 9.16 when the transformer turns ratio is 2:10. Figure 9.16a, b shows the waveforms of the leakage inductance current i_{Lr}, the transformer primary voltage $v_{ab,}$ and the secondary voltage v_{cd} in boost mode, respectively. And the waveforms in buck mode are also given in Fig. 9.16c, d. As seen, the amplitudes of voltage v_{ab} and v_{cd} are matched with DCSR control when the transformer turns ratio is 2:10 in spite of the power capacity transmitted and the operation modes (buck or boost).

Figure 9.17 shows the experimental results when the turns ratio varies from 2:10 to 2:11. Figure 9.17a, b shows the waveforms of the leakage inductance current i_{Lr}, the transformer primary voltage $v_{ab,}$ and secondary voltage v_{cd} with conventional

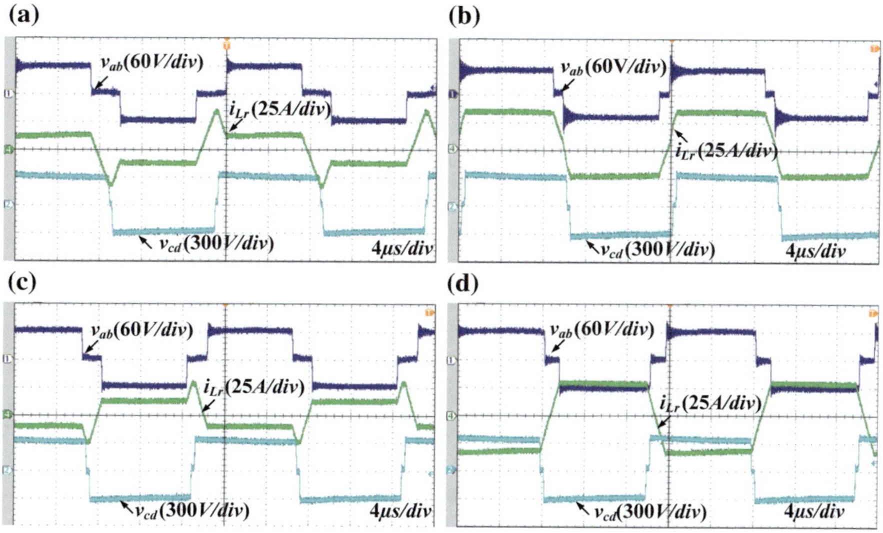

Fig. 9.16 Experimental results with DCSR control (turns ratio is 2:10): **a** boost mode at 24 V and 500 W; **b** boost mode at 24 V and 1200 W; **c** buck mode at 24 V and 500 W; **d** buck mode at 24 V and 1200 W

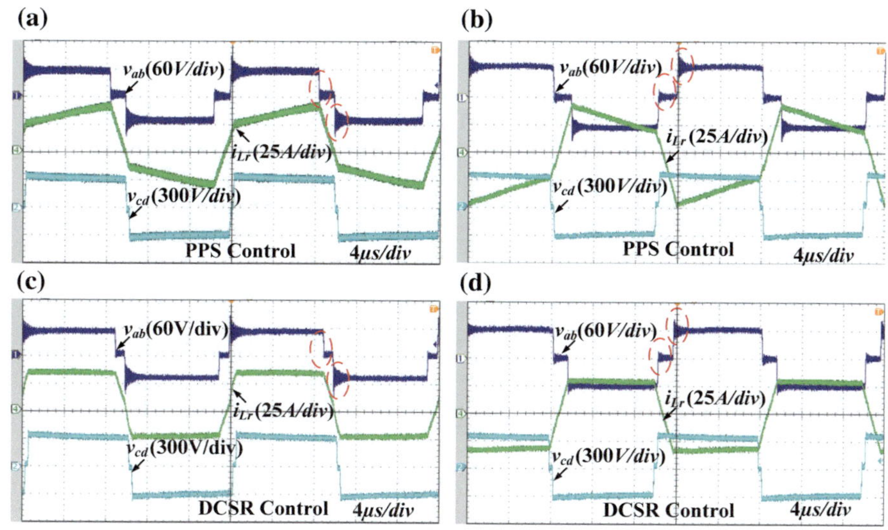

Fig. 9.17 Experimental results (turns ratio is 2:11) at 24 V input and full load: **a** PPS control in boost mode; **b** PPS control in buck mode; **c** DCSR control in boost mode; **d** DCSR control in buck mode

PPS control in boost and buck modes, respectively. Waveforms under DCSR control are given in Fig. 9.17c, d.

As seen, the amplitudes of v_{ab} and v_{cd} cannot be matched well with PPS control because the voltage reference of the PWM control loop is not accurate due to the turns ratio mismatch. Hence, the slew rate of i_{Lr} during the range (θ_7–θ_8) is not zero, leading to higher current stress and circulating current loss. As seen in Fig. 9.17c, d, the amplitudes of v_{ab} and v_{cd} are matched well by using the proposed control not only in buck mode but also in boost mode in facing the turns ratio mismatch. Thus, the DCSR control can adapt to the turns ratio variation automatically, and the slew rate of i_{Lr} during the range (θ_7–θ_8) can be kept zero even when the turns ratios vary. Besides, as addressed previously, with PPS control, the turns ratio mismatch cannot be corrected, the peak current of the leakage inductance, the current stress, and the turn-off voltage spike marked by the dashed circle increases.

Figure 9.18 illustrates the experimental results for the same prototype with conventional PPS control and the proposed DCSR control at different power outputs in facing the same input voltage. The transformer turns ratio is fixed with 2:10. As seen, with the conventional PPS control, although the leakage inductance current can be controlled to be flat during the power transfer stage at light loads as shown in Fig. 9.18a. However, at heavy loads as shown in Fig. 9.18b, the leakage inductance current cannot be regulated to be flat any more due to the larger voltage drop of power switches, line impedance, and ESR of the transformer windings because the active clamp voltage is controlled to be fixed in spite of the transmitted power. Comparatively, with the proposed control, since the slew rate of the leakage

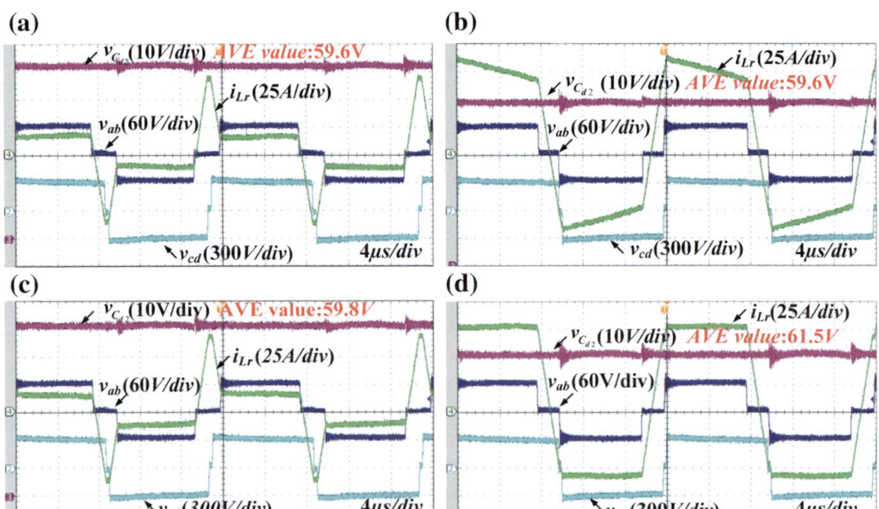

Fig. 9.18 Experimental results in boost mode (turns ratio is 2:10) at 24 V input: **a** PPS control at 200 W output; **b** PPS control at 1.2 kW output; **c** DCSR control at 200 W output; **d** DCSR control at 1.2 kW output

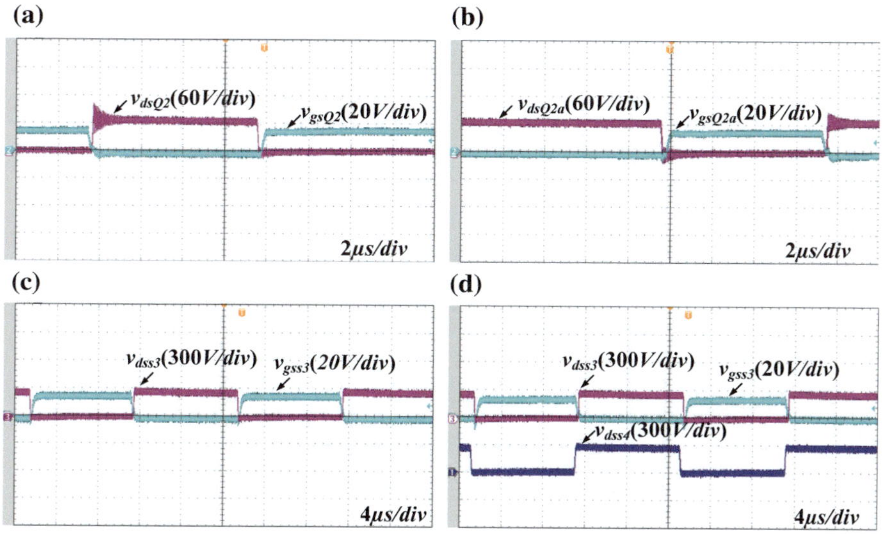

Fig. 9.19 Experimental results with DCSR control (turns ratio is 2:10): gate signals and the drain–source voltage over some switches in boost mode: **a** Q_2; **b** Q_{2a}; **c** S_3; **d** drive signal of S_3 and drain–source voltage of S_3 and S_4

inductance current can be controlled directly, thus the active clamp voltage can be adjusted automatically in facing load variation. As seen in Fig. 9.18c, d, despite the load variation, the slew rate of the leakage inductance current can be controlled to be zero and the active clamp voltage increases a little bit automatically to compensate for the voltage drop at heavy loads.

Figure 9.19 shows the experimental results of ZVS switching in boost mode, where Fig. 9.19a–c represents the gate signal and drain-to-source voltage over switches Q_2, Q_{2a}, and S_3, respectively. As seen, ZVS can be obtained not only for the LVS switches but also for the HVS switches. Figure 9.19d shows the gate signal of S_3 and drain-to-source voltage of S_3 and S_4. As seen, HVS voltage is balanced well by S_3 and S_4.

Figure 9.20 shows the experimental results in buck mode, which is similar to the boost mode. As seen, ZVS can be achieved not only for LVS switches but also for HVS switches. Besides, for the HVS three-level, the HVS high voltage can be balanced well with the proposed gate logic.

The dynamic transition waveforms with DCSR control are shown in Fig. 9.21, where Fig. 9.21a shows the transition from boost mode to buck mode and Fig. 9.21b shows the transition from buck mode to boost mode. If V_2 is higher than 600 V, the working mode will change from boost mode to buck mode automatically. As seen in Fig. 9.21b, if V_2 decreases and is lower than 600 V, the converter transfers from buck mode to boost mode quickly. The transition between the two modes is very smooth with the proposed control.

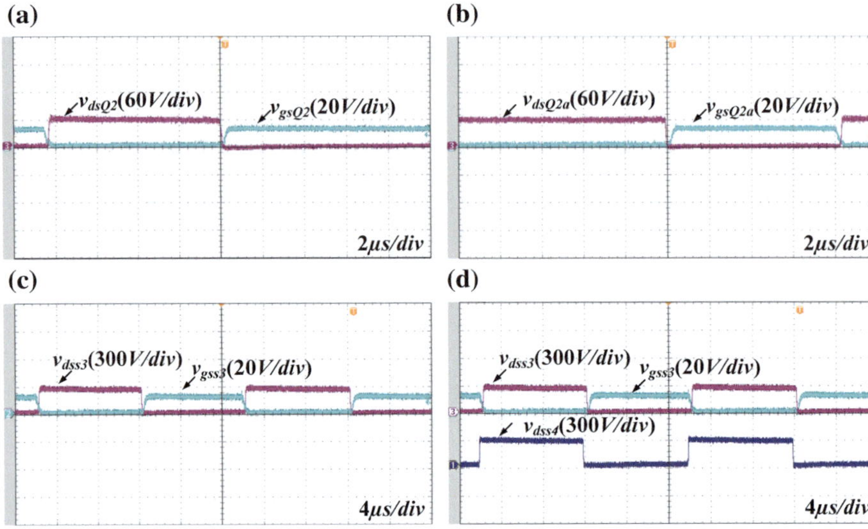

Fig. 9.20 Experimental results of DCSR control (turn ratio is 2:10): drive signal and drain–source voltage in buck mode: **a** Q_2; **b** Q_{2a}; **c** S_3; **d** gate signal of S_3 and drain–source voltage over S_3 and S_4

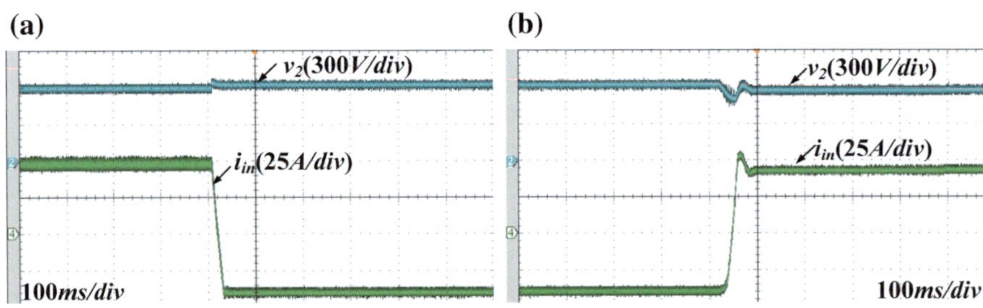

Fig. 9.21 Experimental results of operation mode transition with DCSR control (turns ratio is 2:10): **a** boost to buck; **b** buck to boost

Figure 9.22a shows the conversion efficiency at different loads with the proposed control and PPS control when the turns ratio is fixed as 2:10 at different input voltages. As seen, at light loads, the conversion efficiency differs not much with the aforementioned two control strategies. As the load increases, the conversion efficiency with DCSR control is higher than conventional PPS control. When there is slight mismatch for the turns ratio varying from 2:10 to 2:11, the conversion efficiency comparison at different loads and input voltages can be seen in Fig. 9.22b. As seen, in the same working condition, the conversion efficiency obtained by DCSR control is higher than the PPS control.

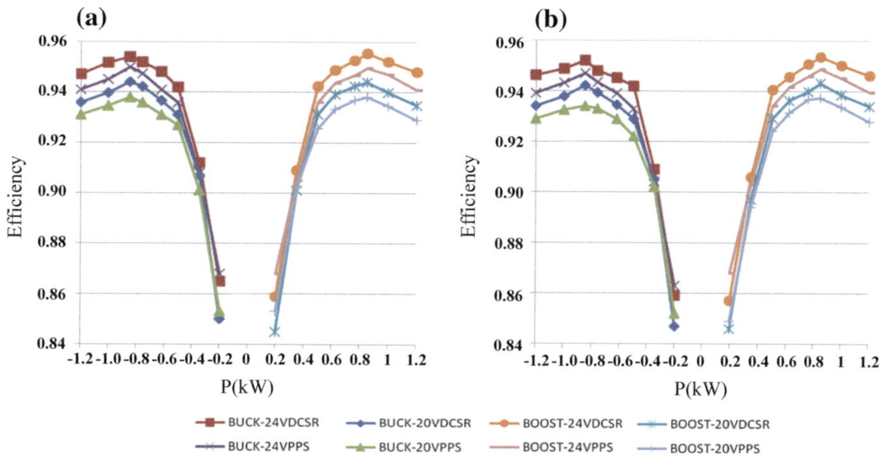

Fig. 9.22 Efficiency: **a** turns ratio: 2:10; **b** turns ratio: 2:11

9.6 Conclusion

The chapter proposes a bidirectional high-frequency isolated three-level DC–DC converter for high-voltage applications. With the proposed gate sequence, for the HVS, the high voltage can be shared perfectly and ZVS can be achieved well. It is feasible to sample the leakage inductance current in order to achieve zero slew rate of the leakage inductance current during the power transfer stage directly. The high-frequency leakage inductance current can be sampled by a low-cost current transformer and the inner A/D converter of the DSP. This avoids the voltage sensing of the clamp voltages by voltage sensors and becomes cost effective.

With the proposed DCSR control, the slew rate of the leakage inductance current during the power transfer stage can be controlled to be zero during the power transfer stage in spite of the variations of turns ratio mismatch, output power variations, and system parasitic parameters. However, the slew rate of leakage inductance current during the power transfer stage using the conventional PPS control cannot be maintained always to be zero in facing those variations. Comparatively, with the DCSR control, conduction loss, peak current, and voltage spike over LVS switches can be reduced more than the PPS control within the designed LVS voltage range. Thus, the conversion efficiency is higher than the PPS control. ZVS can be obtained for all power switches in spite of the power flow direction. Besides, with the proposed control, seamless transition between buck and boost modes can be obtained. The validity of the proposed control strategy and converter has been verified by a simulation and experimental results of a 1.2 kW prototype.

References

1. Zhao T, Wang G, Bhattacharya S, Huang A (2013) Voltage and power balance control for a cascaded H-bridge converter-based solid-state transformer. IEEE Trans Power Electron 28 (4):1523–1532
2. Yu H, Xiang X, Zhao C, Zhao Y, Li W, He X (2013) Performance analysis of a ZVS bidirectional dc-dc converter with reduced voltage stress on high voltage side. Proc IEEE ECCE 2013:44–49
3. Shi Y, Yang X (2013) Zero-voltage switching PWM three-level full-bridge DC–DC converter with wide ZVS load range. IEEE Trans Power Electron 28(10):4511–4524
4. Narimani M, Moschopoulos G (2014) An investigation on the novel use of high-power three-level converter topologies to improve light-load efficiency in low power DC/DC full-bridge converters. IEEE Trans Ind Electron 61(10):5690–5692
5. Jin K, Yang M, Ruan X, Xu M (2010) Three-level bidirectional converter forfuel-cell/battery hybrid power system. IEEE Trans Ind Electron 57(6):1976–1986
6. Grbovi´c PJ, Delarue P, Moigne PL, Bartholomeus P (2010) A bidirectional three-level DC–DC converter for the ultra capacitor applications. IEEE Trans Ind Electron 57(10):3415–3430
7. Xu D, Zhao C, Fan H (2004) A PWM plus phase-shift control bidirectional dc–dc converter. IEEE Trans Power Electron 19(3):666–675
8. Xiao H, Xie S (2008) A ZVS bidirectional DC–DC converter with phase shift plus PWM control scheme. IEEE Trans Power Electron 23(2):813–823

Chapter 10
A Bidirectional Three-Level DC–DC Converter with Reduced Circulating Loss and Fully ZVS Achievement for Battery Charging/Discharging

Abstract A high-frequency isolated bidirectional three-level DC–DC converter is proposed for battery charging/discharging applications. To reduce the circulating loss, a double PWM plus double phase shifted control employing voltage matching control is proposed. ZVS for all power switches can be obtained even at no-load condition by employing the magnetizing inductance. The mode analysis is given to illustrate its working principle. The system parameter design criterion is given in view of the efficiency improvement and ZVS achievement for all switches. A detailed comparison of RMS value and peak value of the leakage inductance current is made with the voltage matching control and mismatching control, respectively. The effectiveness of the proposed control for the DC–DC converter is verified by experimental results of a 1 kW three-level prototype.

Keywords Three-level DC–DC converter · ZVS · Battery charger
Bidirectional power flow

10.1 Converter Mode Analysis with Proposed Control Strategy

To reduce the charging current ripple, current-fed DAB is an attractive option [1]. When the voltage conversion ratio varies, the conduction loss during the power transfer stage is not minimized. To adapt to the wide voltage conversion ratio situations, a PWM plus phase shift (PPS) control was used. With PPS control, the battery side voltage is PWM modulated to ensure the matching between clamp voltage and DC bus voltage [2–4]. However, high current spike can be generated and imposed on switches as the battery voltage is rather low. This causes lower conversion efficiency and even system reliability issues. To suppress the current spike, the two sides of the CF-DAB can be modulated with unequal duty cycles [5]. However, ZVS cannot be ensured for all the switches at the same time.

For all the CF-DAB converters in the literature, the DC bus voltage side is only a half bridge or full bridge, so they cannot be used directly using silicon-based

D. Sha and G. Xu, *High-Frequency Isolated Bidirectional Dual Active Bridge DC–DC Converters with Wide Voltage Gain*, CPSS Power Electronics Series,
https://doi.org/10.1007/978-981-13-0259-6_10

MOSFETs if the DC bus voltage is high. Thus, in this chapter, a three-level CF-DAB DC–DC converter is presented for battery charging/discharging. To reduce the circulation loss and suppress the current spike in spite of the battery voltage, a dual PWM plus dual phase shifted (DPDPS) control with voltage match is used. By proper design of the transformer magnetizing current, ZVS can be achieved for all switches even at no-load condition.

The proposed topology is shown in Fig. 10.1. For the battery side, L_1 and L_2 are DC inductors which can be treated as two constant current sources. Q_1 and Q_2 are the main switches with body diodes. Q_{1a}, Q_{2a}, and C_{c2} constitute the clamp circuit. The AC inductor L_r represents the sum of transformer leakage inductance in battery side and the external AC inductance. In the HVS, S_1–S_4 are the main switches. D_u and D_d are the neutral point clamped diodes. C_u and C_d are voltage-dividing capacitors. C_{c1} is the flying capacitor. V_1 is the DC bus voltage, and V_2 is the battery voltage side. The direction of assumed current flow is highlighted in Fig. 10.1.

The steady-state waveforms in charging mode during one complete period are shown in Fig. 10.2. The duty cycles for battery side bottom switches Q_1 and Q_2 are the same, but their PWM gating signals are interleaved with each other. v_{ab} and v_{cd} are the high-frequency AC three-level waveforms with the same shape, and they are phase shifted according to the transferred power. The leakage inductor current is maintained at zero during the circulation stage, and the circulation loss can be reduced by employing the PWM control for the HVS like the battery side with the same duty cycle.

The typical operation modes are shown in Fig. 10.3. The detailed mode analysis is described as follows:

Stage 1 (Before θ_0): Q_1, Q_{2a}, S_3, and S_4 conduct, and v_{ab}, v_{cd}, i_{Lr}, i_{Lm} are negative. During this stage, the power flows from HVS (V_1) to battery side (V_2).

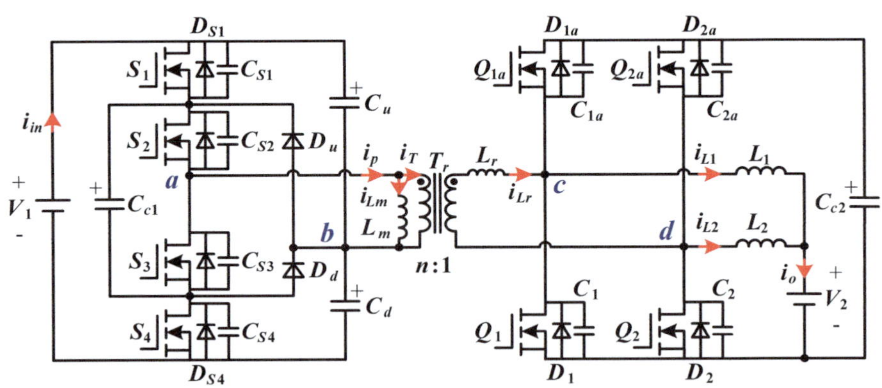

Fig. 10.1 Topology of the bidirectional current-fed three-level DC–DC converter

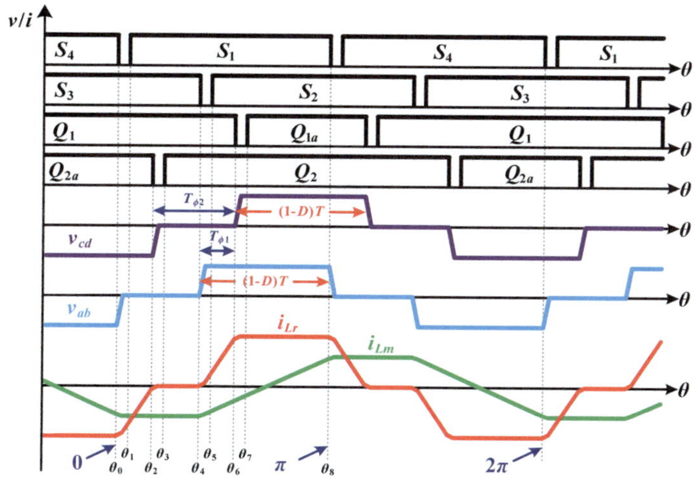

Fig. 10.2 Steady-state waveforms of the proposed control in charging mode

Stage 2 (θ_0–θ_1): At θ_0, S_4 turns off. The sum of magnetizing current i_{Lm} and primary winding i_T charges/discharges the junction capacitors C_{S4}/C_{S1} until the voltage of C_{S1} decays to zero. Then, S_1 turns on under ZVS.

Stage 3 (θ_1–θ_2): At θ_1, S_1 is gated on under ZVS.

Stage 4 (θ_2–θ_3): At θ_2, Q_{2a} turns off. The difference between i_{L2} and i_{Lr} begins to charge/discharge the junction capacitors C_{2a}/C_2 until the body diode of Q_2 conducts. Then, Q_2 turns on under ZVS.

Stage 5 (θ_3–θ_4): At θ_3, Q_2 turns on under ZVS.

Stage 6 (θ_4–θ_5): At θ_4, S_3 turns off. The difference between magnetizing current i_{Lm} and leakage reflected current i_{Lr}/n charges/discharges the junction capacitors C_{S3}/C_{S2} until the body diode of S_2 conducts. Then, S_2 turns on under ZVS.

Stage 7 (θ_5–θ_6): At θ_5, S_2 turns on under ZVS. D_d conducts until the voltage of V_{Cc1} rises to its steady value.

Stage 8 (θ_6–θ_7): At θ_6, Q_1 turns off, the difference between i_{Lr} and i_{L1} charges/discharges the junction capacitors C_1/C_{1a} until the body diode of Q_{1a} conducts.

Stage 9 (θ_7–θ_8): At θ_7, Q_{1a} is gated on under ZVS. i_{Lm} is increasing positively and linearly. At θ_8, S_1 turns off, and the second half period is similar to the first half period.

The steady-state waveforms and the operation modes in discharging mode with the proposed control strategy are similar to those in the charging mode; thus, they are omitted herein for the sake of brevity.

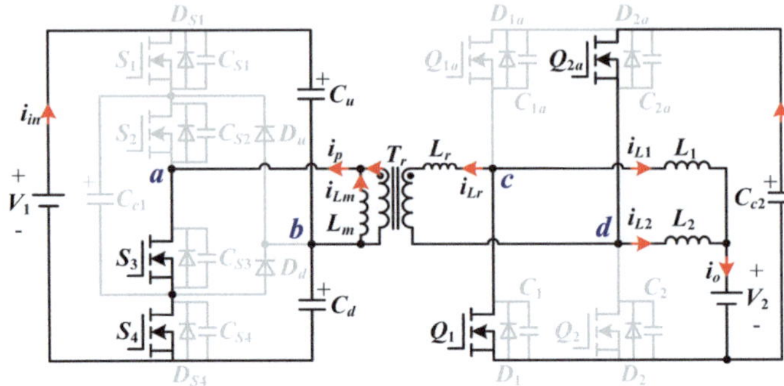

(a) Before θ_0

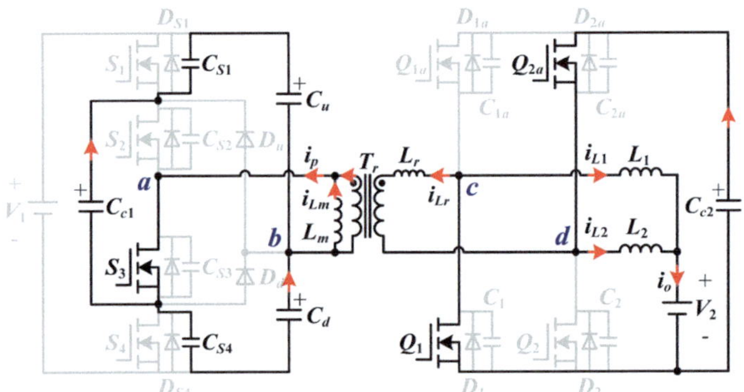

(b) θ_0 - θ_1

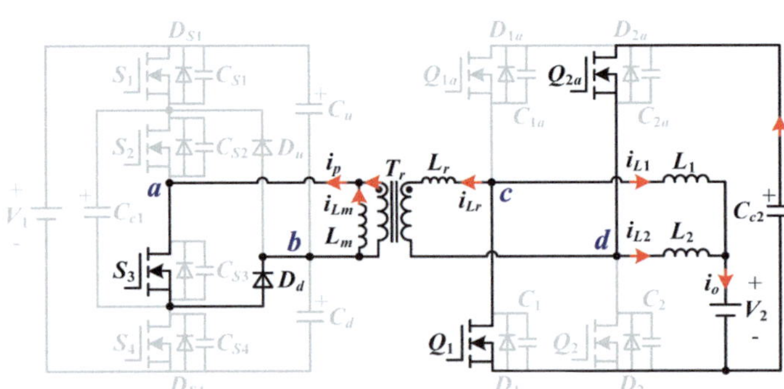

(c) θ_0 - θ_1

Fig. 10.3 Modes of the proposed topology during half period in charging mode

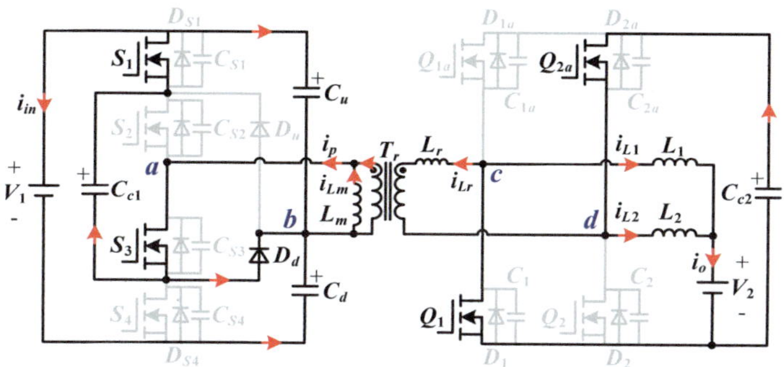

(d) θ_1 - θ_2

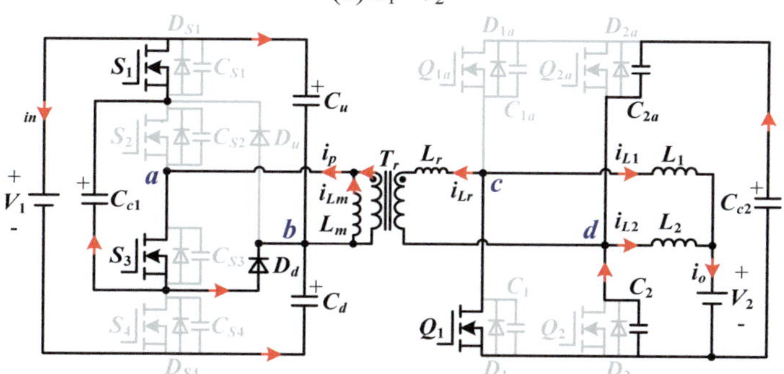

(e) θ_2 - θ_3

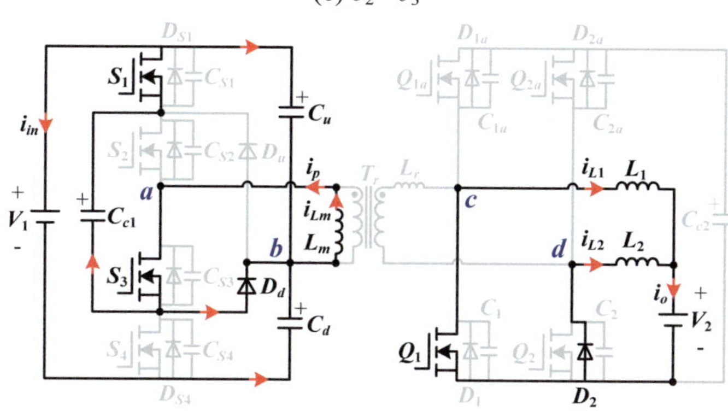

(f) θ_2 - θ_3

Fig. 10.3 (continued)

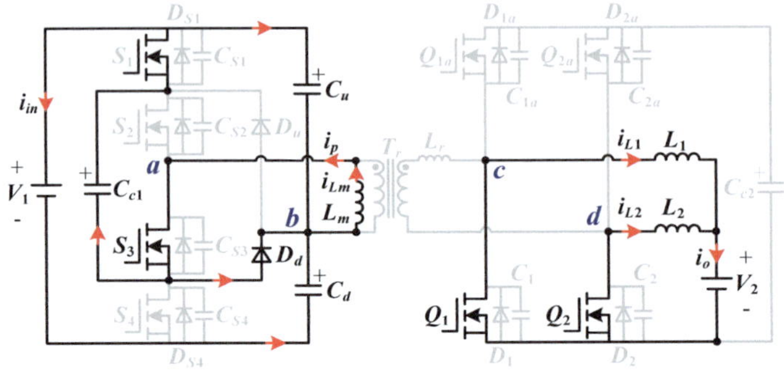

(g) θ_3 - θ_4

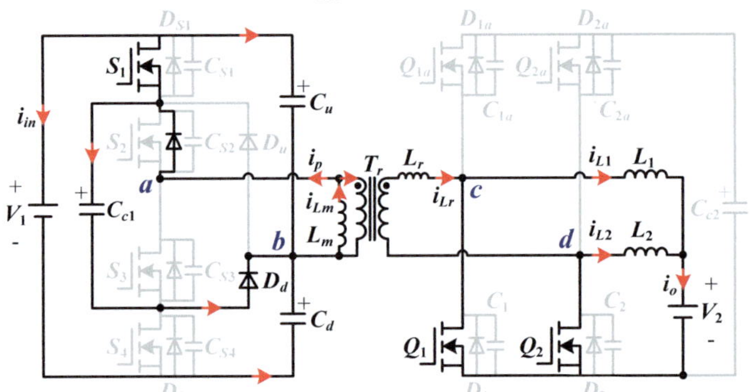

(h) θ_4 - θ_5

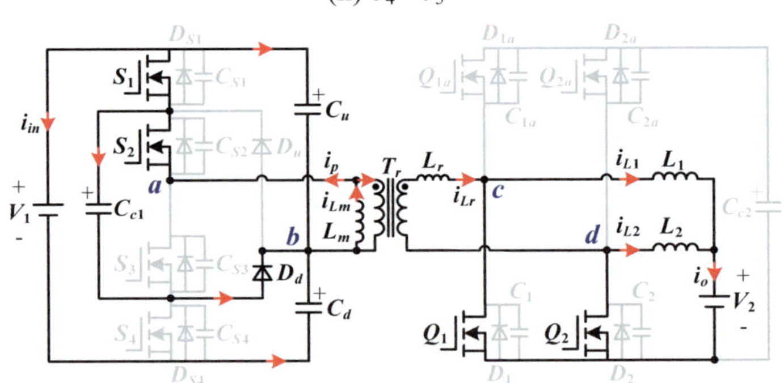

(i) θ_5 - θ_6

Fig. 10.3 (continued)

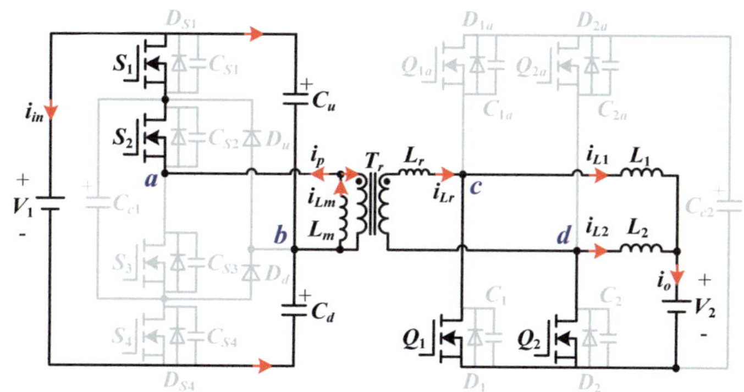

(j) θ_5 - θ_6

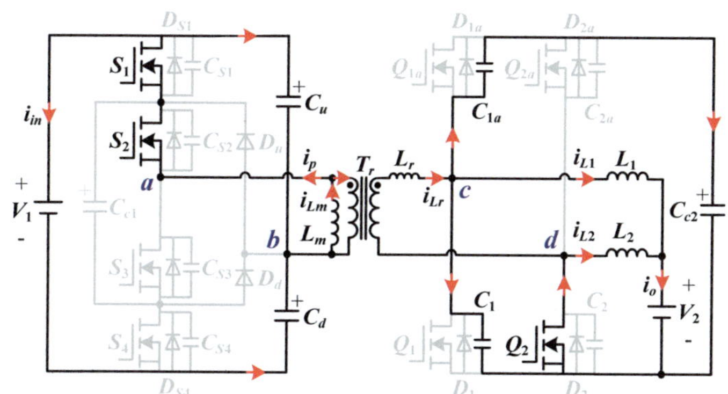

(k) θ_6 - θ_7

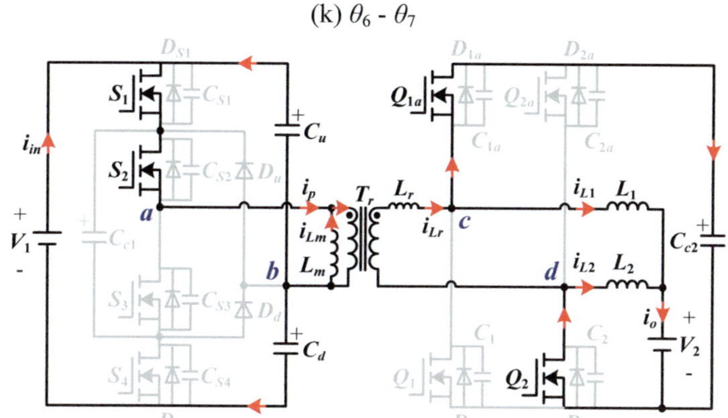

(l) θ_7 - θ_8

Fig. 10.3 (continued)

10.2 Performance Analysis and Discussion

The stages in which the transferred power $P_{o\_cir} = 0$ are defined as circulation stages. It can be proved that the minimum circulation can be implemented only when $T_{\phi 2} = (D - 0.5)T$. Therefore, the same duty cycle control is employed. In order to simplify the analysis, it is assumed that the charging and discharging processes of the junction capacitors are instantaneous when the switches turn off. The analysis is under the condition that $T_{\phi 2} = (D - 0.5)T$, $0 < T_{\phi 1} < T_{\phi 2}$, and $D > 0.5$ [6]. ϕ_1 is the phase shift angle between the two sides, and $T_{\phi 1} = \phi_1 T/2\pi$ is the corresponding time.

10.2.1 Derivation of System Output Power

As shown in Fig. 10.4, the leakage inductor current of battery side can be divided into four stages under the condition $0.5 < D < 0.75$ in half a period. The maximum value of leakage inductor current can be expressed as follows:

$$I_{\text{Lrmax}} = \frac{V_{\text{Cc2}}}{L_r} T_{\phi 1} \tag{10.1}$$

Ignoring the power loss during the power conversion, when $0.5 < D < 0.75$ and $T_{\phi 1} < T_{\phi 2}$, expressions of the transferred energy can be obtained as follows:

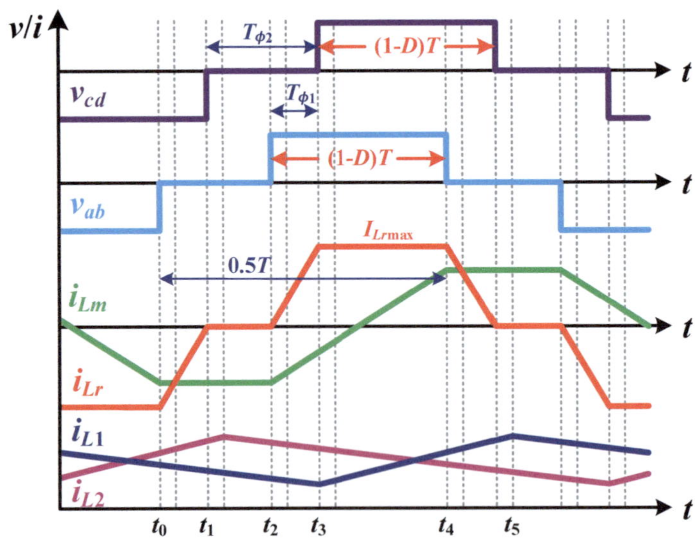

Fig. 10.4 Simplified key waveforms when $0.5 < D < 0.75$ and $T_{\phi 1} < T_{\phi 2}$

$$W_1 = \int_0^{\Delta t_1} v_{cd} i_{\mathrm{Lr\_1}} \mathrm{d}t = \frac{V_{\mathrm{Cc2}}^2}{2L_r} T_{\phi 1}^2 \tag{10.2}$$

$$W_2 = \int_0^{\Delta t_2} v_{cd} i_{\mathrm{Lr\_2}} \mathrm{d}t = 0,\ W_3 = \int_0^{\Delta t_3} v_{cd} i_{\mathrm{Lr\_3}} \mathrm{d}t = 0 \tag{10.3}$$

$$W_4 = \int_0^{\Delta t_4} v_{cd} i_{\mathrm{Lr\_4}} \mathrm{d}t = \frac{V_{\mathrm{Cc2}}^2}{L_r} T_{\phi 1} \left[(1-D)T - T_{\phi 1} \right] \tag{10.4}$$

where $\Delta t_1 = t_1 - t_0 = T_{\phi 1}$, $\Delta t_2 = t_2 - t_1 = T_{\phi 2} - T_{\phi 1}$, $\Delta t_3 = t_3 - t_2 = T_{\phi 1}$, $\Delta t_4 = t_4 - t_3 = (1 - D)T - T_{\phi 1}$, $i_{\mathrm{Lr\_1}} = V_{\mathrm{Cc2}} \cdot t / L_r - I_{\mathrm{Lrmax}}$, $i_{\mathrm{Lr\_2}} = 0$, $i_{\mathrm{Lr\_3}} = V_{\mathrm{Cc2}}(t - t_2) / L_r$, $i_{\mathrm{Lr\_4}} = I_{\mathrm{Lrmax}}$. Therefore, the output power in this case can be derived as follows:

$$P_{\mathrm{o}} = \frac{\sum_{i=1}^{4} W_i}{0.5T} = \frac{V_{\mathrm{Cc2}}^2}{TL_r} T_{\phi 1} \left[2(1-D)T - T_{\phi 1} \right] \tag{10.5}$$

The output power expressions in other cases can be obtained similarly, which are shown below:

$$D \in [0.5, 0.75):$$
$$P_{\mathrm{o}} = \begin{cases} \frac{V_{\mathrm{Cc2}}^2}{TL_r} \left[2(1-D)TT_{\phi 1} - T_{\phi 1}^2 \right] & T_{\phi 1} \in [0, T_{\phi 2}) \\ \frac{V_{\mathrm{Cc2}}^2}{TL_r} \left[2(1-D)TT_{\phi 1} - \left(T_{\phi 1} - \psi \right)^2 - T_{\phi 1}^2 \right] & T_{\phi 1} \in \left[T_{\phi 2}, (1-D)T \right) \\ \frac{V_{\mathrm{Cc2}}^2}{TL_r} \left[(1-D)^2 T^2 - \left(T_{\phi 1} - \psi \right)^2 \right] & T_{\phi 1} \in [(1-D)T, 0.5T] \end{cases} \tag{10.6}$$

$$D \in [0.75, 1]:$$
$$P_{\mathrm{o}} = \begin{cases} \frac{V_{\mathrm{Cc2}}^2}{TL_r} \left[2(1-D)TT_{\phi 1} - T_{\phi 1}^2 \right] & T_{\phi 1} \in [0, (1-D)T) \\ \frac{V_{\mathrm{Cc2}}^2}{TL_r} (1-D)^2 T^2 & T_{\phi 1} \in [(1-D)T, T_{\phi 2}) \\ \frac{V_{\mathrm{Cc2}}^2}{TL_r} \left(0.5T - T_{\phi 1} \right) \left[(1.5 - 2D)T + T_{\phi 1} \right] & T_{\phi 1} \in \left[T_{\phi 2}, 0.5T \right] \end{cases} \tag{10.7}$$

The output power expressions in discharging mode can also be obtained in the same way, which are omitted due to the simplicity of analysis. According to (10.6) and (10.7), the output power curves versus the phase shift angle ϕ_1 with different duty cycle D are plotted in Fig. 10.5.

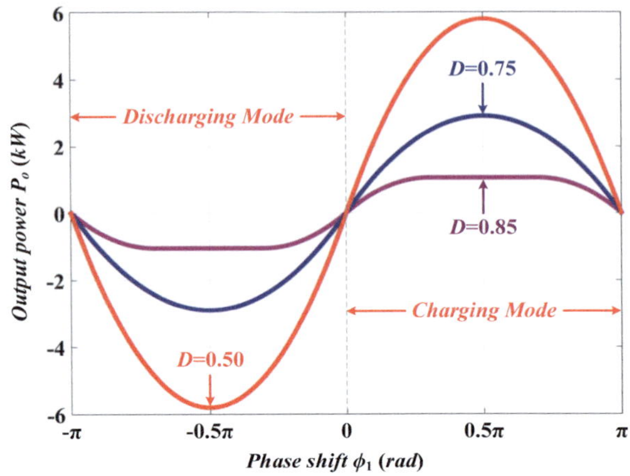

Fig. 10.5 Output power curves versus phase shift angle ϕ_1

As shown in Fig. 10.5, it can be seen that the maximum output power in charging mode occurs when $\phi_1 = 0.5\pi$ while $\phi_1 = -0.5\pi$ in discharging mode. The maximal transferred power capability takes place in $D = 0.5$. With the increase of the duty cycle D, the maximum power transferred capability is limited. There is a stage that the output power P_o does not change with the phase shift angle ϕ_1 when $D > 0.75$.

10.2.2 Clamp Voltage and Voltage Gain of Converter

According to the DPDPS control principle, the voltage across the clamp capacitance C_{c2} can be obtained as follows:

$$V_{Cc2} = \frac{V_2}{1 - D} \tag{10.8}$$

The circulating loss and peak/RMS values of leakage inductor currents can be reduced by achieving amplitude matching between voltage v_{ab} and v_{cd}. Therefore, the following relationship between voltage V_{Cc2} on the capacitor C_{c2} and the DC bus voltage V_1 is expressed as follows:

$$V_{Cc2} = \frac{V_1}{2n} \tag{10.9}$$

Voltage gain G is defined as follows:

$$G = \frac{V_1}{V_2} \tag{10.10}$$

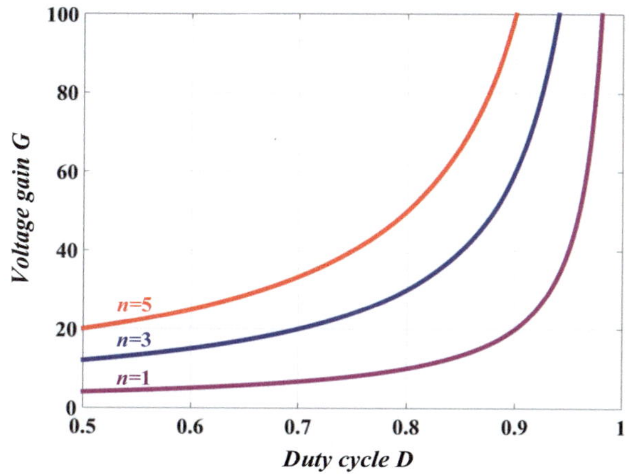

Fig. 10.6 Voltage gain curves versus duty cycle D

Substituting (10.8) and (10.9) into (10.10) yields the voltage gain

$$G = \frac{2n}{1 - D} \tag{10.11}$$

The voltage gain curves versus the duty cycle D with different turns ratio n are plotted in Fig. 10.6. It can be seen that high voltage gain can be obtained by employing the proposed topology even when turns ratio n = 1 and duty cycle is very low.

10.2.3 Design Considerations

A 1 kW prototype is built according to the example of parameter design procedure. V_1 is designed as 600 V, and the battery voltage range is 18–28 V. The switching frequency is selected as 50 kHz.

(1) ***Determination of transformer turns ratio***

According to the mode analysis mentioned above, larger DC inductance and larger output capacitance of battery side switches make the ZVS turn on hard to realize. In order to achieve ZVS for bottom switches of battery side, the duty cycle D must be greater than 0.5 or equal to 0.5. Considering the clamp voltage of battery side and the battery side voltage, the following relationship stands:

$$V_{Cc2} > \frac{V_2}{1 - 0.5} \tag{10.12}$$

In this chapter, the output voltage V_2 is designed within the range 18–28 V; thus, the V_{Cc2} must greater or equal to 56 V. V_{Cc2} is determined as 60 so as to simplify the transformer design, and the turns ratio n can be calculated as follows:

$$n = \frac{V_2}{2V_{Cc2}} = 5 \tag{10.13}$$

(2) ***Leakage inductance design***

Leakage inductor L_r limits the power transfer capability and has a great impact on the RMS value of i_{Lr}. It is feasible to discuss the i_{Lr} to reduce the loss of whole system. To analyze the leakage inductor current waveforms shown in Fig. 10.4, the leakage inductor current expression during half period is easy to be derived which is as follows:

$$i_{Lr} = \begin{cases} \frac{V_{Cc2}}{L_r} t - I_{Lrmax} & t \in [t_0, t_1) \\ 0 & t \in [t_1, t_2) \\ \frac{V_{Cc2}}{L_r}(t - t_2) & t \in [t_2, t_3) \\ I_{Lrmax} & t \in [t_3, t_4] \end{cases} \tag{10.14}$$

$T_{\phi 1}$ can be deduced according to the first expression of (10.6) which is as follows:

$$T_{\phi 1} = \frac{V_2 T - \sqrt{T^2 V_2^2 - T L_r P_o}}{V_{Cc2}} \tag{10.15}$$

Then, the leakage inductor RMS current can be obtained as follows:

$$I_{LrRMS} = \sqrt{\frac{2}{3 L_r^2 V_{Cc2}} \left(2T^2 V_2^3 - L_r P_o H - 2TV_2^2 H\right)} \tag{10.16}$$

where $H = \sqrt{T^2 V_2^2 - L_r P_o T}$. In view of the limitation of the maximum output power, the leakage inductance must be smaller than $\frac{TV_{2min}^2}{P_{omax}}$. Based on (10.16) and (10.1), the RMS current and peak current of the leakage inductor versus leakage inductance value in different battery voltage are plotted in Fig. 10.7.

As is demonstrated in Fig. 10.7, the RMS and peak current will increase if leakage inductance increases. It means that the larger leakage inductance will cause more loss and current stress. On the other hand, too small leakage inductance will cause rush current during the start-up process.

(3) ***DC inductance design***

Ignoring the power loss during the power conversion and assuming the inductors L_1 and L_2 have good current-sharing performance, the relationship between output power and the average DC inductor current I_{L1} can be expressed as follows:

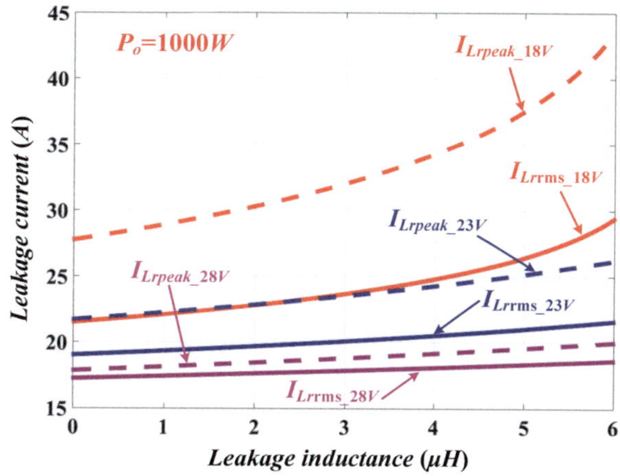

Fig. 10.7 RMS leakage current and peak current in 1 kW versus different leakage inductance in different output voltage

$$P_o = P_{in} = 2V_2 I_{L1} \tag{10.17}$$

where I_{L1} is the average DC inductor current, and V_2 can be written by

$$I_{L1} = \frac{1}{2}(I_{L1\max} + I_{L1\min}) \tag{10.18}$$

$$V_2 = L_1 \frac{I_{L1\max} - I_{L1\min}}{DT} \tag{10.19}$$

Based on (10.17) to (10.19), the minimum and the maximum DC inductor current can be, respectively, expressed as follows:

$$I_{L1\min} = \frac{P_o}{2V_2} - \frac{V_2 DT}{2L_1} \tag{10.20}$$

$$I_{L1\max} = \frac{P_o}{2V_2} + \frac{V_2 DT}{2L_1} \tag{10.21}$$

As shown in Fig. 10.4, marking t_3 as 0, the equations of DC inductor current during whole period can be derived as follows:

$$i_{L1} = \begin{cases} \frac{V_2}{L_1} t + I_{L1\min} & t \in [0, DT) \\ -\frac{V_2}{L_1}(t - DT) + I_{L1\max} & t \in [DT, T] \end{cases} \tag{10.22}$$

By substituting (10.20) and (10.21) into (10.22), the DC inductor RMS current can be obtained as follows:

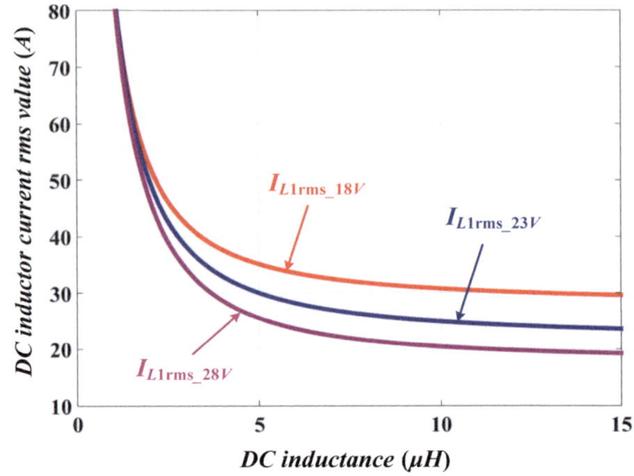

Fig. 10.8 DC inductor RMS current versus DC inductance value

$$I_{\mathrm{L1RMS}} = \sqrt{\frac{P_{\mathrm{o}}^2}{4V_2^2} - \frac{P_{\mathrm{o}}TV_2^2}{L_1V_{\mathrm{Cc2}}^2} + \frac{P_{\mathrm{o}}TV_2}{2L_1V_{\mathrm{Cc2}}} + \frac{T^2V_2^2}{12L_1^2} - \frac{3T^2V_2^4}{4L_1^2V_{\mathrm{Cc2}}^2} + \frac{T^2V_2^5}{L_1^2V_{\mathrm{Cc2}}^3}} \quad (10.23)$$

The DC inductor RMS current curves versus DC inductance value in different output voltage on account of (10.23) are plotted in Fig. 10.8.

As shown in Fig. 10.8, DC inductor RMS current decreases as DC inductance value increases. The RMS current decreases dramatically when the DC inductance value increases from 0 to 2.5 μH. And the RMS current reduces slightly when the DC inductance value is larger than 5 μH. It reveals that the larger DC inductance value will reduce the RMS current and then reduce the DC copper loss. Meanwhile, larger DC inductance will also reduce the current ripple. This reduces the DC inductor core loss as well. It is worth noting that too large DC inductance value results in the loss of ZVS for battery side switches, and the limiting condition of DC inductance value will be discussed in detail later in this chapter.

10.2.4 Comparison of Voltage Matching Mode and Mismatching Mode

Figure 10.9 shows the mismatched waveforms, where $T_{\phi 1\mathrm{Mis}}$, $T_{\phi 2\mathrm{Mis}}$, D_{Mis} are the corresponding parameters under mismatch. $T_{\phi 1\mathrm{Mis}} < T_{\phi 2\mathrm{Mis}}$ is always satisfied in order to achieve minimum circulation loss. The leakage inductor current $i_{\mathrm{Lr{\cdot}Mis}}$ can be written owing to Fig. 10.9, and it can be written as follows:

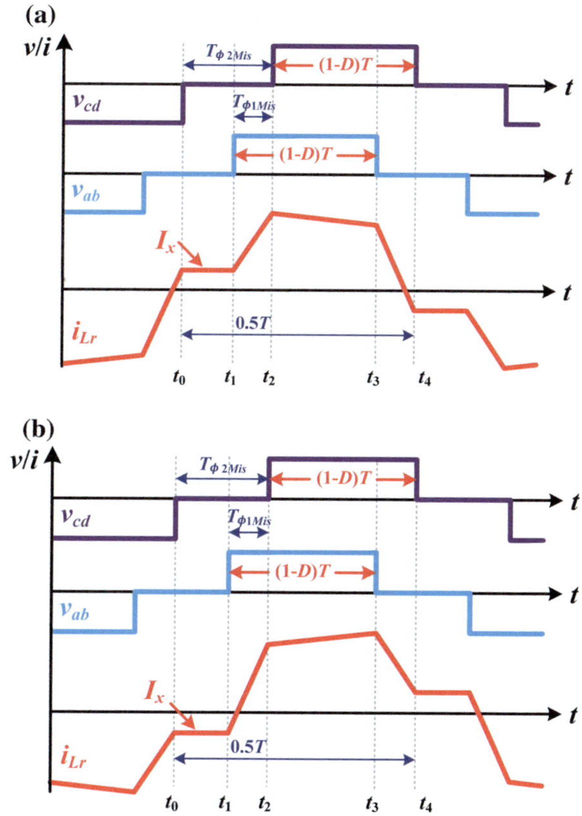

Fig. 10.9 Mismatching mode waveforms when **a** V_{Cc2_Mis} is higher than V_{Cc2}; **b** V_{Cc2_Mis} is lower than V_{Cc2}

$$i_{\mathrm{Lr\cdot Mis}} = \begin{cases} I_x & [t_0, t_1) \\ \frac{V_{\mathrm{Cc1}}}{nL_r}(t - t_1) + I_x & [t_1, t_2) \\ i_{\mathrm{Lr\cdot Mis}}(t_2) - \frac{nV_{\mathrm{Cc2\cdot Mis}} - V_{\mathrm{Cc1}}}{nL_r}(t - t_2) & [t_2, t_3) \\ i_{\mathrm{Lr\cdot Mis}}(t_3) - \frac{V_{\mathrm{Cc2\cdot Mis}}}{L_r}(t - t_3) & [t_3, t_4] \end{cases} \tag{10.24}$$

I_x can be deduced as follows:

$$I_x = \frac{V_2 T}{2L_r}\left(1 - \frac{V_{\mathrm{Cc1}}}{nV_{\mathrm{Cc2\cdot Mis}}}\right) \tag{10.25}$$

The output power is derived as follows:

$$P_{\mathrm{o\cdot Mis}} = \frac{T_{\phi 1\mathrm{Mis}} V_{\mathrm{Cc1}}}{nL_r}\left(2V_2 - \frac{T_{\phi 1\mathrm{Mis}} V_{\mathrm{Cc2\cdot Mis}}}{T}\right) \tag{10.26}$$

The relationship between ϕ_{TMis} and ϕ_T can be obtained based on the fact that $P_{o\cdot Mis} = P_o$

$$T_{\phi 1Mis} = \frac{TV_2V_{Cc1} - \sqrt{nV_{Cc1}V_{Cc2}V_{Cc2\cdot Mis}T_{\phi 1}\left(T_{\phi 1}V_{Cc2} - 2TV_2\right) + T^2V_2^2V_{Cc1}^2}}{V_{Cc1}V_{Cc2\cdot Mis}} \tag{10.27}$$

Thus, the maximum value of $T_{\phi 1Mis}$ can be obtained by assuming $T_{\phi 1Mis} = T_{\phi 2Mis}$; that is, $T_{\phi 1Mis\_max} = T(1/2 - V_2/V_{Cc2\cdot Mis})$. Figure 10.10 shows the curves $T_{\phi 1Mis}$ versus V_2 and $V_{Cc2\cdot Mis}$.

The upper surface in Fig. 10.10 is the limited phase shift angle $\phi_{1Mis\cdot Lim}$, and the bottom one is the phase shift angle ϕ_{1Mis}. It is clear that these two surfaces have an intersection when the mismatched value $V_{Cc2\cdot Mis}$ is lower than 58.8 V, which indicates that the system cannot output enough power if the clamp voltage is rather low.

The RMS value of the leakage inductor current can be derived as follows:

$$\begin{aligned} i_{Lr\cdot Mis\cdot RMS}^2 = & \frac{2}{nL_r^2V_{Cc2\cdot Mis}^2}\left[\frac{A_2}{8n} + V_2^2T\left(TV_2V_{Cc1} - \frac{2A_1}{3}\right)\right] - \frac{T^2V_2^2}{L_r^2V_{Cc2\cdot Mis}}\left(\frac{V_2}{3} + \frac{V_{Cc1}}{2n}\right) \\ & + \frac{T^2V_2^2}{4L_r^2} - \frac{2P_oA_1}{3L_rV_{Cc1}V_{Cc2\cdot Mis}} - \frac{T^2V_2^3V_{Cc1}^2}{3n^2L_r^2V_{Cc2}^3} \end{aligned} \tag{10.28}$$

where $A_1 = \sqrt{A_2 - nTL_rP_oV_{Cc1}V_{Cc2\cdot Mis}}$, $A_2 = (TV_2V_{Cc1})^2$. The value of $i_{Lr\cdot Mis\cdot RMS}$ versus V_2 and $V_{Cc2\cdot Mis}$ which is plotted in Fig. 10.11a, b is the projection on

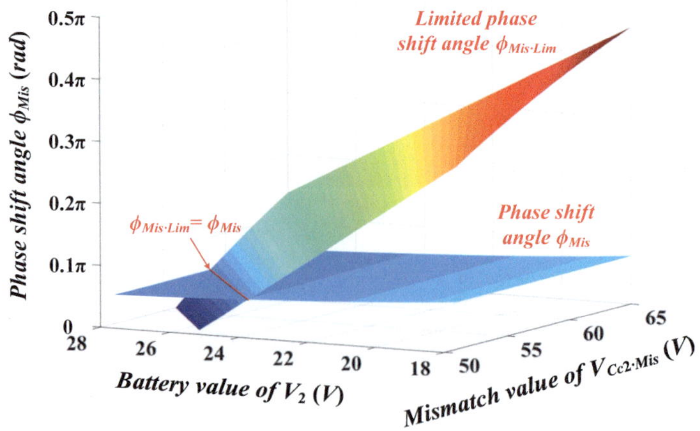

Fig. 10.10 ϕ_{1Mis} versus V_2 and $V_{Cc2\cdot Mis}$

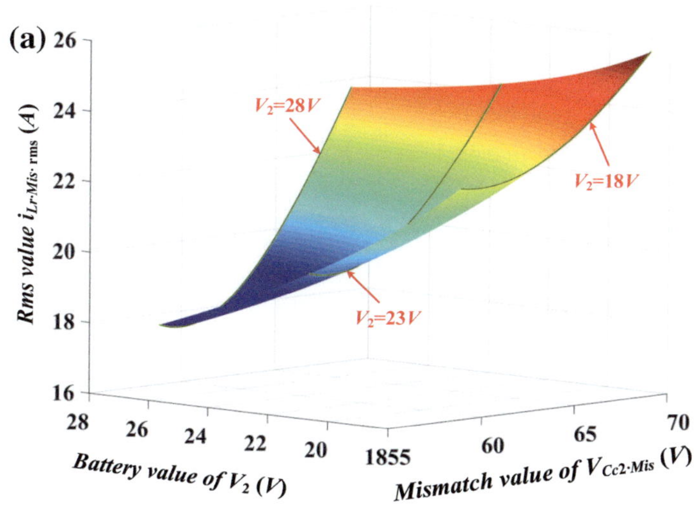

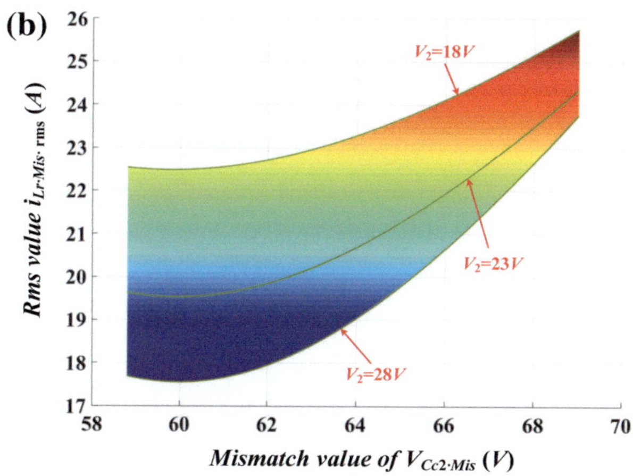

Fig. 10.11 $i_{\mathrm{Lr\cdot Mis\cdot RMS}}$ versus V_2 and $V_{\mathrm{Cc2\cdot Mis}}$: **a** three-dimensional graph; **b** projection graph

$V_{\mathrm{Cc2\cdot Mis}} - i_{\mathrm{Lr\cdot Mis\cdot RMS}}$ side. From Fig. 10.11, it can be seen that leakage inductor RMS value will increase when $V_{\mathrm{Cc2\cdot Mis}}$ is larger than V_{Cc2} or $V_{\mathrm{Cc2\cdot Mis}}$ is smaller than V_{Cc2}. The minimum RMS value occurs at $V_{\mathrm{Cc2\cdot Mis}} = V_{\mathrm{Cc2}}$, which means that the voltage of battery side matches with HVS.

And the peak current $i_{\mathrm{Lr\cdot Mis\ peak}}$ can be deduced as follows:

$$i_{\mathrm{Lr\cdot Mis\cdot peak}} = \begin{cases} \frac{V_{\mathrm{Cc1}}}{nL_r} T_{\phi 1\mathrm{Mis}} + I_x & V_{\mathrm{Cc2\cdot Mis}} \geq V_{\mathrm{Cc2}} \\ \frac{V_{\mathrm{Cc2\cdot Mis}}}{L_r} T_{\phi 1\mathrm{Mis}} - I_x & V_{\mathrm{Cc2\cdot Mis}} < V_{\mathrm{Cc2}} \end{cases} \tag{10.29}$$

The $i_{Lr \cdot Mis \cdot peak}$ surface versus $V_{Cc2 \cdot Mis}$ and V_2 which is plotted in Fig. 10.12a, b is the projection on $i_{Lr \cdot Mis \cdot peak} - V_{Cc2 \cdot Mis}$ side.

Figure 10.12 indicates that the leakage inductor current peak value would increase regardless of the fact that $V_{Cc2 \cdot Mis}$ is larger than or smaller than V_{Cc2}. The minimum peak value occurs at $V_{Cc2 \cdot Mis} = V_{Cc2}$, which means that the voltage of battery side matches with HVS. As stated previously, ZVS achievement for HVS switches depends on the magnetizing current i_{Lm} and the leakage current i_{Lr}. The ZVS achievement for battery side switches is determined by the DC inductor current and leakage current i_{Lr}. The ZVS conditions of battery side switches are quite different and asymmetric between the charging mode and discharging mode if

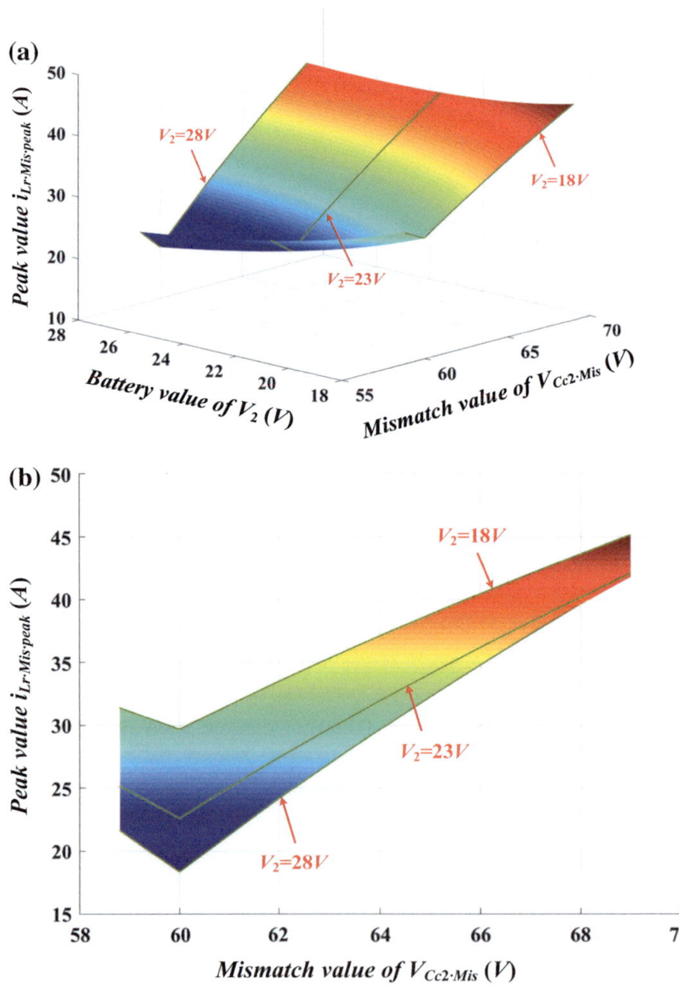

Fig. 10.12 $i_{Lr \cdot Mis \cdot peak}$ versus V_2 and $V_{Cc2 \cdot Mis}$: **a** three-dimensional graph; **b** projection graph

the voltages are mismatched. This issue makes main circuit parameter optimization difficult. Even worse, ZVS may be lost completely if the voltage mismatch is too much.

10.2.5 Soft-Switching Condition

(1) ***High-voltage side switches***

Switches S_1 and S_2 are taken as an example to analyze the ZVS condition for HVS active switches. The key waveforms of HVS are shown in Fig. 10.13.

According to the mode analysis, the sum of magnetizing current i_{Lm} and battery side winding current i_T is shown in Fig. 10.1 to charge/discharge the junction capacitors during the dead time. Thus, the ZVS condition for HVS four switches in charging and discharging modes can be calculated as follows:

$$\int_{0}^{T_{\mathrm{dH}}} |i_T + i_{\mathrm{Lm}}|\mathrm{d}t \geq 2nC_{\mathrm{oss}\cdot H}V_{\mathrm{Cc2}} \quad (10.30)$$

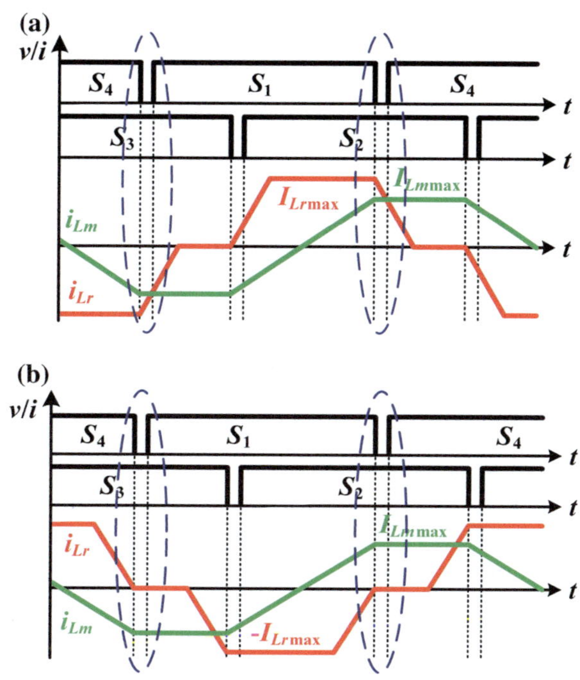

Fig. 10.13 Key waveforms of HVS switches **a** in charging mode; **b** in discharging mode

For the lagging leg switches in charging mode, before S_1 turns on, the magnetizing inductor current i_{Lm} is the maximum and negative $-I_{Lmmax}$ and the leakage inductor current i_{Lr} begins to increase. Based on (10.30), the ZVS condition for the lagging leg switches in charging mode is shown as follows:

$$\begin{cases} I_{Lmmax}T_{dH} + \frac{T_{dH}}{2n}\left(2I_{Lrmax} - \Delta i_{Lr}\right) \geq 2nC_{oss\cdot H}V_{Cc2} & T_{\phi1} \geq T_{dH} \\ I_{Lmmax}T_{dH} + \frac{1}{2n}I_{Lrmax}T_{\phi1} \geq 2nC_{oss\cdot H}V_{Cc2} & T_{\phi1} < T_{dH} \end{cases} \tag{10.31}$$

Simplifying (10.31) leads to

$$\begin{cases} \frac{V_2T}{L_m} + \frac{V_{Cc2}}{n^2L_r}\left(2T_{\phi1} - T_{dH}\right) \geq \frac{4C_{oss\cdot H}V_{Cc2}}{T_{dH}} & T_{\phi1} \geq T_{dH} \\ \frac{V_2T}{L_m} + \frac{V_{Cc2}}{n^2L_rT_{dH}}T_{\phi1}^2 \geq \frac{4C_{oss\cdot H}V_{Cc2}}{T_{dH}} & T_{\phi1} < T_{dH} \end{cases} \tag{10.32}$$

It can be proved that in (10.32), the ZVS condition of lagging leg switches in HVS becomes more difficult to achieve with the decrease of $T_{\phi1}$. And the most difficult condition occurs at $T_{\phi1} = 0$, where there is no power transmission. Thus, the magnetizing inductance must be designed as follows to achieve ZVS in this case:

$$L_m \leq \frac{V_2TT_{dH}}{4V_{Cc2}C_{oss\cdot H}} \tag{10.33}$$

It can be seen from (10.33) that the maximum magnetizing inductance L_m is independent of the output power. Actually, in discharging mode, the ZVS condition for lagging leg is the same as (10.33). Likewise, the ZVS condition for leading lag switches in charging mode and discharging mode can be expressed as follows:

$$L_m \leq \frac{T_{dH}(V_2T - V_{Cc2}T_{dH})}{4C_{oss\cdot H}V_{Cc2}} \tag{10.34}$$

From the (10.33) and (10.34), it is obvious that with the decrease of battery voltage, the maximum magnetizing inductance value decreases as well, which implies that the most difficult ZVS occurs at the lowest battery voltage in the same situation. The comparison of (10.33) and (10.34) concludes that (10.34) is more difficult than (10.33). In other words, once the magnetizing inductance is designed as shown in (10.34), the ZVS condition for HVS can be guaranteed in spite of the working modes. Based on (10.34), the maximum magnetizing inductance L_m curve versus the dead time T_{dH} to guarantee ZVS for HVS switches in 18 V battery voltage is plotted in Fig. 10.14.

Figure 10.14 indicates that if the dead time T_{dH} is within 1 μs, the maximum magnetizing inductance value is increasing with the increasing of the dead time of HVS. To minimize the magnetizing loss, L_m must be designed as large as possible. But too large dead time causes more switching body diode conduction loss. So the magnetizing inductor must be designed with a proper value.

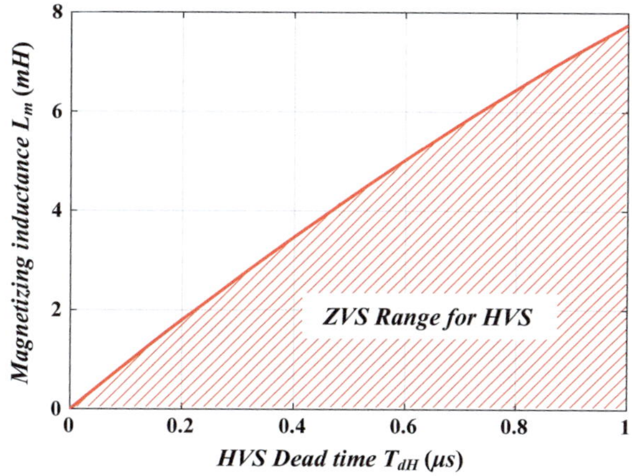

Fig. 10.14 Maximum magnetizing inductance L_{m} curve versus the dead time T_{dH}

(2) ***Low-voltage side switches***

The battery side DC inductors must be designed carefully to satisfy the ZVS condition for the battery side switches. The key waveforms before Q_1 and Q_{1a} turning on are shown in Fig. 10.15. The analysis for the other two switches is similar.

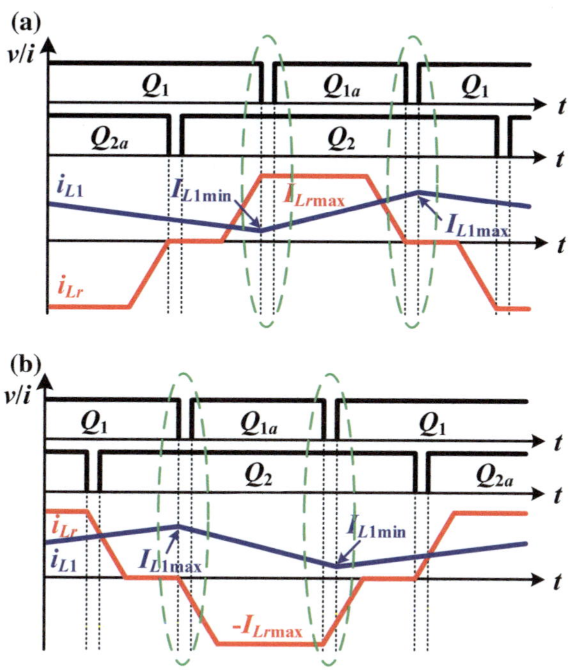

Fig. 10.15 Key waveforms of battery side switches **a** in charging mode; **b** in discharging mode

As mentioned previously, the absolute value of the difference between leakage inductor i_{Lk} and DC inductor i_{L1} must be large enough to charge/discharge the junction capacitors of battery side switches during the dead time. Thus, the ZVS condition for four battery side switches can be roughly obtained as shown below:

$$\int_{0}^{T_d} |i_{\mathrm{Lr}} - i_{L1}|\mathrm{d}t \geq 2C_{\mathrm{oss}\cdot L}V_{\mathrm{Cc2}} \tag{10.35}$$

From Fig. 10.15a, the ZVS condition for upper switches in charging mode can be derived as follows:

$$I_{\mathrm{Lrmax}}T_{\mathrm{dL}} - \frac{T_{\mathrm{dL}}}{2}(2I_{L1\mathrm{min}} + \Delta i_{L1}) \geq 2C_{\mathrm{oss}\cdot L}V_{\mathrm{Cc2}} \tag{10.36}$$

Simplifying the (10.36) yields

$$\frac{2V_{\mathrm{Cc2}}}{L_r}T_{\phi1} - \frac{V_{\mathrm{Cc2}}^2}{V_2TL_r}\left[2(1-D)TT_{\phi1} - T_{\phi1}^2\right] - \frac{V_2}{L_1}(T_{\mathrm{dL}} - DT) \geq \frac{4C_{\mathrm{oss}\cdot L}V_{\mathrm{Cc2}}}{T_{\mathrm{dL}}} \tag{10.37}$$

Derivation of the left side equation versus $T_{\phi1}$ and setting it to be zero will lead to $T_{\phi1} = 0$. It indicates that the ZVS is more difficult to achieve with the reduction of output power. The most difficult condition occurs at the output power which is zero, and the DC inductance value must be designed as follows to achieve ZVS condition of upper switches in this case:

$$L_1 \leq \frac{V_2T_{\mathrm{dL}}(DT - T_{\mathrm{dL}})}{4V_{\mathrm{Cc2}}C_{\mathrm{oss}\cdot L}} \tag{10.38}$$

It can be seen that the value of DC inductance is independent of the output power. Actually, the ZVS condition of upper switches in discharging mode and the bottom switches in charging mode is the same as shown in (10.38). Similarly, the ZVS condition of bottom switches in discharging mode can be obtained as follows:

$$L_1 \leq \frac{V_2(DT - T_{\mathrm{dL}})}{-\frac{V_{\mathrm{Cc2}}}{L_r}\left(\frac{V_{\mathrm{Cc2}}}{V_2T} + \frac{1}{T_{\mathrm{dL}}}\right)T_{\phi1a}^2 + \frac{2V_{\mathrm{Cc2}}}{L_r}T_{\phi1a} + \frac{4C_{\mathrm{oss}\cdot L}V_{\mathrm{Cc2}}}{T_{\mathrm{dL}}}} \tag{10.39}$$

where $T_{\phi1a} = \frac{V_2T}{V_2T + T_{\mathrm{dL}}V_{\mathrm{Cc2}}}T_{\mathrm{dL}}$. As shown in (10.38) and (10.39), it can be inferred that the most difficult condition to achieve ZVS is illustrated in (10.39) under the minimum battery voltage 18 V. Thus, the boundary DC inductance value versus dead time of battery side T_{dL} is plotted in Fig. 10.16.

The figure reaches its peak point at the dead time T_{dL} around 135 ns. As aforementioned, the larger DC inductance value is beneficial to reduce the system

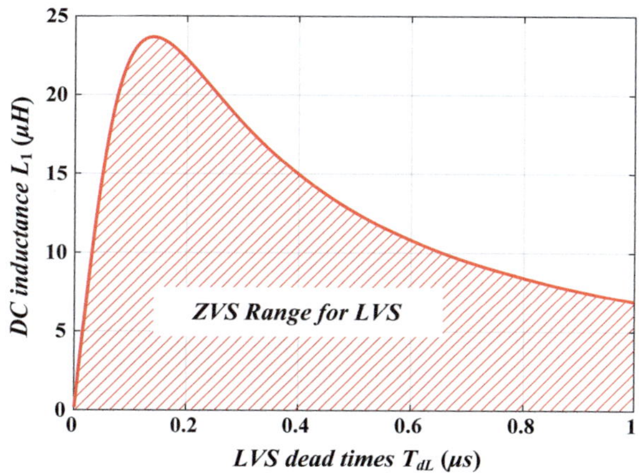

Fig. 10.16 Maximum DC inductance L_1 versus dead time T_{dL}

loss and minimize the current stress. However, the volume of the DC inductors is also an issue worth considering. Hence, it is necessary to design the DC inductor legitimately.

10.3 Experimental Results

10.3.1 Prototype

A 1 kW experimental prototype has been built in order to verify the effectiveness of the current-fed bidirectional DC–DC converter with the proposed DPDPS control. The prototype photograph is shown in Fig. 10.17. The system specifications are given in Table 10.1. Besides, battery side switch type is IXFN360N15T2 for Q_1 and Q_2, and HVS switch type is FDA50N50 for S_1–S_4. The high-frequency transformer core is EE55. The entire control of the system is implemented on a Texas Instruments TMS320F28335 DSP.

10.3.2 Operation Waveforms of Charging Mode and Discharging Mode

The steady-state waveforms of the proposed converter under 1 kW are shown in Fig. 10.18. The waveforms in Fig. 10.18a–c are tested in charging mode in 18, 23, and 28 V battery voltage, respectively, while Fig. 10.18d–f are tested in discharging mode in 18, 23, and 28 V battery voltage, respectively. As seen, the leakage inductance current i_{Lr} keeps constant during the power-transferring stage

Fig. 10.17 Photograph of the test prototype

Table 10.1 System specifications

Components	Parameters
Voltage of HVS (V_1)	600 V
Voltage of battery side (V_2)	18–28 V
Output power (P_{omax})	1 kW
Switching frequency (f_S)	50 kHz
Leakage inductance (L_r)	1.5 μH
DC inductance (L_1 and L_2)	11 μH
Battery side MOSFETs (Q)	IXFN360N15T2
HVS MOSFETs (S)	FDA50N50

due to the voltage-matched control and the circulation loss is suppressed by employing DPDPS modulation.

10.3.3 *Soft-Switching Waveforms of Discharging Mode and Charging Mode*

Figures 10.19, 10.20, 10.21, and 10.22 illustrate the ZVS achievement for HVS switches at the maximum load and no-load conditions. It can be seen that the proposed topology can achieve soft switching within full load range as long as the parameters are designed reasonable. It is clear to see that all HVS switches can achieve ZVS, which indicates that the proposed topology can work at full load range with a good performance.

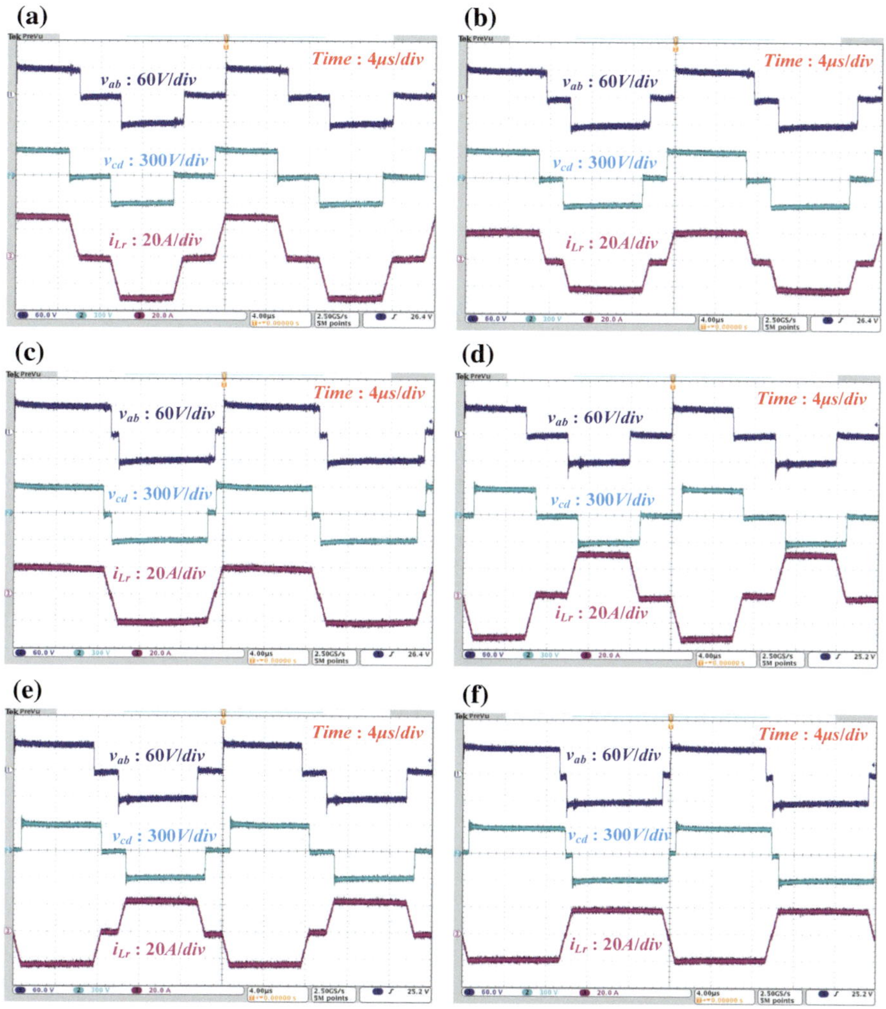

Fig. 10.18 Steady-state waveforms at 1 kW output: **a** V_2 = 18 V in charging mode; **b** V_2 = 23 V in charging mode; **c** V_2 = 28 V in charging mode; **d** V_2 = 18 V in discharging mode; **e** V_2 = 23 V in discharging mode; **f** V_2 = 28 V in discharging mode

The efficiency curves of the proposed converter are plotted in Fig. 10.23. The left curves represent the discharging mode, while the right curves show charging mode. As seen, the high conversion efficiency is achieved over wide battery voltage and load range owing to the soft-switching operation, the utilization of low-voltage rating power devices, and low peak/RMS currents. The highest efficiency can be obtained at full load when the battery voltage is 28 V in both modes. The maximum efficiency in charging mode is 96.71% while 96.79% in discharging mode.

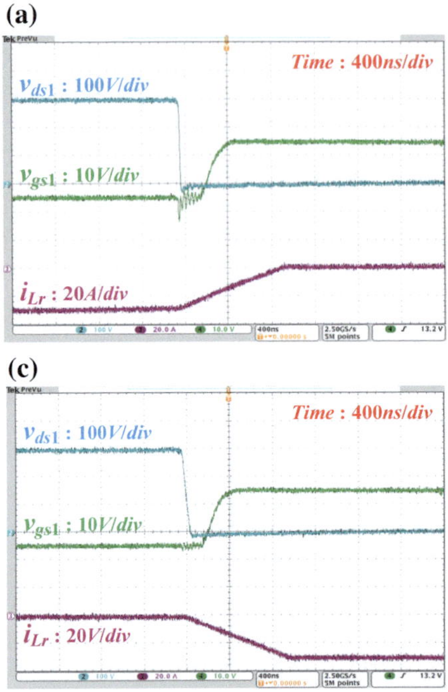

Fig. 10.19 ZVS waveforms in charging mode at 1 kW output for HVS switches in $V_2 = 18$ V: **a** S_1; **b** S_2; **c** S_3; **d** S_4

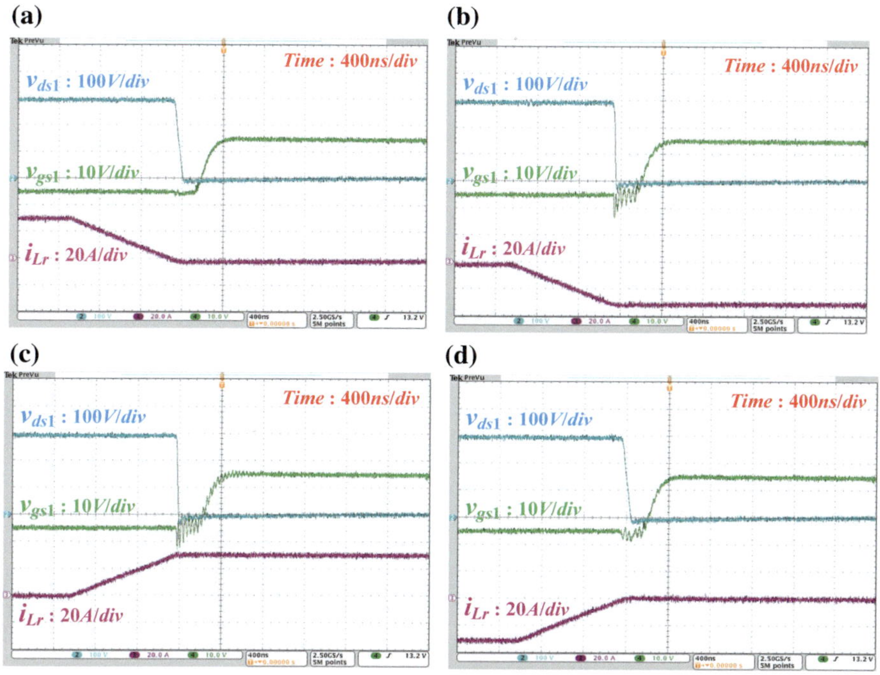

Fig. 10.20 ZVS waveforms in discharging mode at 1 kW output for HVS switches in V_2 = 18 V: **a** S_1; **b** S_2; **c** S_3; **d** S_4

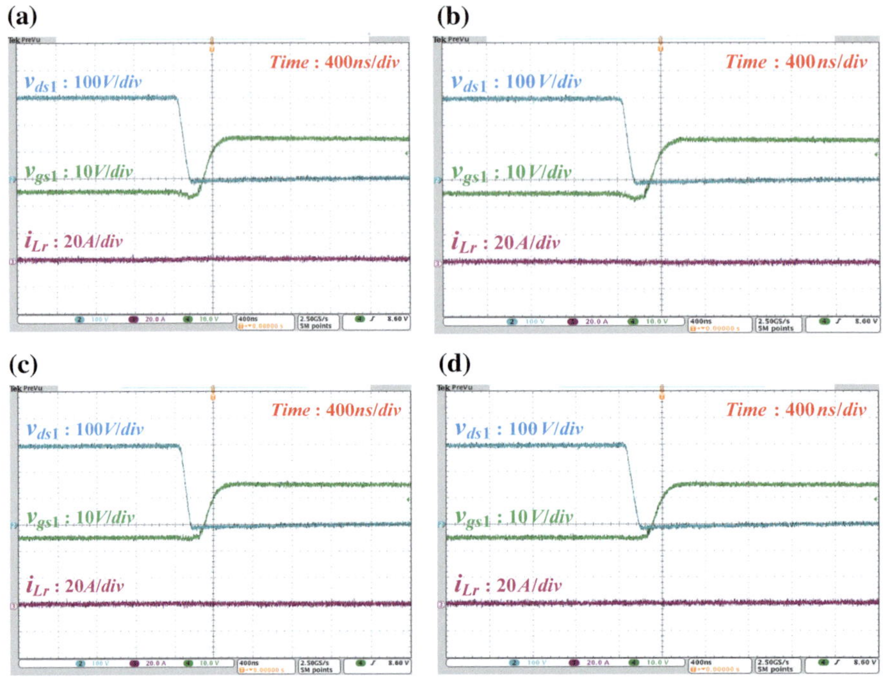

Fig. 10.21 ZVS waveforms in charging mode at no-load for HVS switches in V_2 = 18 V: **a** S_1; **b** S_2; **c** S_3; **d** S_4

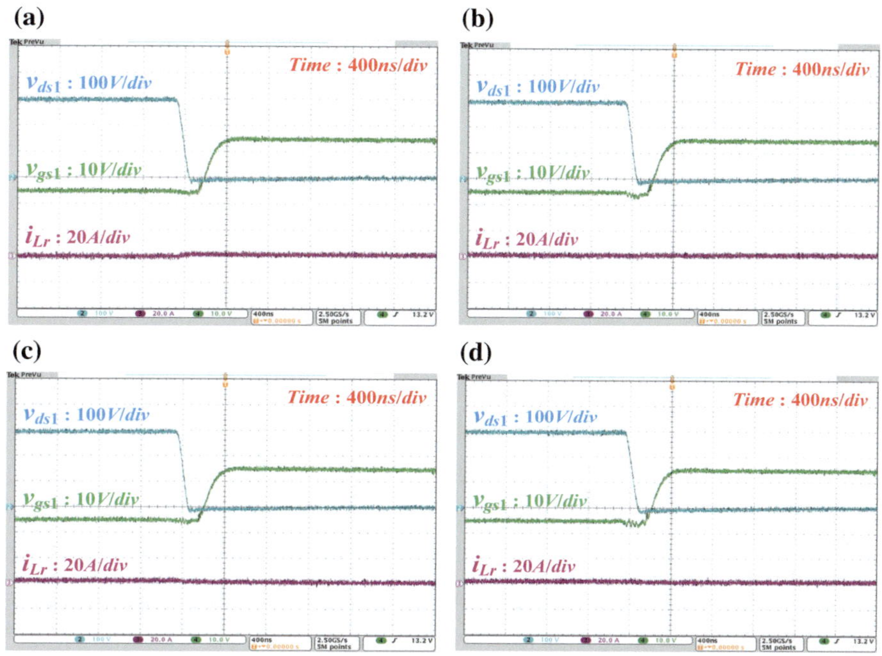

Fig. 10.22 ZVS waveforms in discharging mode at no-load for HVS switches in $V_2 = 18$ V: **a** S_1; **b** S_2; **c** S_3; **d** S_4

Fig. 10.23 Conversion efficiency

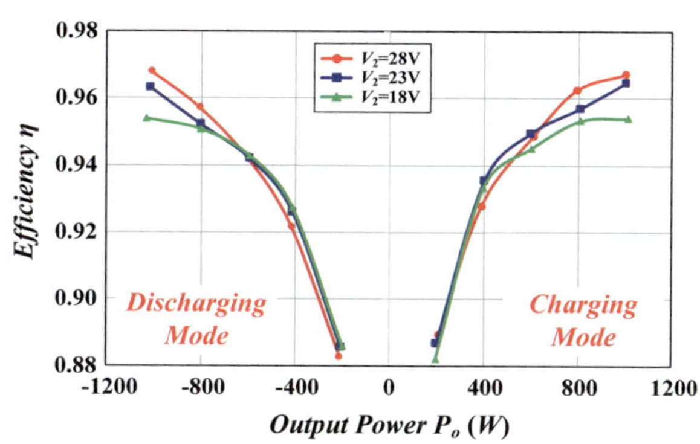

10.4 Conclusion

This chapter proposes a DPDPS control strategy with equal duty cycles for current-fed three-level DC–DC converters for battery charging/discharging. With the proposed control strategy, the conduction loss can be reduced and the system reliability can be improved. The comparison between voltage matching and mismatching modes has been made in detail. Besides, the relationship between the maximum DC inductance and dead time of battery side, and the relationship between the maximum magnetizing inductance and dead time of HVS are investigated to ensure the ZVS. ZVS can be achieved for all switches within full load range with proper parameter design. The validity of the proposed control strategy has been verified by experimental results of a 1 kW prototype.

References

1. Peng FZ, Li H, Su G, Lawler JS (2004) A new ZVS bidirectional DC–DC converter for fuel cell and battery application. IEEE Trans Power Electron 19(1):54–65
2. Xu D, Zhao C, Fan H (2004) A PWM plus phase-shift control bidirectional DC–DC converter. IEEE Trans Power Electron 19(3):666–675
3. Xiao H, Xie S (2008) A ZVS bidirectional DC–DC converter with phase shift plus PWM control scheme. IEEE Trans Power Electron 23(2):813–823
4. Sha D, Lin Q, You F, Wang X, Xu G (2016) A ZVS bidirectional three-level DC–DC converter with direct current slew rate control of leakage inductance current. IEEE Trans Ind Appl 52 (3):2368–2377
5. Ding Z, Yang C, Zhang Z, Wang C, Xie S (2014) A novel soft-switching multiport bidirectional DC–DC converter for hybrid energy storage system. IEEE Trans Power Electron 29(4):1595–1609
6. Sha D, You F, Wang X (2016) A high efficiency current-fed semi dual active bridge DC–DC converter for low input voltage applications. IEEE Trans Ind Electron 63(4):2155–2164

Chapter 11
A Current-Fed Hybrid Dual Active Bridge DC–DC Converter for Fuel Cell Power Conditioning System with Reduced Input Current Ripple

Abstract In this chapter, a novel current-fed hybrid dual active bridge DC–DC converter is proposed, which is suitable for low-voltage fuel cell power conditioning system. The high-frequency input current ripple can be reduced to minimum because the input-side switches are always switched at 50% duty cycles in spite of the fuel cell voltage and the load variation. Notch filter is used in the voltage feedback path to reduce the low-frequency input current ripple when interfaced with a single-phase inverter load. All of the power devices can achieve zero-voltage switching on by the proposed control strategy. The mode analysis, the operation principle, ZVS conditions, and parameter design are given in this chapter. A 1 kW prototype has been fabricated to verify the effectiveness of the proposed converter and control strategy.

Keywords DC–DC converter · Fuel cell · Zero-voltage switching
Current ripple

11.1 Converter Topology and Operating Principles

In power conditioning systems, not only low-frequency current ripple, but also high-frequency current ripple can be a threat for devices [1, 2]. It is, therefore, necessary to suppress the high-frequency current ripple [3, 4].

For voltage-fed DAB topologies [5, 6], the high-frequency current ripple is rather large, so a large capacitance is needed to snub the high-frequency component. As to conventional current-fed DAB converters [7, 8], the high-frequency current ripple can be reduced by interleaving operation of the two buck/boost circuits. But the current ripple cannot be minimized to be zero, since PWM control is still used for the current-fed side to adjust the clamp voltage. In order to minimize the high-frequency current ripple more effective, current-fed hybrid DAB converter is proposed.

D. Sha and G. Xu, *High-Frequency Isolated Bidirectional Dual Active Bridge DC–DC Converters with Wide Voltage Gain*, CPSS Power Electronics Series,
https://doi.org/10.1007/978-981-13-0259-6_11

11.1.1 Proposed Converter

The proposed converter topology is shown in Fig. 11.1. The input side (also the fuel cell side) is dual-boost half bridges consisting of four power MOSFETs (Q_1, Q_{1a}, Q_2, Q_{2a}) and two DC inductors (L_1, L_2). The output side is composed of a full bridge and an auxiliary half bridge. The full bridge consists of four MOSFETs (S_1–S_4); the auxiliary half bridge consists of two MOSFETs (S_5, S_6) and two capacitors C_u and C_d. The two sides are connected by a high-frequency transformer T, whose turns ratio is 1:n. L_k denotes the leakage inductance of the transformer. m is the voltage conversion ratio defined as

$$m = V_o/(2nV_{in}) \tag{11.1}$$

11.1.2 Modulation Strategy

The typical operating waveforms are shown in Fig. 11.2. For the input-side power switches, Q_1 and Q_{1a}, Q_2 and Q_{2a} are operated complementarily, respectively, with

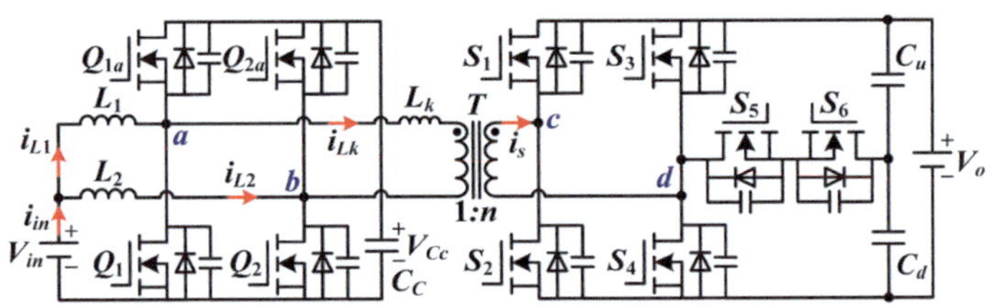

Fig. 11.1 Topology of the proposed converter

Fig. 11.2 Typical operating waveforms

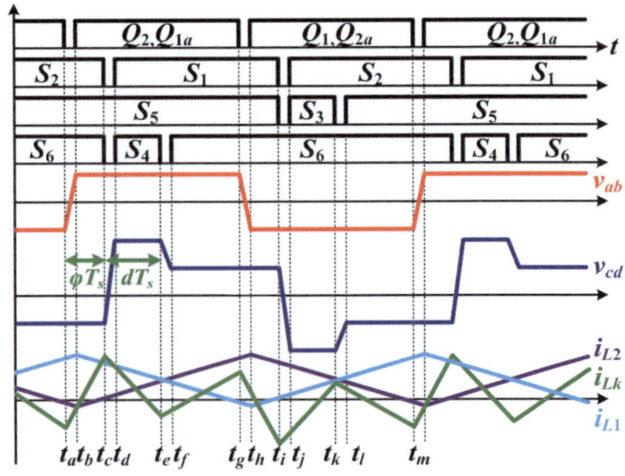

duty cycles fixed as 50%; Q_1 and Q_2 are phase shifted by $T_s/2$, where T_s is the switching period. For the output-side switches, S_1 and S_2, S_3 and S_5, S_4 and S_6 also operate complementarily. The duty cycles of S_1 and S_2 are fixed as 50%, and the gating signal of S_1 lags behind that of Q_{1a} for phase shift ratio φ. The duty cycles of S_3 and S_4 are both denoted as d. S_3 and S_4 are turned on simultaneously with S_2 and S_1, respectively. The proper selection of d can ensure ZVS for all switches in wide operation range, which will be studied in later sections. v_{ab} denotes the voltage across the points "a" and "b" shown in Fig. 11.1, and v_{cd} is the voltage across "c" and "d". The current flowing through the leakage inductor is denoted as i_{Lk}.

11.1.3 Typical Operating Periods

According to the operating waveforms in Fig. 11.2, an entire switching period can be divided into twelve stages. Since the operation mode is symmetrical, only half of a switching cycle will be illustrated in Fig. 11.3 and described as follows. The modes in the other half cycle are similar.

Period 1 (t_a, t_b): Prior to t_a, S_2, S_5, S_6, Q_1 and Q_{2a} are on. The leakage inductance current i_{Lk} is decreasing linearly.

At the instant t_a, Q_1 and Q_{2a} are turned off. Then, the output junction capacitors of the input-side switches begin to resonate with the L_k and the input-side DC inductances. The difference between i_{L1} and i_{Lk} begins to charge the junction capacitor of Q_1 and discharge that of Q_{1a} until the drain–source voltage of Q_{1a} falls to zero; then, the body diode of Q_{1a} begins to conduct, waiting for the gating signal for Q_{1a}. Thus, ZVS can be obtained. Likewise, the sum of $-i_{L2}$ and $-i_{Lk}$ begins to charge the junction capacitor of Q_{2a} and discharge that of Q_2 until the drain–source voltage of Q_2 falls to zero.

Period 2 (t_b, t_c): At t_b, Q_{1a} and Q_2 are turned on. By proper design of L_1, L_2, and L_k, Q_{1a} and Q_2 can achieve ZVS. When the resonance process is completed, v_{ab} is equal to $+V_{Cc}$ and v_{cd} equals $-V_o/2$, so i_{Lk} will increase linearly with a higher slew rate. During this period, the following equations can be obtained

$$\begin{cases} v_{ab} = +V_{Cc},\ v_{cd} = -V_o/2 \\ i_{Lk}(t) = i_{Lk}(t_b) + \frac{v_{ab} - v_{cd}/n}{L_k}(t - t_b) \end{cases} \tag{11.2}$$

It should be noticed that the voltages across S_3 and S_4 during this period are both $V_o/2$.

Period 3 (t_c, t_d): At t_c, S_2 and S_6 are turned off at the same time. The current i_s, the transformer current referred to the secondary side, will charge the junction capacitors of S_2, S_3, and S_6 while discharge those of S_1 and S_4, until the drain–source voltage of S_1 and S_4 falls to zero. Simultaneously, both the drain–source voltages of S_2 and S_3 increase to V_o and that of S_6 increases to $V_o/2$. Then, the body diodes of S_1 and S_4 begin to conduct until their gating signals come.

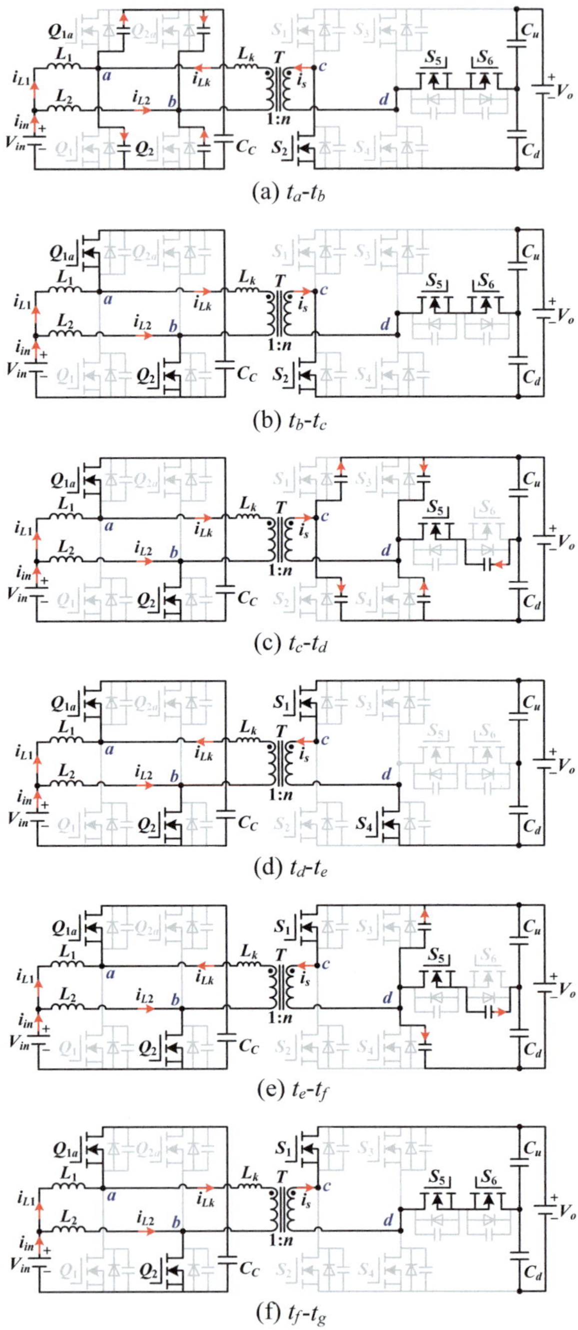

Fig. 11.3 Typical operating periods

Period 4 (t_d, t_e): At t_d, S_1 and S_4 are turned on with ZVS. During this stage, v_{ab} is still equal to $+V_{Cc}$ but v_{cd} changes to $+V_o$. In our configuration, V_{Cc} is lower than V_o/n, so i_{Lk} will decrease linearly according to the following relationship

$$\begin{cases} v_{ab} = +V_{Cc}, \; v_{cd} = +V_o \\ i_{Lk}(t) = i_{Lk}(t_d) + \frac{v_{ab} - v_{cd}/n}{L_k}(t - t_d) \end{cases} \tag{11.3}$$

During this stage, the secondary side current i_s will not flow through the bidirectional switch composed by S_5 and S_6.

Period 5 (t_e, t_f): At t_e, S_4 is turned off. i_s will charge the junction capacitor of S_4 and discharge that of S_3 and S_6, until the drain–source voltage of S_4 increases to $V_o/2$ while the voltage across S_3 and S_6 falls to $V_o/2$ and zero, respectively. Then, the body diode of S_6 begins to conduct since S_5 is still on.

This stage will occur only if the $i_{Lk}(t_e)$ is negative. In order to achieve this, compensation for d may be needed; this will be explained in detail in the later sections.

Period 6 (t_f, t_g): At t_f, S_6 can be turned on with ZVS if the value of $i_{Lk}(t_f)$ is ensured negative, and then i_{Lk} starts to increase linearly according to the following relationship

$$\begin{cases} v_{ab} = +V_{Cc}, \; v_{cd} = +V_o/2 \\ i_{Lk}(t) = i_{Lk}(t_f) + \frac{v_{ab} - v_{cd}/n}{L_k}(t - t_f) \end{cases} \tag{11.4}$$

11.2 ZVS Conditions and Control Strategy

11.2.1 ZVS Conditions

As illustrated in previous sections, the power transferred by the proposed converter is controlled by two independent variables, including the phase shift ratio φ and the output-side duty cycle d. Therefore, for a certain operating point with fixed input voltage and transferred power, different waveforms of i_{Lk} can be generated by changing the combination possibility of the two control variables, leading to different ZVS conditions.

Figure 11.4 shows the theoretical operating waveforms of the proposed converter. t_0–t_5 are the moments when switches are turned on. Since the values of the leakage inductance current at these points determine the achievement of ZVS for the switches, calculation for the expression of these currents is necessary.

Based on Fig. 11.4, (11.2) and (11.4), the differences of current values between these moments can be calculated as follows.

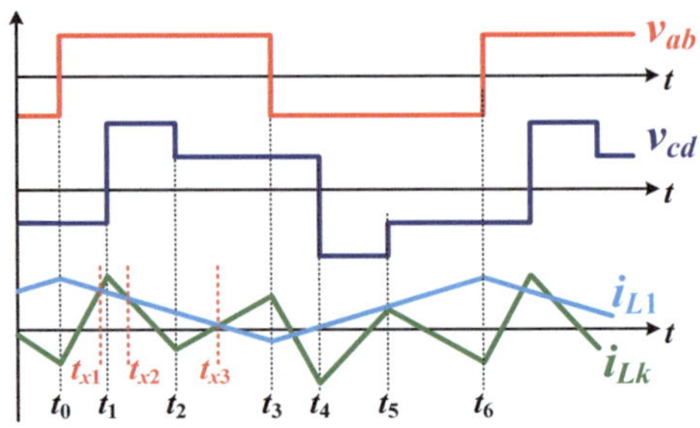

Fig. 11.4 Theoretical operating waveforms

$$\begin{cases} i_{\mathrm{Lk}}(t_1) = -i_{\mathrm{Lk}}(t_4) = i_{\mathrm{Lk}}(t_0) + (2V_{\mathrm{in}} + V_{\mathrm{o}}/2n)\varphi T_{\mathrm{s}}/L_k \\ i_{\mathrm{Lk}}(t_2) = -i_{\mathrm{Lk}}(t_5) = i_{\mathrm{Lk}}(t_1) + (2V_{\mathrm{in}} - V_{\mathrm{o}}/n)dT_{\mathrm{s}}/L_k \\ i_{\mathrm{Lk}}(t_3) = -i_{\mathrm{Lk}}(t_0) = i_{\mathrm{Lk}}(t_2) + (2V_{\mathrm{in}} - V_{\mathrm{o}}/2n)(0.5 - \varphi - d)T_{\mathrm{s}}/L_k \end{cases} \tag{11.5}$$

Solving (11.5) yields current values at the switching moments

$$\begin{cases} i_{\mathrm{Lk}}(t_0) = (-0.5nV_{\mathrm{in}} + V_{\mathrm{o}}(0.125 + 0.25d - 0.5\varphi))T_{\mathrm{s}}/nL_k \\ i_{\mathrm{Lk}}(t_1) = ((0.125 + 0.25d)V_{\mathrm{o}} + nV_{\mathrm{in}}(2\varphi - 0.5))T_{\mathrm{s}}/nL_k \\ i_{\mathrm{Lk}}(t_2) = ((0.125 - 0.75d)V_{\mathrm{o}} + nV_{\mathrm{in}}(2d + 2\varphi - 0.5))T_{\mathrm{s}}/nL_k \end{cases} \tag{11.6}$$

Meanwhile, the expressions of the input DC inductor currents are written by

$$\begin{cases} i_{L1}(t_0) = i_{L2}(t_3) = P_{\mathrm{o}}/(2V_{\mathrm{in}}) + (V_{\mathrm{in}}T_{\mathrm{s}})/(4L_{\mathrm{in}}) \\ i_{L1}(t_3) = i_{L2}(t_0) = P_{\mathrm{o}}/(2V_{\mathrm{in}}) - (V_{\mathrm{in}}T_{\mathrm{s}})/(4L_{\mathrm{in}}) \end{cases} \tag{11.7}$$

P_{o} is the output power. The ZVS conditions for switches can be summarized as shown in Table 11.1.

11.2.2 Control Strategy

From Table 11.1, for output-side full-bridge switches S_1–S_4, ZVS can be obtained once $i_{\mathrm{Lk}}(t_1)$ is ensured positive. If the output-side duty cycle d is selected as follows

$$d = 1/m - 0.5 \tag{11.8}$$

Table 11.1 ZVS conditions for switches

Switches	Current
S_1, S_2, S_3, and S_4	$i_{\mathrm{Lk}}(t_1) > 0$
S_5 and S_6	$i_{\mathrm{Lk}}(t_2) < 0$
Q_{1a} and Q_{2a}	$i_{L1}(t_0) > i_{\mathrm{Lk}}(t_0)$
Q_1 and Q_2	$-i_{\mathrm{Lk}}(t_0) > i_{L2}(t_0)$

then the formula of $i_{Lk}(t_1)$ in (11.6) can be simplified as

$$i_{Lk}(t_1) = 2V_{in}\varphi T_s/L_k \tag{11.9}$$

Since the phase shift ratio φ is positive for the unidirectional power flow feature in the fuel cell system, $i_{Lk}(t_1)$ can be always kept positive. Hence, ZVS for S_1–S_4 can be obtained. Meanwhile, ZVS for Q_1, Q_{1a}, Q_2, and Q_{2a} can be achieved by proper design of the input DC inductance values and L_k.

In this situation, d is only decided by the voltage conversion gain and will not be affected by the load condition. However, ZVS for the auxiliary switches S_5 and S_6 is related to the combination of φ and d. If d is selected only according to (11.8), ZVS for S_5 and S_6 cannot be achieved when transferred power is high. Thus, to ensure ZVS for S_5 and S_6, d cannot be decided only by (11.8), and the control needs to be improved.

According to Table 11.1, ZVS condition for S_5 and S_6 is given by

$$i_{Lk}(t_2) < 0 \tag{11.10}$$

Based on (11.1) and (11.6), it can be written as

$$(1 - 6d)m + (8d + 8\varphi - 2) < 0 \tag{11.11}$$

In view of the value $i_{Lk}(t_2)$, it can be assumed to be zero, and then (11.12) can be derived. If (11.12) is satisfied, $i_{Lk}(t_2)$ equals zero at any conditions, which is the boundary condition for ZVS of S_5 and S_6.

$$d = (8\varphi + m - 2)/(6m - 8) \tag{11.12}$$

In this situation, since i_{Lk} decreases during the time interval $[t_1, t_2]$ and $i_{Lk}(t_2)$ is zero, it is obvious that $i_{Lk}(t_1)$ can still be ensured positive and so S_1–S_4 can still achieve ZVS.

But this equation leads to another problem. The power transferred by the converter in this case can be calculated as

$$\begin{aligned} P_o &= \frac{\int_0^{T_s} 2V_{in}(t) i_{Lk}(t)\mathrm{d}t}{T_s} \\ &= \frac{2[d^2 + \varphi(1 - 2\varphi) + d(2\varphi - 0.5)]mV_{in}^2 T_s}{L_k} \end{aligned} \tag{11.13}$$

Substituting (11.12) into (11.13) yields

$$\begin{aligned} P_o = \frac{mT_s V_{in}^2}{L_k(3m - 4)^2}&[-4(9m^2 - 36m + 40)\varphi^2 \\ &+ 4(6m^2 - 19m + 12)\varphi - (m - 1)(m - 2)] \end{aligned} \tag{11.14}$$

(11.14) indicates that the power transferred has a minimum limitation. The minimum power $P_{\min}$ can be expressed as

$$P_{\min} = \frac{mT_s V_{\text{in}}^2 (m-1)(m-2)}{L_k(3m^2 - 12m + 8)} \tag{11.15}$$

To sum up, based on the relationship of (11.8), ZVS for the auxiliary switches can be achieved at light load, but cannot be ensured at heavy load. In comparison, employing control strategy of (11.12), ZVS for the auxiliary switches can be achieved over certain kind of loads since the minimum output power is limited. Therefore, the two control laws can be combined together. So, the proposed control strategy is given as (11.16).

$$d = \begin{cases} \max\left\{\frac{1}{m} - 0.5, \frac{8\varphi + m - 2}{6m-8}\right\}, & 6m - 8 > 0 \\ \min\left\{\frac{1}{m} - 0.5, \frac{8\varphi + m - 2}{6m-8}\right\}, & 6m - 8 < 0 \end{cases} \tag{11.16}$$

11.2.3 Control Diagram Implementation

The control block diagram is shown in Fig. 11.5. It consists of phase shift loop and duty cycle loop. The phase shift loop is composed of output voltage loop and input current loop and aimed to regulate the output voltage by phase shifting. The duty cycle loop changes the duty cycle d for S_5 and S_6 according to (11.16). $G_v(s)$ and $G_i(s)$ are the transfer functions of the output voltage regulator and the input current regulator, respectively. H_v and H_i are sampling coefficients for v_o and i_{in}.

If the input voltage is fixed and the output voltage v_o is lower than its reference, the phase shift ratio φ increases and the transferred power increases as well. Since the load is fixed, v_o will rise until it is equal to the reference value.

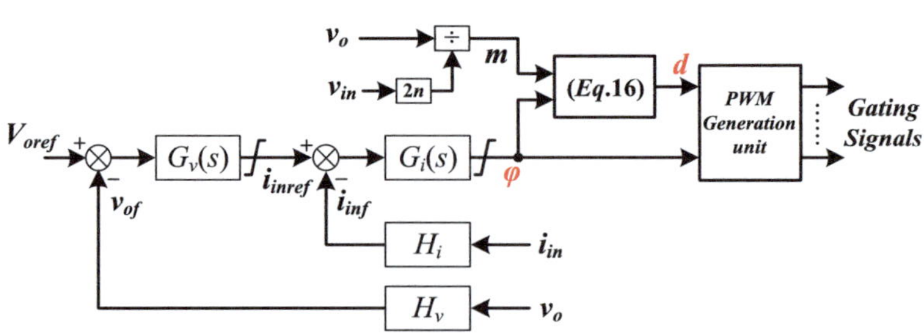

Fig. 11.5 Control block diagram

11.3 Characteristic Analysis and Parameter Design

11.3.1 Power Transfer Characteristics

The system specifications are given in Table 11.2.

Based on (11.16), the power transfer characteristics curve of the proposed converter is plotted in Fig. 11.6.

As shown in Fig. 11.6, the relationship between P_o and φ is piecewise. For the left piece, P_o always increases monotonically with φ increasing. For the right piece, P_o also increases monotonically with φ increasing when $V_{in} < 30$ V ($m > 1.5$). However, when $V_{in} > 30$ V ($m < 1.5$), the relationship between P_o and φ is not monotonic and P_o exists a minimum point, of which the coordinate is given as follows.

$$\begin{cases} \varphi = \frac{6m^2 - 19m + 12}{2(9m^2 - 36m + 40)} \\ P_o = \frac{mT_s V_{in}^2 (m-1)(m-2)}{L_k(3m^2 - 12m + 8)} \end{cases} \tag{11.17}$$

In order to avoid the possible instability caused by the non-monotonicity, the minimum power point should be avoided and thus need to be higher than the rated power. Following this principle, the leakage inductance L_k needs to satisfy the following relationship

Table 11.2 System specifications

Symbol	Value	Symbol	Value
V_{in}	24–36 V	L_1, L_2	8 μH
V_o	400 V	C_c	60 μF
P_o	1 kW	C_u, C_d	18.8 μF
f_s	50 kHz	Q_1, Q_2, Q_{1a}, Q_{2a}	IPT015N10N5
1:n	1:5	S_1–S_4	FDA50N50
L_k	1.6 μH	S_5, S_6	IXFH69N30P

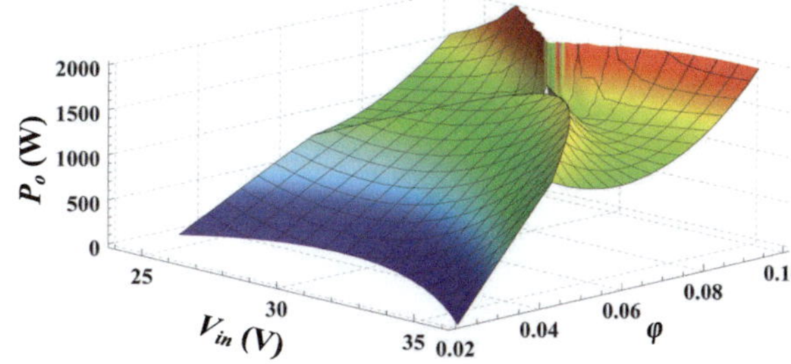

Fig. 11.6 Power transfer characteristics curve

$$L_k < \frac{mT_s V_{in}^2 (m-1)(m-2)}{P_{rated}(3m^2 - 12m + 8)} \tag{11.18}$$

Under the specifications shown in Table 11.2, the design for L_k should satisfy the following relationship

$$L_k < 1.745\ \mu\text{H} \tag{11.19}$$

11.3.2 Input Inductance Design

According to Table 11.2, ZVS condition for primary side switches is given by

$$-i_{Lk}(t_0) > i_{L2}(t_0) \tag{11.20}$$

Substituting (11.6) and (11.7) into (11.20) leads to

$$L_{in}(-8m\varphi^2 + 8md\varphi + m - 2) < L_k \tag{11.21}$$

The range of φ, d, and m can be summarized as the following

$$\begin{cases} 0 < \varphi < 0.5 \\ 0 < d < 0.5 \\ \varphi + d < 0.5 \\ 1.1 < m < 1.8 \end{cases} \tag{11.22}$$

Based on (11.22), the following can be obtained from (11.21)

$$\begin{aligned} &L_{in}(-8m\varphi^2 + 8md\varphi + m - 2) \\ &\quad < L_{in}(2md^2 + m - 2) < L_{in}(11m/9 - 2) < 0.2L_{in} \end{aligned} \tag{11.23}$$

So if $0.2L_{in} < L_k$, (11.21) will be satisfied at any condition. Therefore, in order to ensure ZVS for the primary side switches, the input-side inductance can be designed as

$$L_{in} < 5L_k \tag{11.24}$$

Since L_k is designed as 1.6 μH, the input inductance can be designed as 8 μH.

11.3.3 Clamping Capacitor Design

The design of the primary side clamping capacitor C_c should ensure that the voltage ripple across C_c is confined within a narrow range. Due to the symmetry of the primary side modulation, only half a switching cycle $[t_0, t_3)$ is analyzed. The clamping capacitor voltage $v_C(t)$ can be calculated by

$$v_C(t) = \left(\int_{t_0}^{t} (i_{L1}(t) - i_{Lk}(t))\mathrm{d}t \right) / C_c + V_0 \tag{11.25}$$

The three instants t_{x1}, t_{x2}, and t_{x3}, denoted in Fig. 11.4, are the instants when i_{L1} intersects with i_{Lk}. The current $i_{L1} - i_{Lk}$ will charge/discharge C_c. During $[t_0, t_{x1})$ and (t_{x2}, t_{x3}), i_{L1} is larger than i_{Lk}, so C_c will be charged and v_C will rise; during (t_{x1}, t_{x2}) and (t_{x3}, t_3), i_{L1} is smaller than i_{Lk}, so C_c is discharged and v_C will fall. So, the maximum of v_C is either at t_{x1} or at t_{x3}, and minimum one is either at t_0 or at t_{x2}. Letting i_{L1} equal to i_{Lk} yields

$$\begin{aligned}
t_{x1} &= t_0 + \left(\frac{P_o}{2V_{in}} + \frac{V_{in}T_s}{4L_1} + \frac{V_o\varphi T_s}{2nL_k} \right) \Big/ \left(\frac{V_{in}}{L_1} + \frac{2V_{in} + V_o/(2n)}{L_k} \right) \\
t_{x2} &= t_0 + \left(\frac{P_o}{2V_{in}} + \frac{V_{in}T_s}{4L_1} - \frac{(4nV_{in} - V_o)\varphi T_s}{nL_k} \right) \Big/ \left(\frac{V_{in}}{L_1} + \frac{2V_{in} - V_o/n}{L_k} \right) \\
t_{x3} &= t_0 + \frac{\left(\frac{P_o}{2V_{in}} + \frac{V_{in}T_s}{4L_1} - \frac{(2nV_{in}(2d+2\varphi-1) + (V_o/2)(1-d-\varphi))T_s}{nL_k} \right)}{\left(\frac{V_{in}}{L_1} + \frac{2V_{in} - V_o/n}{L_k} \right)}
\end{aligned} \tag{11.26}$$

So, the voltage ripple ΔV_C across C_c can be illustrated as

$$\Delta V_C = \max\{v_C(t_{x1}), v_C(t_{x3})\} - \min\{v_C(t_0), v_C(t_{x2})\} \tag{11.27}$$

In order that ΔV_C in not too large, the following criteria need to be satisfied

$$\frac{\Delta V_C}{V_C} = \frac{\Delta V_C}{2V_{in}} \leq 5\% \tag{11.28}$$

However, due to the nonlinearity of (11.27), no analytical solution can be obtained. Calculation for different input voltage and load conditions by the software Mathcad shows that the maximum voltage ripple occurs for the case of 24 V input and 1 kW load; so, we choose this case to design the clamping capacitor. Substituting V_{in} = 24 V and P_o = 1 kW into relevant equations and the critical value for C_c can be obtained

$$C_c \geq 28.96\,\mu\text{F} \tag{11.29}$$

Since input-side switches are open-loop controlled with duty cycles fixed as 50%, margin should be considered. Thus, the clamping capacitor C_c can be designed as 60 μF.

11.3.4 High-Frequency Current Ripple Analysis

Figure 11.7 shows the input-side inductor current waveforms if the dead time is ignored and all the devices are ideal. Duty cycles of all input-side switches are 50%, so the average clamping capacitor voltage V_C is twice the input voltage and expressed as

$$V_C = 2V_{\text{in}} \tag{11.30}$$

During $[T_0, T_1]$, the voltage across L_1 is $V_{\text{in}} - V_C$ and that of L_2 is V_{in}. During $[T_1, T_2]$, the voltage across L_1 is V_{in} and that of L_2 is $V_{\text{in}} - V_C$. Hence, i_{L1} and i_{L2} can be described as

$$\begin{aligned} i_{L1}(t) &= \begin{cases} I_1 + \frac{V_{\text{in}}T_s}{4L_1} - \frac{V_{\text{in}}}{L_1}(t - T_0), & T_0 \leq t < T_1 \\ I_1 - \frac{V_{\text{in}}T_s}{4L_1} + \frac{V_{\text{in}}}{L_1}(t - T_1), & T_1 \leq t < T_2 \end{cases} \\ i_{L2}(t) &= \begin{cases} I_2 - \frac{V_{\text{in}}T_s}{4L_2} + \frac{V_{\text{in}}}{L_2}(t - T_0), & T_0 \leq t < T_1 \\ I_2 + \frac{V_{\text{in}}T_s}{4L_2} - \frac{V_{\text{in}}}{L_2}(t - T_1), & T_1 \leq t < T_2 \end{cases} \end{aligned} \tag{11.31}$$

I_1 and I_2 are the average currents of L_1 and L_2, respectively. Assuming that the values of L_1 and L_2 are identical, the input current i_{in} is obtained as

$$i_{\text{in}}(t) = i_{L1}(t) + i_{L2}(t) = I_1 + I_2 \tag{11.32}$$

It can be seen from (11.32) that input current i_{in} is a constant DC value and there is no high-frequency current ripple if L_1 and L_2 are identical. So, the proposed converter can achieve high-frequency ripple-free input current theoretically.

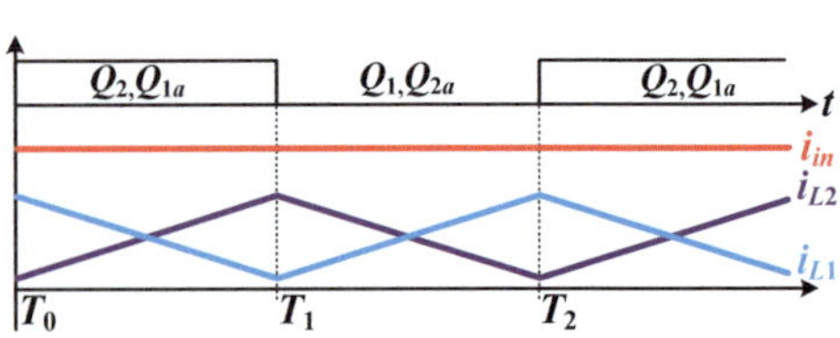

Fig. 11.7 Input inductor current waveforms neglecting the dead time effect

11.4 Simulation Results

Simulation is carried out by PSIM to verify the effectiveness of the proposed topology and control strategy. It must be noted that the DC inductors are ideal for the simulation. The system specifications are listed in Table 11.2.

Figure 11.8a illustrates the steady-state waveforms for the proposed converter with fixed 50% duty cycle interleaving operation when the input voltage is 30 V and the load is 1 kW. Figure 11.8b shows those for the conventional current-fed DAB converter by using conventional interleaving operation of varying duty cycle control.

As seen from Fig. 11.8, for the proposed converter and the proposed fixed 50% duty cycle interleaving operation, both i_{L1} and i_{L2} rise linearly in a half switching cycle and fall linearly in another half cycle. The absolute values of the rising and falling slew rates are identical. Besides, i_{L1} and i_{L2} are 180° phase shifted. Thus, i_{in} is a constant DC value and contains no ripple at all. However, for the conventional current-fed DAB, the high current ripple is still quite large even with interleaving technology, because the duty cycle for input-side bottom switches is not fixed as 50%. This verifies the effectiveness of the proposed control in high-frequency ripple reduction.

Figure 11.9 shows turning-on process of the Q_1, Q_{1a}, S_1, and S_6 when the input voltage is 24 V; the output power is 800 W, and d is decided only by (11.8). Figure 11.9a shows the turning-on process of Q_{1a} and S_1. Figure 11.9b, c shows that of Q_1 and S_6, respectively. For Q_1, Q_{1a}, and S_1, the drain–source voltage falls to zero before the gating signal is given, so ZVS for them is achieved. While for S_6, ZVS turn-on is lost. Actually, in this condition, the leakage inductor current is positive when S_6 is turned on, so it cannot discharge the junction capacitor of S_6.

In comparison, Fig. 11.10 shows the ZVS achievement of S_6 with the same working condition in Fig. 11.9. It is worth noting that d is obtained in (11.12). d and d_0 are the duty cycle signals for S_3 obtained by (11.8) and (11.12), respectively. $v_{\text{GS\_S6}}$ and $v_{\text{DS\_S6}}$ are the gating signal and drain–source voltage of S_6.

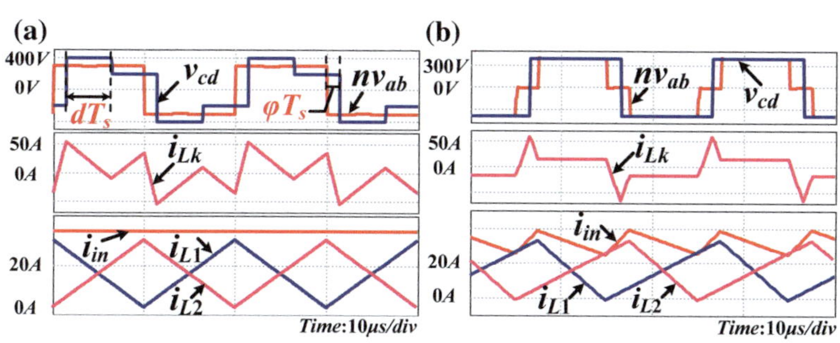

Fig. 11.8 Steady-state waveforms with $V_{\text{in}} = 30$ V, $P_o = 1$ kW. **a** Proposed converter with fixed 50% duty cycle interleaving control for the input side; **b** conventional current-fed DAB with varying duty cycle interleaving control for the input side

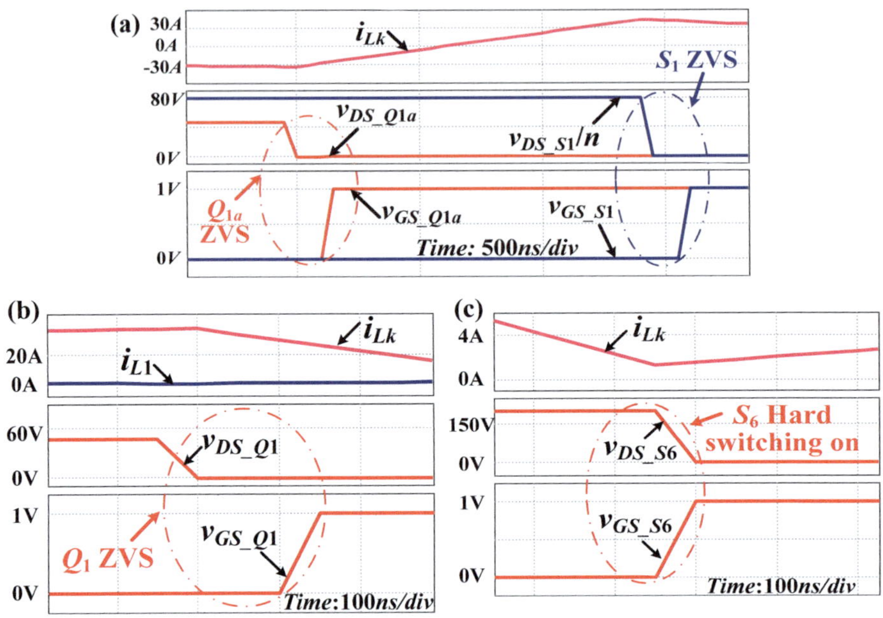

Fig. 11.9 Turning-on process for V_{in} = 24 V, P_o = 800 W with control relationship (11.8): **a** Q_{1a} and S_1, **b** Q_1, **c** S_6

Fig. 11.10 Turning-on process of S_6 with V_{in} = 24 V, P_o = 800 W, controlled by (11.12)

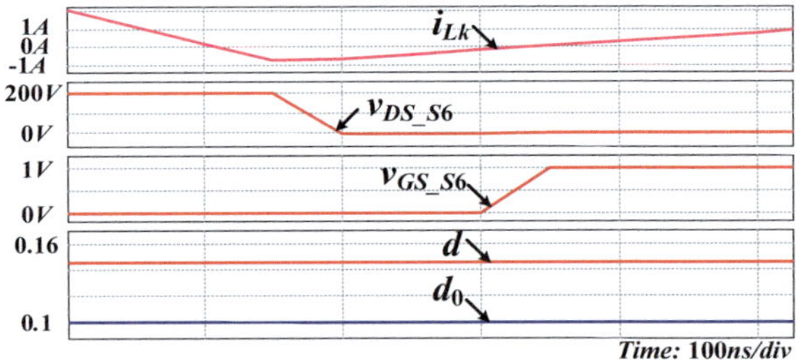

Actually, in order to verify the control strategy accurately in the simulation, the switches junction capacitor is set as small as 1 pF, so only a very small negative i_{Lk} is needed to discharge the junction capacitor of S_6. For the aforementioned discussion of the ZVS conditions in Sect. 11.2, the energy required for achieving ZVS is not considered and only the polarity of i_{Lk} is taken into account. So, the real compensation is a little bit higher than that in theory. It can be found from Fig. 11.10 that i_{Lk} turns negative before S_6 is turned on, so S_6 can achieve ZVS.

11.5 Experimental Results

11.5.1 Prototype

A 1 kW prototype was fabricated to verify the proposed converter and control strategy. The converter specifications are the same as those in Table 11.2. The photograph of the prototype is shown in Fig. 11.11.

11.5.2 Experimental Waveforms for Positive Power Flow

Figure 11.11 shows the steady-state waveforms under different input voltage and load conditions when the output-side duty cycle d is controlled only by (11.8). The converter works well in wide input voltage range and wide load range.

It can be seen from Fig. 11.12 that the polarity of $i_{\mathrm{Lk}}(t_2)$, denoted in Fig. 11.4, may vary with the input voltage and load. Hence, if d is only decided by (11.8) and no compensation is done, ZVS for S_5 and S_6 cannot be achieved in the whole operation range. For example, when the input voltage is 24 V and the load is 1 kW, $i_{\mathrm{Lk}}(t_2)$ is positive and ZVS for the S_5 and S_6 will be lost.

Just as the proposed control in (11.12) shows, d needs to be compensated in certain cases to achieve ZVS for the auxiliary switches S_5 and S_6.

Figure 11.13 shows steady-state waveforms under 24 V input and 1 kW load condition using the proposed control. The duty cycle compensation is 0.1. As seen, $i_{\mathrm{Lk}}(t_2)$ is negative with proper compensation. Actually, the practical compensation

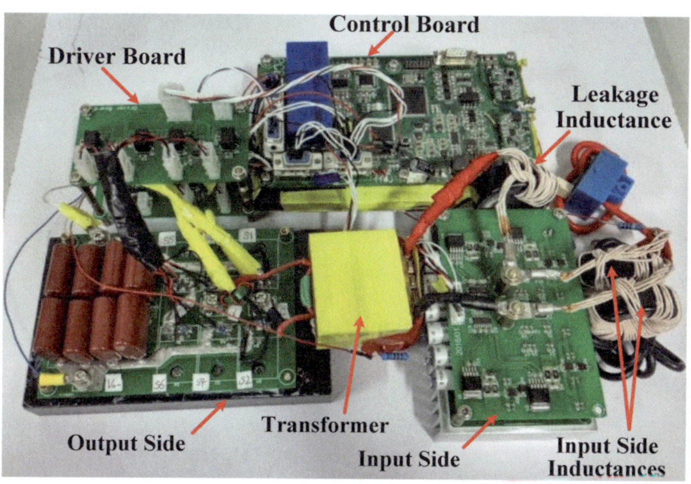

Fig. 11.11 Photograph of the prototype

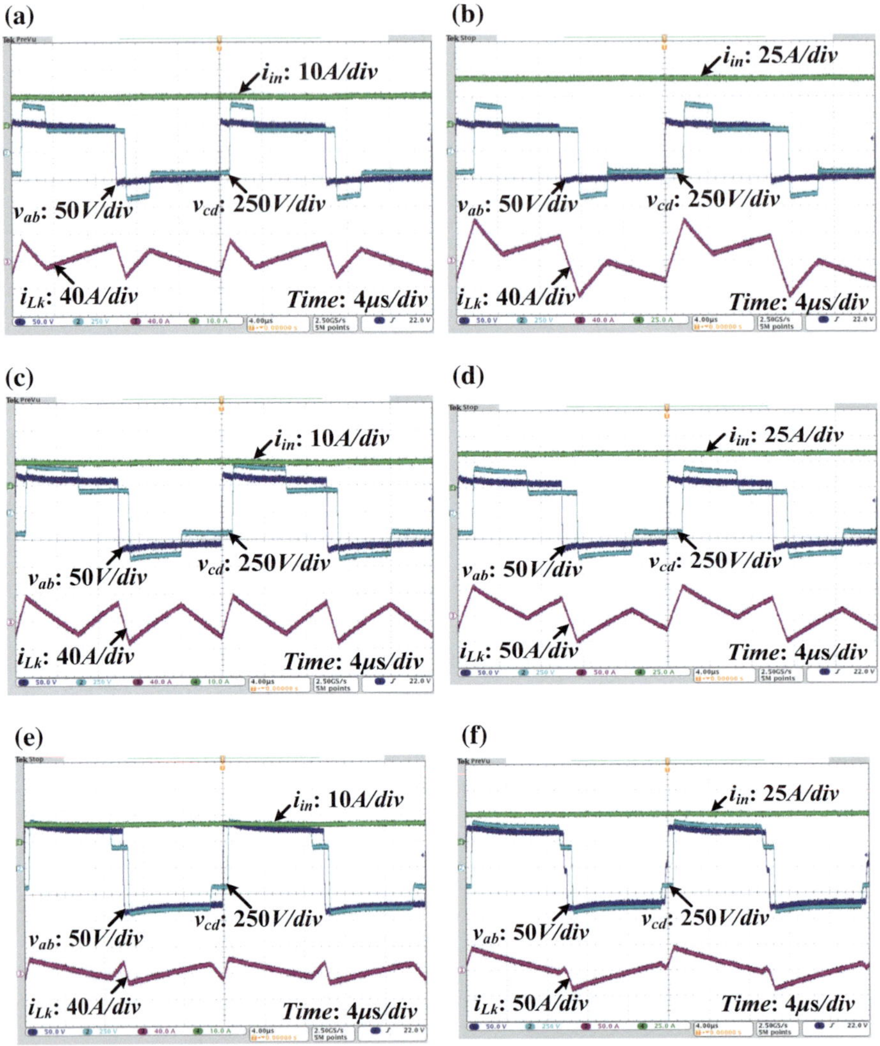

Fig. 11.12 Steady-state waveforms when controlled by (11.8): **a** $V_{in} = 24$ V, $P_o = 200$ W. **b** $V_{in} = 24$ V, $P_o = 1$ kW. **c** $V_{in} = 30$ V, $P_o = 200$ W. **d** $V_{in} = 30$ V, $P_o = 1$ kW. **e** $V_{in} = 36$ V, $P_o = 200$ W. **f** $V_{in} = 36$ V, $P_o = 1$ kW

is a little bit larger than the theoretical value, due to the necessary energy to discharge junction capacitors of MOSFETs completely.

From the experimental results above, it can be seen that the proposed converter can work in wide operation range with the proposed control.

Figure 11.14 shows the input current and its high-frequency AC component for the conventional current-fed DAB employing conventional interleaving operation with 1 kW load and 30 V input voltage. In comparison, Fig. 11.15 gives those

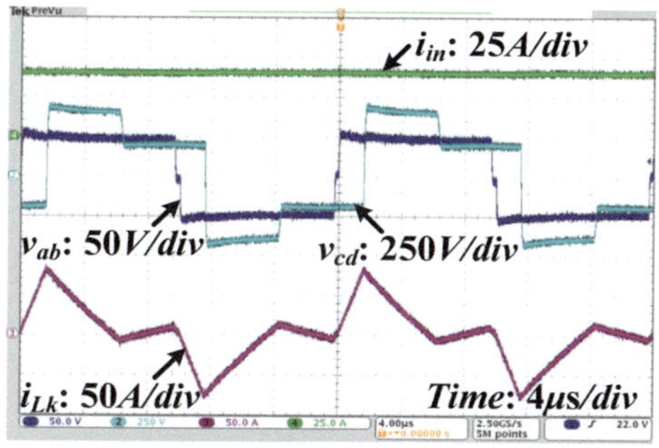

Fig. 11.13 Steady-state for proposed control with $V_{\rm in}$ = 24 V, $P_{\rm o}$ = 1 kW

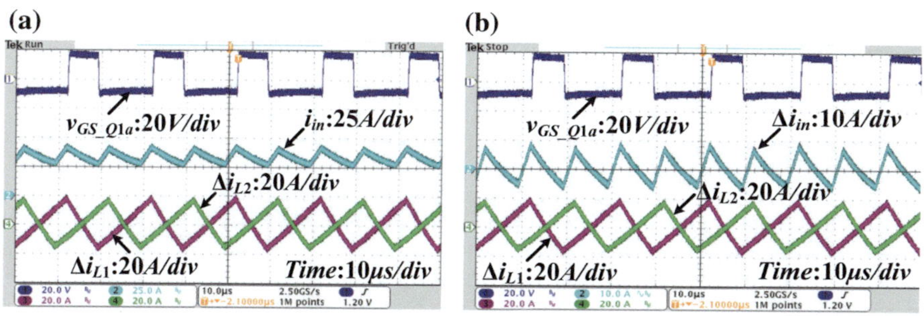

Fig. 11.14 Experiment results of high-frequency current ripple for conventional current-fed DAB using conventional interleaving technology; **a** input current, **b** AC component of input current

based on the proposed converter and the proposed interleaving control with 1 kW load and different input voltage conditions, in which $i_{\rm in}$ is the input current and $\Delta i_{\rm in}$ is its AC component. Δi_{L1} and Δi_{L2} represent the AC components of i_{L1} and i_{L2}, respectively. As seen in Fig. 11.15, duty cycles for input-side switches are fixed as 50% in spite of input voltage variation, and thus it can achieve minimal high-frequency input current ripple.

In Fig. 11.14, the peak–peak value of the high-frequency input current ripple can reach 16.4 A when the input voltage is 30 V and the load is 1 kW; while in Fig. 11.15, it is only 1.5, 1.7, and 2 A in case of 24, 30, and 36 V input voltage, respectively. As shown, the high-frequency input current ripple for the proposed converter and control can be suppressed to be minimum while that for the conventional current-fed DAB is relatively high. Unlike the ideal simulation results as shown in Fig. 11.8a, the rather small ripple in Fig. 11.15 may come from the mismatch of L_1 and L_2 inductance values, resonance process during the dead time, parasitic resistors of inductors, and parasitic parameters of power semiconductors.

Figure 11.16 illustrates the turning-on process of input-side bottom switch Q_1 under different input voltage and load conditions. As seen, Q_1 can achieve ZVS within wide input voltage range and wide load range. From analysis in Sect. 11.2, it

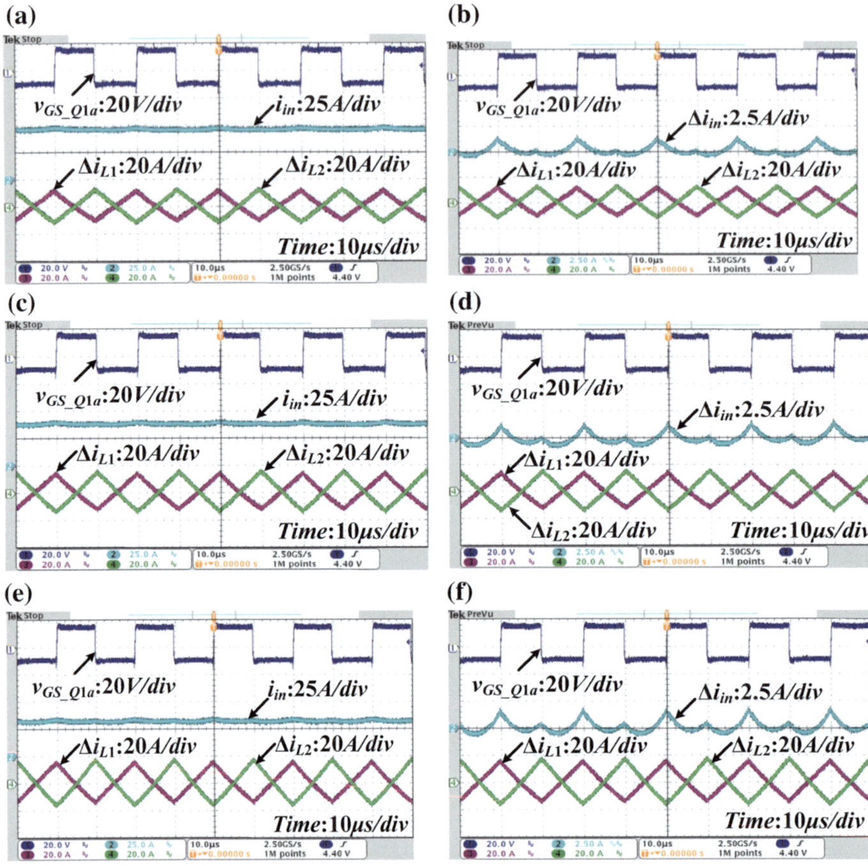

Fig. 11.15 Experiment results of high-frequency current ripple for proposed converter using 50% duty cycle interleaving operation: **a** V_{in} = 24 V input current, **b** V_{in} = 24 V AC component of input current, **c** V_{in} = 30 V input current, **d** V_{in} = 30 V AC component of input current, **e** V_{in} = 36 V input current, **f** V_{in} = 36 V AC component of input current

is easier for input-side top switches than bottom switches to achieve ZVS. Hence, Q_{1a} and Q_{2a} can achieve ZVS too.

Figure 11.17 illustrates the ZVS process of S_1 under different input voltage and load cases. It can be found out that S_1 can achieve ZVS within wide input voltage range and wide load range. Actually, S_1 and S_2 are relatively easy to achieve ZVS.

For S_4, it is turned on at the same time with S_1, so the resonance current for ZVS achieving of S_1 and S_4 is also the same. While the voltage across S_4 in off state is $V_o/2$, and that across S_1 is V_o. Hence, S_4 can obtain ZVS as well once ZVS of S_1 is achieved.

Figure 11.18 illustrates ZVS process of the auxiliary switch S_5. It can be seen that S_5 can achieve ZVS with the proposed control strategy in wide operation range.

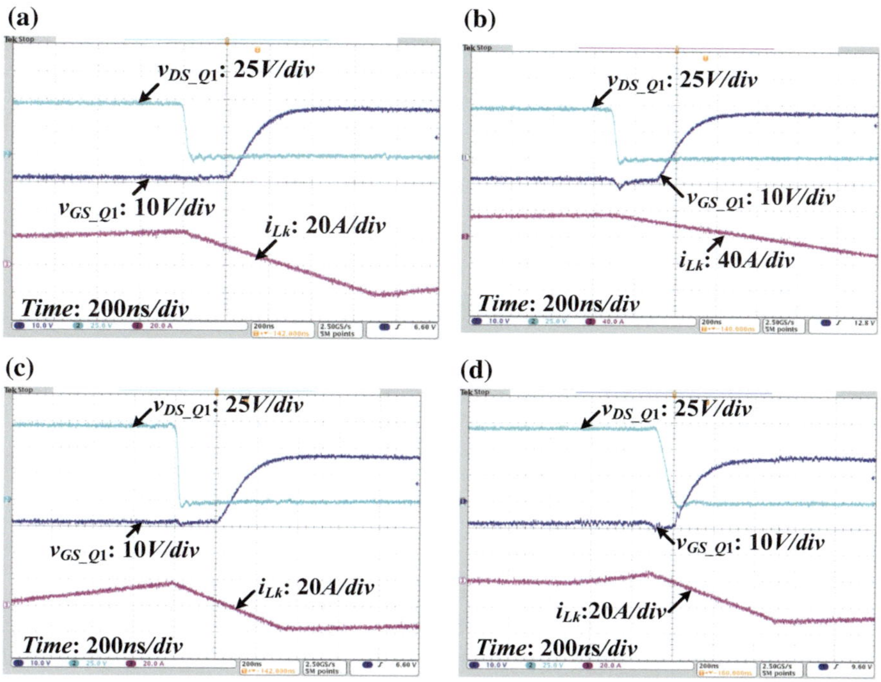

Fig. 11.16 Soft switching waveforms for Q_1: **a** V_{in} = 24 V, P_o = 200 W. **b** V_{in} = 24 V, P_o = 1 kW. **c** V_{in} = 36 V, P_o = 200 W. **d** V_{in} = 36 V, P_o = 1 kW

The experimental waveforms above show that the proposed converter works well within wide input voltage range and wide load range. Meanwhile, all the switches can achieve ZVS by the proposed control and proper inductance design. Besides, the input current high-frequency ripple can be suppressed to nearly zero in theory; low-frequency current ripple can also be reduced significantly when powering a single-phase inverter load. This verifies the effectiveness of the proposed converter and control.

For the proposed converter, efficiency of the main circuit under different input voltage and different load conditions for positive power flow was measured. The efficiency curves are illustrated in Fig. 11.19. As seen, for the same input voltage, the efficiency of the main circuit increases with the transferred power P_o increasing until the efficiency reaches its maximum value. The maximum conversion efficiency can exceed 95%.

For the proposed converter, the conduction loss can be reduced without using rectifying diodes for the output side since high-voltage rating diodes normally have relatively higher voltage drop compared with voltage drop over power MOSFETs. ZVS achievement for all switches in wide operation range helps to reduce switching loss. Comparing with conventional current-fed DAB converters employing varying duty cycle control at the input side [7, 8], duty cycles for input-side bottom switches

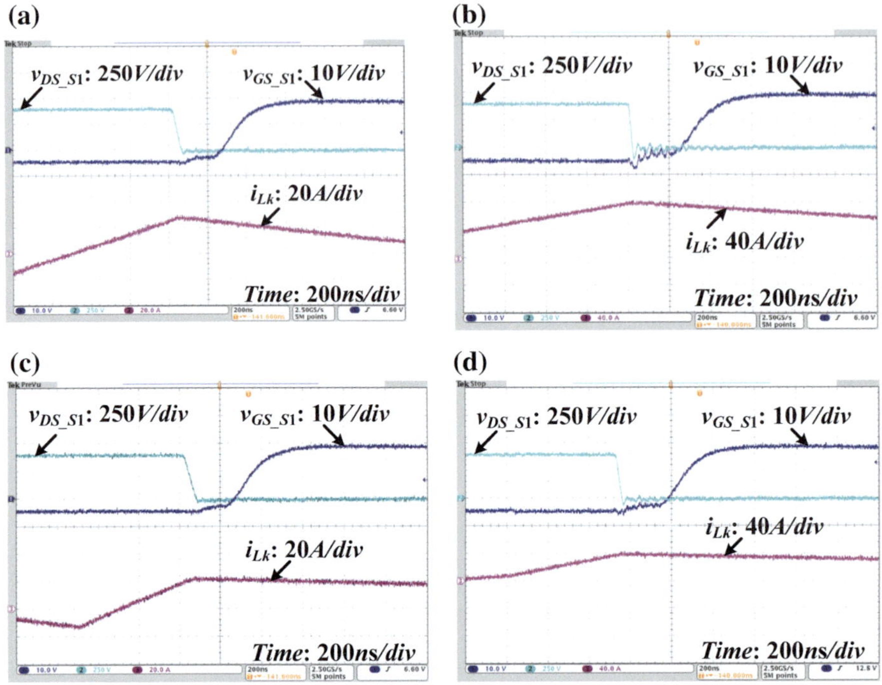

Fig. 11.17 Soft switching waveforms for S_1: **a** $V_{in} = 24$ V, $P_o = 200$ W. **b** $V_{in} = 24$ V, $P_o = 1$ kW. **c** $V_{in} = 36$ V, $P_o = 200$ W. **d** $V_{in} = 36$ V, $P_o = 1$ kW

are fixed as 50%, so both the ripple and the RMS values of the individual DC inductor currents are reduced for the same working condition. This causes the reduction of both the core loss and copper loss for DC inductors. The aforementioned attributes contribute to the conversion efficiency improvement.

11.5.3 Experimental Waveforms for Negative Power Flow

Figure 11.20 shows the steady-state waveforms when the power flow direction is reversed. As seen, the converter can also work stably and effectively. One thing needed to be pointed out is that the phase shift ratio φ may be positive for some negative power flow cases. This is because that when d is neither 0 nor 50%, the zero-crossing point of fundamental for v_{cd} leads to the zero-crossing point of v_{cd} in phase. Hence, the fundamental component of v_{ab} may lag behind that of v_{cd} even for a positive value of φ, resulting in negative power flow.

Figure 11.21 shows the high-frequency input current ripple for the proposed converter with fixed 50% modulation and −1 kW load. It is obvious that the

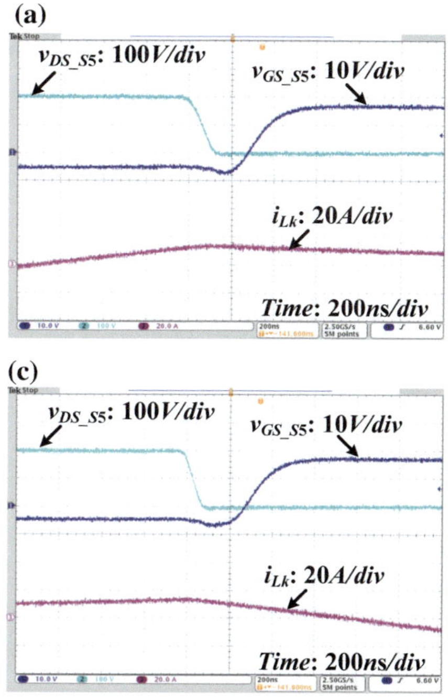

Fig. 11.18 Soft switching waveforms for S_5: **a** V_{in} = 24 V, P_o = 200 W. **b** V_{in} = 24 V, P_o = 1 kW. **c** V_{in} = 36 V, P_o = 200 W. **d** V_{in} = 36 V, P_o = 1 kW

Fig. 11.19 Efficiency curves of the converter for positive power flow

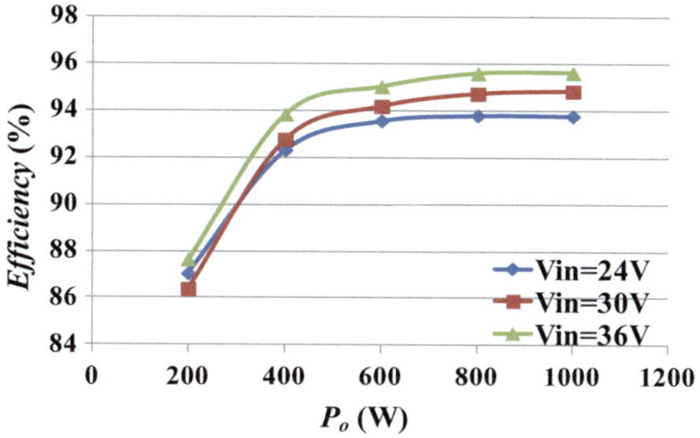

low-voltage side duty cycles are fixed as 50% in spite of input voltage and load variation. As a result, the high-frequency current ripple can be reduced to be nearly zero.

Figure 11.22 illustrates the turning-on process of low-voltage side upper switch Q_{1a} for different voltage and load conditions. As seen, the drain–source voltage of Q_{1a} has fallen to zero before the driving signal turns to high, so ZVS is achieved.

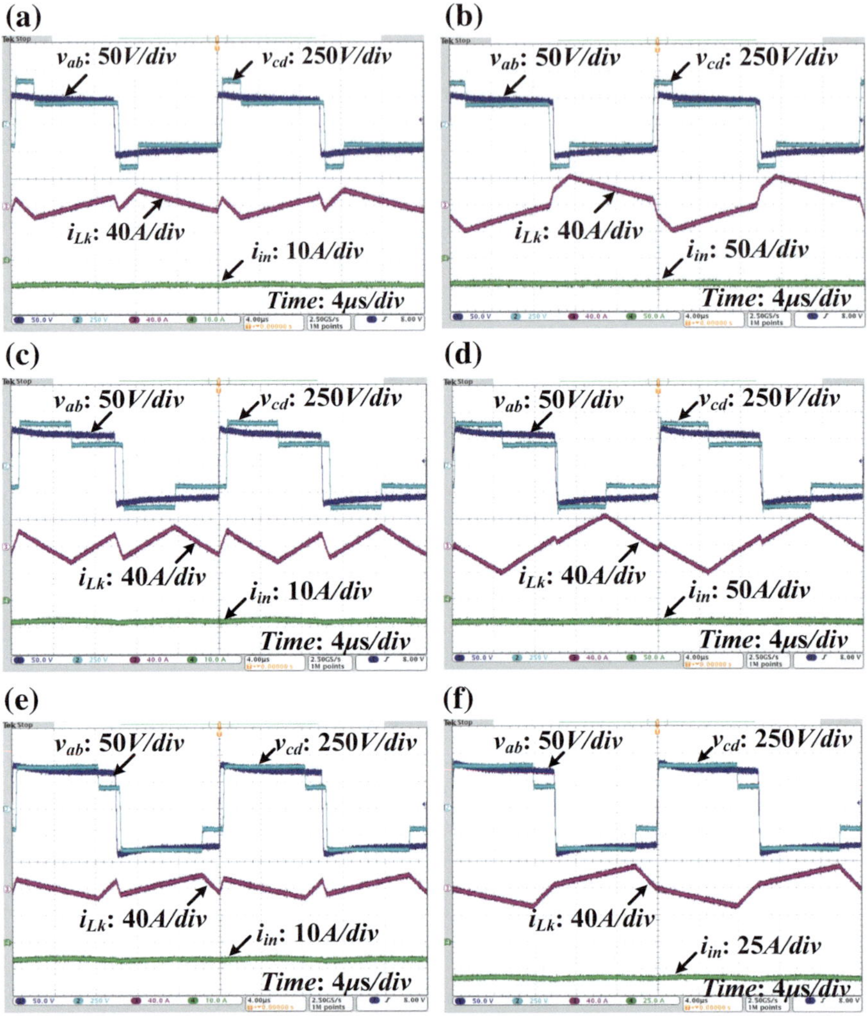

Fig. 11.20 Steady-state waveforms for negative power flow: **a** $V_{in} = 24$ V, $P_o = -200$ W. **b** $V_{in} = 24$ V, $P_o = -1$ kW. **c** $V_{in} = 30$ V, $P_o = -200$ W. **d** $V_{in} = 30$ V, $P_o = -1$ kW. **e** $V_{in} = 36$ V, $P_o = -200$ W. **f** $V_{in} = 36$ V, $P_o = -1$ kW

Besides, Q_1 can also achieve ZVS since Q_1 is easier than Q_{1a} to achieve ZVS under negative power flow conditions.

Figure 11.23 illustrates the ZVS process of S_1 under different voltage and load cases. It can be found out that S_1 can achieve ZVS within wide input voltage range and wide load range. It should be pointed out that the phase shift ratio φ is nearly

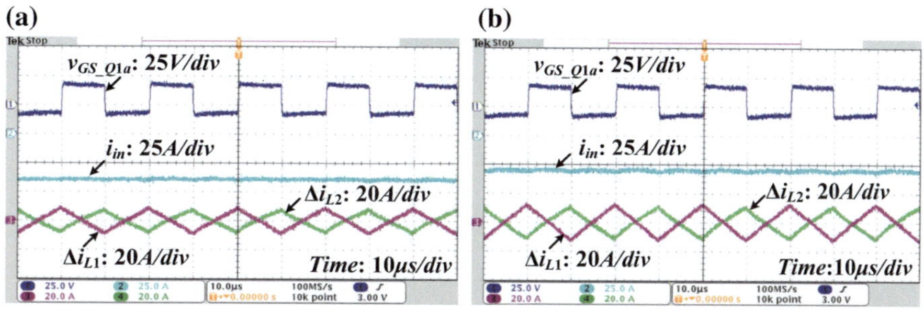

Fig. 11.21 Experiment results of high-frequency current ripple for proposed converter using 50% duty cycle interleaving operation: **a** V_{in} = 24 V, P_o = −1 kW. **b** V_{in} = 36 V, P_o = −1 kW

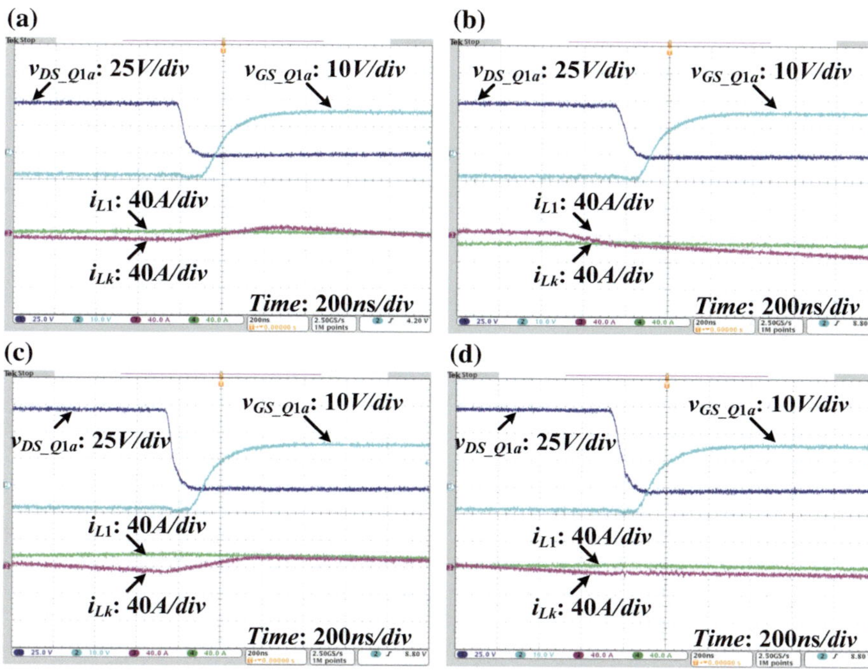

Fig. 11.22 Turning-on process for Q_{1a}: **a** V_{in} = 24 V, P_o = −200 W. **b** V_{in} = 24 V, P_o = −1 kW. **c** V_{in} = 36 V, P_o = −200 W. **d** V_{in} = 36 V, P_o = −1 kW

zero in case of 36 V and −1 kW, so the current i_{Lk} is very small when S_1 is turned on, which cannot discharge the junction capacitor of S_1 completely and oscillation occurs. This can be seen in Fig. 11.23d, but ZVS for S_1 is still achieved if neglecting the energy stored in the junction capacitor.

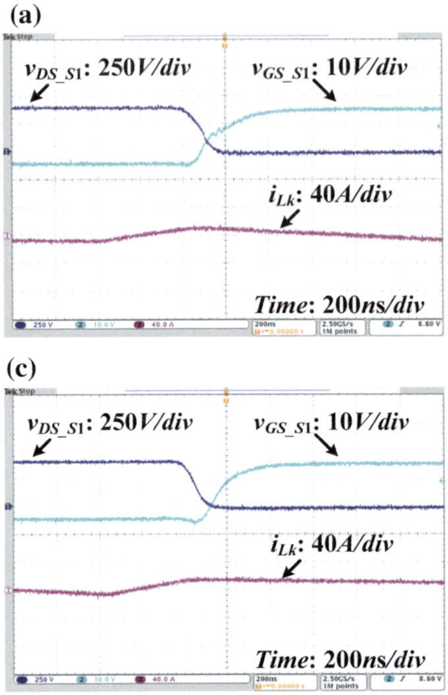

Fig. 11.23 Turning-on process for S_1: **a** $V_{in} = 24$ V, $P_o = -200$ W. **b** $V_{in} = 24$ V, $P_o = -1$ kW. **c** $V_{in} = 36$ V, $P_o = -200$ W. **d** $V_{in} = 36$ V, $P_o = -1$ kW

Figure 11.24 illustrates ZVS process of S_5. When S_5 is turned on, the leakage inductance current i_{Lk} is positive, which will discharge its output capacitance, so S_5 can achieve ZVS by the proposed control strategy within wide operation range.

The experimental results in this section show that the proposed converter can also work well within operation range for the negative power flow. ZVS for all switches can be ensured by the proposed control strategy and proper parameter design. Besides, high-frequency input current ripple can be reduced significantly.

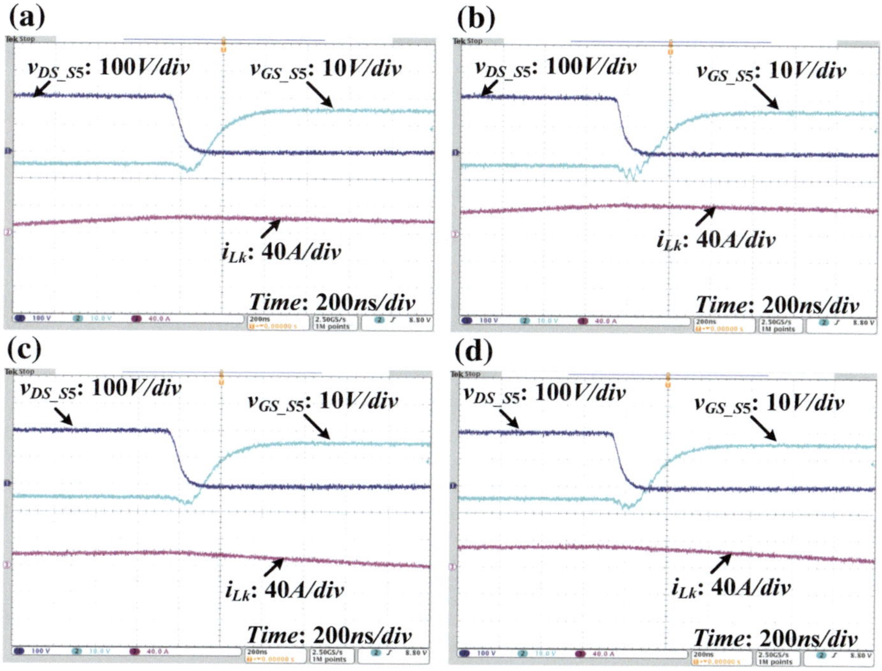

Fig. 11.24 Turning-on process for S_5: **a** V_{in} = 24 V, P_o = −200 W. **b** V_{in} = 24 V, P_o = −1 kW. **c** V_{in} = 36 V, P_o = −200 W. **d** V_{in} = 36 V, P_o = −1 kW

11.6 Conclusion

In this chapter, a novel current-fed hybrid dual active bridge DC–DC converter is proposed for fuel cell power conditioning system. For the proposed converter, duty cycles for all the input-side switches are fixed as 50% and gating signals of the dual interleaved boost circuits are 180° phase shifted. Thus, the converter with the proposed control can achieve less high-frequency input current ripple than that by using conventional interleaving technology with variable duty cycle control. In addition, by using the proposed control strategy, all the power switches can achieve ZVS within wide input voltage range and wide load range. The effectiveness of the proposed converter and control strategy has been verified by simulation and experimental results.

References

1. Wang S, Kenarangui Y, Fahimi B (2006) Impact of boost converter switching frequency on optimal operation of fuel cell systems. Paper presented at the 2006 IEEE vehicle power and propulsion conference, Windsor, 1–5 Sept 2006
2. Mazumder SK, Acharya K, Haynes CL (2004) Solid-oxide-fuel-cell performance and durability: resolution of the effects of power-conditioning systems and application loads. IEEE Trans Power Electron 19(5):1263–1278
3. Mazumder SK, Burra RK, Acharya K (2007) A ripple-mitigating and energy-efficient fuel cell power-conditioning system. IEEE Trans Power Electron 22(4):1437–1452
4. Rathore AK (2011) High-frequency soft-switching current-fed inverter for off-grid micro-generation: fuel cell application for rural electrification/development. In: Proceedings of the 14th European conference on power electronics and applications, Birmingham, 1–10 Aug 2011
5. Demetriades GD, Nee HP (2008) Characterization of the dual-active bridge topology for high-power applications employing a duty-cycle modulation. Paper presented at the 2008 IEEE power electronics specialists conference, Rhodes, June 2008, pp 2791–2798
6. Bai H, Mi C (2008) Eliminate reactive power and increase system efficiency of isolated bidirectional dual-active-bridge DC–DC converters using novel dual-phase-shift control. IEEE Trans Power Electron 23(6):2905–2914
7. Xiao H, Xie S (2008) A ZVS bidirectional DC–DC converter with phase-shift plus PWM control scheme. IEEE Trans Power Electron 23(2):813–823
8. Sha D, You F, Wang X (2016) A high efficiency current fed semi dual active bridge DC–DC converter for low input voltage applications. IEEE Trans Ind Electron 63(4):2155–2164

Chapter 12
Dynamic Response Improvements of Parallel-Connected Bidirectional DC–DC Converters

Abstract Parallel-connected modular current-fed bidirectional DC–DC converters are used for the AC motor drive system powered by batteries with low voltage and wide voltage range. The input current ripple can be reduced significantly by employing interleaving technology not only for individual module but also for all the modules. A current-sharing control strategy is applied for the constituent modules. Double pulse width modulation plus double phase shifted control with equal duty cycles for one module can minimize the circulation loss and avoid non-active power issue. Factors affecting dynamic performance are investigated based on the small-signal modeling. The leakage inductance value is optimized in view of system reliability and better dynamic performance. Besides, to improve the dynamic performance further, feed-forward control employing optimized feed-forward coefficient based on the small-signal analysis is implemented. A 4 kW prototype composed of two bidirectional DC–DC converters is built to verify the effectiveness for the proposed control strategy in AC motor drive application with fast regenerative braking.

Keywords Parallel-connected · DC–DC converter · Current-sharing Dynamic performance · Feed-forward control · Motor drive

12.1 The Drive System Overview and DPDPS Control

In the low-voltage and high-power rating applications, it is not good for one module to handle such large current at LVS. Parallel-connected modular power converters can be a good solution [1]. A bidirectional interleaved current-fed converter was proposed for energy storage system [2]. Current sharing can be achieved by common phase shift [3] or common duty cycle [4] control for parallel-connected bidirectional DC–DC converters [5] or unidirectional DC–DC converters. The unidirectional is not really modular configuration due to the connection of the rectifier stage. The dynamic performance may be limited due to the lack of current loops.

D. Sha and G. Xu, *High-Frequency Isolated Bidirectional Dual Active Bridge DC–DC Converters with Wide Voltage Gain*, CPSS Power Electronics Series,
https://doi.org/10.1007/978-981-13-0259-6_12

Besides, steady-state performances of the DAB, such as circulation loss and non-active current reduction, soft switching for LVS side switches, have been discussed extensively in previous works [6, 7]. The equivalent leakage inductance value has been designed in an optimal way to achieve ZVS, but its effect on the dynamic performance is seldom addressed. Actually, dynamic performance is also important for DAB converters especially for electrical motor drive application which requires fast dynamic response. During the motor acceleration stage, the voltage of the DC bus will drop and may cause under voltage protection for the inverter. On the other hand, the DC bus voltage may increase sharply during the braking stage. Thus, fast dynamic performance must be obtained in electrical drive application where the motor acceleration and braking occur frequently.

To obtain fast dynamic performance, the small-signal analysis is widely used [8–10]. State-space averaging technique can be used to derive the small-signal model, and a closed two-loop controller is designed for current-fed DAB to obtain good transient performance [11]. To improve the dynamic response and suppress the second harmonics further, feed-forward control is used for unidirectional DC–DC converters employing PWM control [12–16]. In comparison, for bidirectional DAB DC–DC converters, the switches are not only PWM modulated but also phase shifted, and its small-signal model is very complicated since there are more subintervals during one entire switching cycle. The output impedance expression of the DAB DC–DC is more complicated as well.

The basic block diagram of the motor drive system is shown in Fig. 12.1. The system is composed of a battery pack, a DC–DC converter, a DC–AC inverter, and a motor. The DC–DC converter consists of two parallel-connected CF-DAB DC–DC modules. The LVS is connected to the battery pack whose voltage range is 18–28 V. The HVS is connected to the DC–AC inverter, and its rated voltage is 300 V/DC.

The proposed topology is shown in Fig. 12.2. For the HVS, a full bridge is made of S_1–S_4. In the LVS, L_1 and L_2 are the DC inductors. Q_1 and Q_2 are bottom MOSFETs while Q_{1a} and Q_{2a} are top switches. The equivalent leakage inductance L_r represents the sum of the external AC inductance and the primary-referred transformer leakage inductance.

To minimize the non-active power and peak current, PWM control can be employed just like the LVS with the same duty cycle for the HVS. Thus, both v_{ab} and v_{de} have the same shape but they are phase shifted according to the transferred

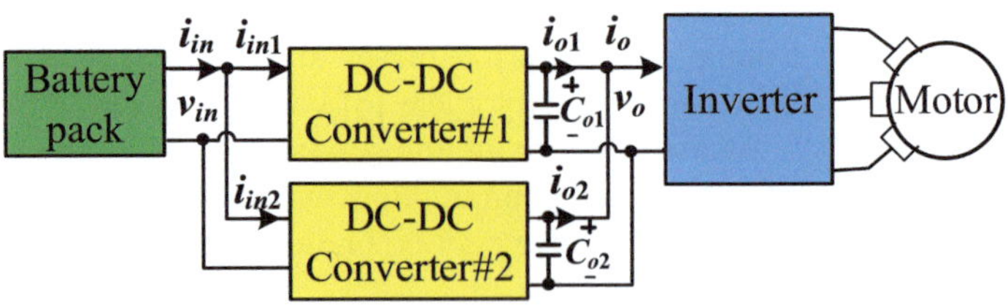

Fig. 12.1 Basic block diagram of the motor drive system

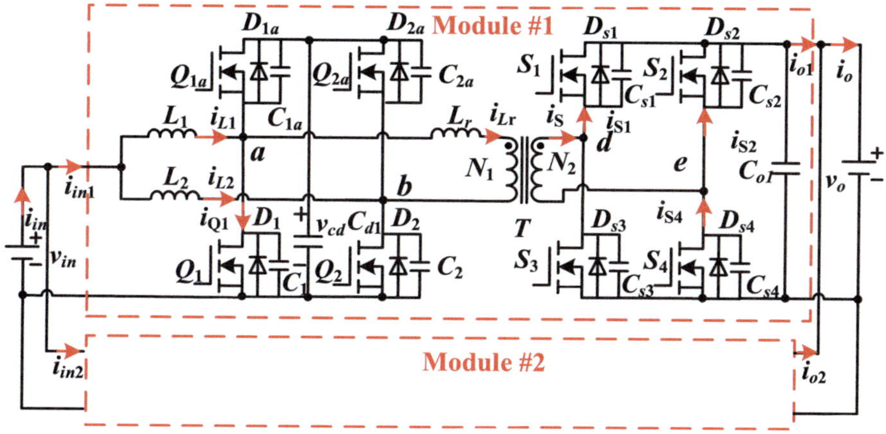

Fig. 12.2 Parallel connection of CF-DAB DC–DC converters

power. The steady-state waveforms in boost and buck mode of module #1 during one complete period can be seen in Fig. 12.3. With the interleaving PWM modulation not only for one module but also for two modules, the LVS current ripple can be minimized, which can be seen in Fig. 12.3c. Figure 12.3a illustrates the operating waveforms in boost mode. The LVS top and bottom switches are gated with complementary PWM signals. The PWM duty cycles for LVS bottom switches Q_1 and Q_2 are equal, but their PWM gating signals are interleaved with each other. φ is defined as the phase shift angle between rising edges of Q_2 and S_1. The voltage v_{ab} is high-frequency alternating voltage. The voltage v_{ab} leads v_{de}. During the circulation stage [θ_3, θ_4], the leakage current drops to zero and stays at zero within the circulation stage range [θ_3, θ_4] until the power transfer stage begins. Thus, during the circulation stage, the conduction loss can be reduced. ZVS can be ensured for the LVS switches in spite of the power flow direction [5]. For the HVS, the lagging leg S2 and S4 can achieve ZVS on while the leading leg switches S1 and S3 can obtain ZCS on. In buck mode as shown in Fig. 12.3b, for the HVS switches, the lagging leg switches S2 and S4 can achieve ZCS while the leading leg switches S1 and S3 can obtain ZVS on. It is worth noting, during one complete cycle, there is no non-active power by neglecting the effect of the magnetizing inductance. Figure 12.3c plots the battery side current curves by using the interleaving technology not only for each module but also between the two modules. In this way, the battery side current ripple frequency is four times the switching frequency and the current ripple can be reduced significantly.

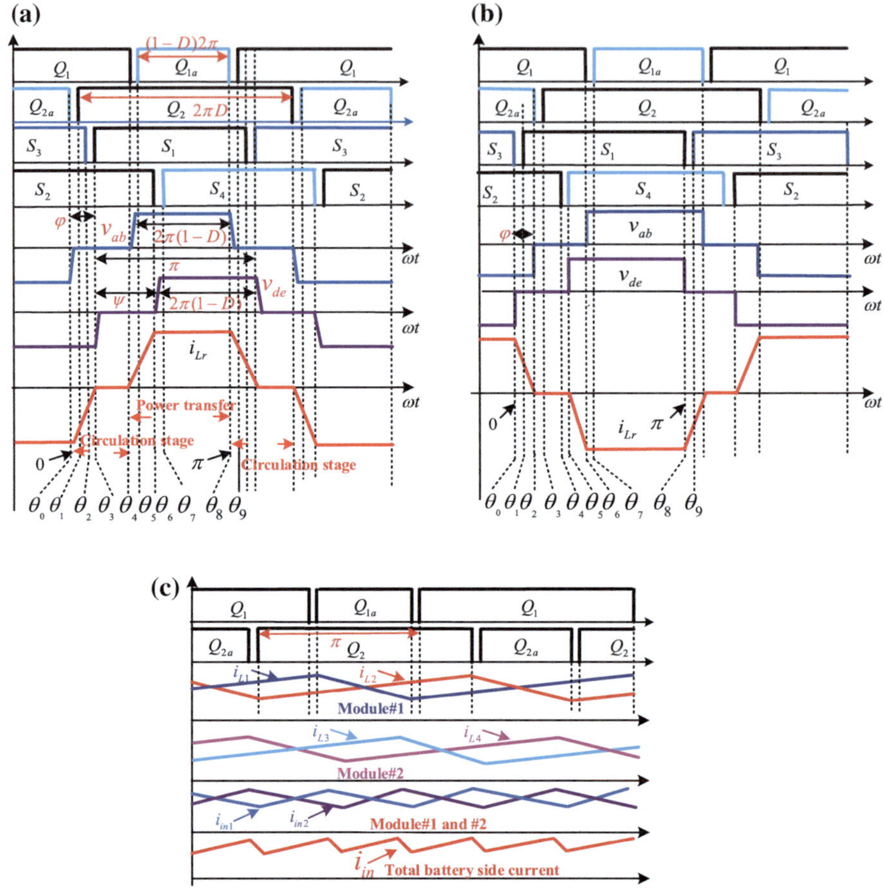

Fig. 12.3 Steady-state waveforms of DPDPS with identical duty cycles in boost and buck mode: **a** boost mode for module #1, **b** buck mode for module #1, **c** battery side current reduction by interleaving technology

12.2 Current-Sharing and Small-Signal Modeling

12.2.1 Implementation of the Current Sharing

The control block diagram for two modules in parallel connection is shown in Fig. 12.4. As can be seen in Fig. 12.4, the current-sharing control scheme consists of one common output voltage loop and two inner current loops. The sum of the voltage compensation loop output and the feed-forward current provides the common reference for both the individual inner current loops. Therefore, current sharing between the two modules can be obtained. For individual modules, the

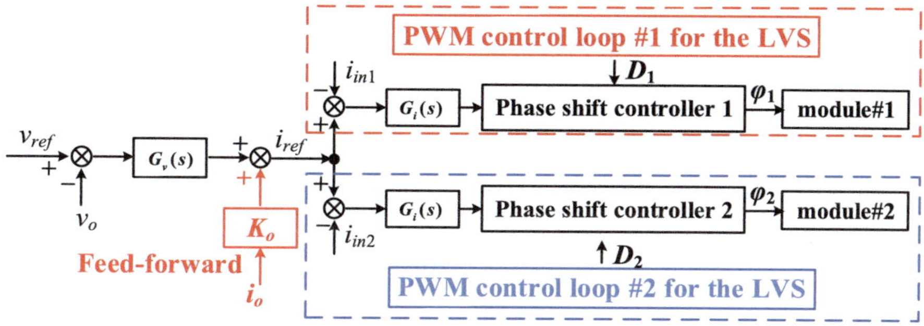

Fig. 12.4 Current-sharing control for the two modules

control is composed of a phase shift control and a PWM control loop [5]. It should be noted the high-voltage current i_o is sensed for the feed-forward control.

12.2.2 Small-Signal Modeling

To simplify the analysis, the dead time effect is ignored. There are eight operation modes in a switching cycle. Figure 12.5 shows the simplified key waveforms of all the modes in a complete switching cycle for module #1.

Fig. 12.5 Simplified waveforms in a switching cycle

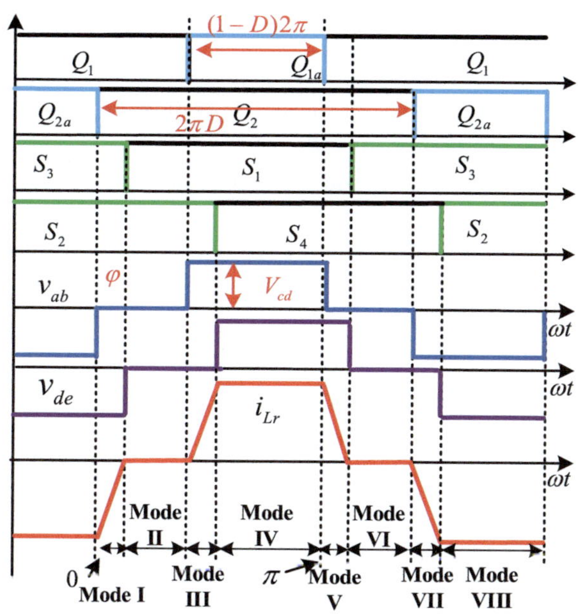

As can be seen in Fig. 12.2, the values i_{L1}, i_{L2}, and v_{cd} are viewed as state variables; d is the duty cycle for bottom switches Q_1 and Q_2. φ is the phase shift angle; v_{in} is the LVS voltage; v_o is the HVS voltage; $\varphi = \Phi + \hat{\varphi}$; $d = D + \hat{d}$; the initial currents in L_r, L_1 and L_2 are I_s, I_{L1} and I_{L2}, respectively. The turns ratio of the transformer is $N_1{:}N_2 = N$.

The small-signal model of each module can be derived as follows

Mode I $[0, \varphi]$	$\begin{cases} c_d \frac{dv_{cd}}{d_t} = 0, L_1 \frac{di_{L1}}{d_t} = v_{\text{in}}, L_2 \frac{di_{L2}}{d_t} = v_{\text{in}} \\ i_{Lr} = \frac{\langle v_{cd}\rangle_{Ts}}{\omega L_r}(\theta - \varphi) \end{cases}$
Mode II $[\varphi, (2d-1)\pi]$	$c_d \frac{dv_{cd}}{d_t} = 0, L_1 \frac{di_{L1}}{d_t} = v_{\text{in}}, L_2 \frac{di_{L2}}{d_t} = v_{\text{in}}, i_{Lr} = 0$
Mode III $[(2d-1)\pi, (2d-1)\pi + \varphi]$	$\begin{cases} c_d \frac{dv_{cd}}{d_t} = i_{L1} - i_{Lr}, L_1 \frac{di_{L1}}{d_t} = v_{\text{in}} - v_{cd}, \\ L_2 \frac{di_{L2}}{d_t} = v_{\text{in}}, i_{Lr} = \frac{\langle v_{cd}\rangle_{T_s}}{\omega L_r}[\theta - (2d-1)\pi] \end{cases}$
Mode IV $[(2d-1)\pi + \varphi, \pi]$	$\begin{cases} c_d \frac{dv_{cd}}{d_t} = i_{L1} - i_{Lr}, L_1 \frac{di_{L1}}{d_t} = v_{\text{in}} - v_{cd} \\ L_2 \frac{di_{L2}}{d_t} = v_{\text{in}}, i_{Lr} = \frac{\langle v_{cd}\rangle_{Ts}}{\omega L_r}\varphi \end{cases}$
Mode V $[\pi, \pi + \varphi]$	$\begin{cases} c_d \frac{dv_{cd}}{d_t} = 0, L_1 \frac{di_{L1}}{d_t} = v_{\text{in}}, L_2 \frac{di_{L2}}{d_t} = v_{\text{in}}, \\ i_{Lr} = \frac{\langle v_{cd}\rangle_{Ts}}{\omega L_r}[\pi - \theta + \varphi] \end{cases}$
Mode VI $[\pi + \varphi, 2d\pi]$	$c_d \frac{dv_{cd}}{d_t} = 0, L_1 \frac{di_{L1}}{d_t} = v_{\text{in}}, L_2 \frac{di_{L2}}{d_t} = v_{\text{in}}, i_{Lr} = 0$
Mode VII $[2d\pi, 2d\pi + \varphi]$	$\begin{cases} c_d \frac{dv_{cd}}{d_t} = i_{L2} + i_{Lr}, L_1 \frac{di_{L1}}{d_t} = v_{\text{in}}, \\ L_2 \frac{di_{L2}}{d_t} = v_{\text{in}} - v_{cd}, i_{Lr} = -\frac{\langle v_{cd}\rangle_{T_s}}{\omega L_r}(\theta - 2d\pi) \end{cases}$
Mode VIII $[2d\pi + \varphi, 2\pi]$	$\begin{cases} c_d \frac{dv_{cd}}{d_t} = i_{L2} + i_{Lr}, L_1 \frac{di_{L1}}{d_t} = v_{\text{in}}, \\ L_2 \frac{di_{L2}}{d_t} = v_{\text{in}} - v_{cd}, i_{Lr} = -\frac{\langle v_{cd}\rangle_{Ts}}{\omega L_r}\varphi \end{cases}$

where $\langle v_{cd}\rangle$ Ts denotes the average of v_{cd} over an interval of length T_s: $\langle v_{cd}\rangle_{Ts} = \left(\int_t^{t+T_s} v_{cd}\,d\tau\right)/T_s$.

Then, the average model can be derived as follows

$$\begin{cases} c_d \frac{d\hat{v}_{cd}}{d_t} = (1-D)\hat{i}_{L1} + (1-D)\hat{i}_{L2} - \frac{2V_{cd}}{\omega L_r}\left(1 - D - \frac{\Phi}{2\pi}\right)\hat{\varphi} \\ \quad -\left(I_{L1} + I_{L2} - \frac{2V_{cd}\Phi}{\omega L_r}\right)\hat{d} - \frac{2\left[(1-D)\Phi - \frac{\Phi^2}{4\pi}\right]}{\omega L_r}\hat{v}_{cd} \\ L_1 \frac{d\hat{i}_{L1}}{d_t} = \hat{v}_{\text{in}} - (1-D)\hat{v}_{cd} + V_{cd}\hat{d} \\ L_2 \frac{d\hat{i}_{L2}}{d_t} = \hat{v}_{\text{in}} - (1-D)\hat{v}_{cd} + V_{cd}\hat{d} \end{cases} \tag{12.1}$$

The transfer function of the phase angle perturbation $\hat{\varphi}$ to the two DC inductor current perturbations $\hat{i}_{L1}$ and $\hat{i}_{L2}$ can be expressed as the following, respectively,

$$G_{i_{L1}\varphi}(s) = \left.\frac{\hat{i}_{L1}(s)}{\hat{\varphi}(s)}\right|_{\substack{\hat{d}(s)=0\\ \hat{v}_{\text{in}}(s)=0\\ \hat{v}_{\text{o}}(s)=0\\ \hat{i}_{L2}(s)=0}} = \frac{2(1-D)(1-D-\frac{\Phi}{2\pi})V_{cd}}{\omega L_1 L_r C_d s^2 + 2[(1-D)\Phi - \frac{\Phi^2}{4\pi}]L_1 s + (1-D)^2\omega L_r} \tag{12.2}$$

$$G_{i_{L2}\varphi}(s) = \left.\frac{\hat{i}_{L2}(s)}{\hat{\varphi}(s)}\right|_{\substack{\hat{d}(s)=0\\ \hat{v}_{\text{in}}(s)=0\\ \hat{v}_{\text{o}}(s)=0\\ \hat{i}_{L1}(s)=0}} = \frac{2(1-D)(1-D-\frac{\Phi}{2\pi})V_{cd}}{\omega L_2 L_r C_d s^2 + 2[(1-D)\Phi - \frac{\Phi^2}{4\pi}]L_2 s + (1-D)^2\omega L_r} \tag{12.3}$$

Supposing that $L_1 = L_2 = L$, $C_{d1} = C_d = C$, $V_{cd} = NV_{\text{o}}$, the transfer function from $\hat{\varphi}$ to $\hat{i}_{\text{in}}$ (input current perturbation of module #1) can be written by

$$G_{i\_\varphi\#1}(s) = \frac{\hat{i}_{\text{in}1}(s)}{\hat{\varphi}_1(s)} = \frac{4(1-D)(1-D-\frac{\Phi}{2\pi})NV_{\text{o}}}{\omega L L_r C s^2 + 2[(1-D)\Phi - \frac{\Phi^2}{4\pi}]Ls + (1-D)^2\omega L_r} \tag{12.4}$$

According to the power conservation $v_{\text{o}}i_{\text{o}} = v_{\text{in}}i_{\text{in}}$, the perturbation of the output voltage is written by

$$\hat{v}_{\text{o}} = \frac{V_{\text{in}}\hat{i}_{\text{in}} + \hat{v}_{\text{in}}I_{\text{in}} - V_{\text{o}}\hat{i}_{\text{o}}}{I_{\text{o}}} \tag{12.5}$$

The complete block diagram of the regulator system on the small-signal model of module #1 is shown in Fig. 12.6. K_o is the feed-forward control coefficient. The output voltage variations $\hat{v}_{\text{o}}$ can therefore be expressed as a linear combination of three independent inputs: the input voltage variations $\hat{v}_{\text{in}}$, the input current variations $\hat{i}_{\text{in}}$, and the load current variations $\hat{i}_{\text{o}}$, where

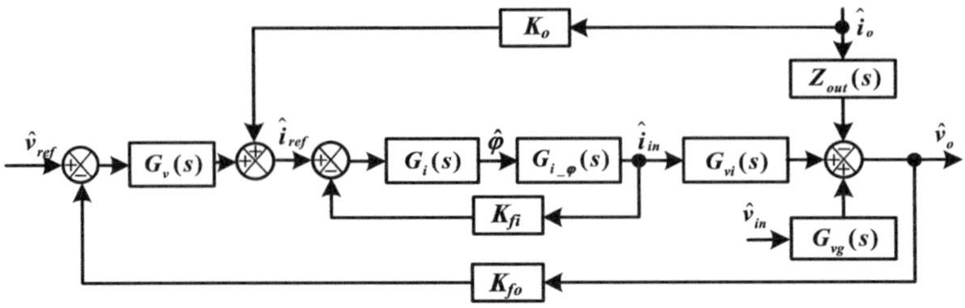

Fig. 12.6 Complete block diagrams on the small-signal model

$$\begin{aligned} G_v(s) = K_{p\_v} + \frac{K_{i\_v}}{s}, G_i(s) = K_{p\_i} + \frac{K_{i\_i}}{s}, \\ G_{vi}(s) = \frac{V_{\text{in}} Z_{\text{out}}}{V_{\text{o}}}, G_{vg}(s) = \frac{I_{\text{in}} Z_{\text{out}}}{V_{\text{o}}} \end{aligned} \tag{12.6}$$

As can be seen in the control system block diagram shown in Fig. 12.6, when K_o is 0, the converter output impedance of the closed loop is written by

$$\begin{aligned} Z_{\text{o1}}(s) &= \frac{\hat{v}_{\text{o}}(s)}{-\hat{i}_{\text{o}}(s)}\bigg|_{\substack{\hat{v}_{\text{in}}=0 \\ \hat{v}_{\text{ref}}=0}} = \frac{Z_{\text{out}}(s)}{1+T(s)}, T(s) = \frac{K_{fo} G_v(s) G_{vi}(s) G_i(s) G_{i\_\varphi}(s)}{1+K_{fi} G_i(s) G_{i\_\varphi}(s)} \\ &= \text{“loop gain''} \end{aligned} \tag{12.7}$$

where the loop gain $T(s)$ is defined in general as the product of the gains around the forward and feedback paths of the loop. If K_o is not zero, the feed-forward control works and the output impedance of the closed loop can be expressed as

$$\left.\begin{aligned} Z_{\text{o2}}(s) &= \frac{\hat{v}_{\text{o}}(s)}{-\hat{i}_{\text{o}}(s)}\bigg|_{\substack{\hat{v}_{\text{in}}=0 \\ \hat{v}_{\text{ref}}=0}} = \frac{Z_{\text{out}}(s) - \frac{K_o T(s)}{K_{fo} G_v(s)}}{1+T(s)} = \frac{Z_{\text{out}}(s)(1-M(s))}{1+T(s)} \\ M(s) &= \frac{K_o T(s)}{K_{fo} G_v(s) Z_{\text{out}}(s)} = \frac{K_o G_i(s) G_{i\_\varphi}(s)}{(V_{\text{o}}/V_{\text{in}})[1+K_{fi} G_i(s) G_{i\_\varphi}(s)]} \end{aligned}\right\} \tag{12.8}$$

12.2.3 Analysis of the Current Sharing

According to (12.4), for module #2, this transfer function can be expressed as

$$\begin{aligned} G_{i\_\varphi\#2}(s) &= \frac{\hat{i}_{\text{in2}}(s)}{\hat{\varphi}_2(s)} \\ &= \frac{4(1-D_s)(1-D_s-\frac{\Phi_s}{2\pi})N_s V_{\text{o}}}{\omega L_s L_{rs} C_s s^2 + 2[(1-D_s)\Phi_s - \frac{\Phi_s^2}{4\pi}]L_s s + (1-D_s)^2 \omega L_{rs}} \end{aligned} \tag{12.9}$$

And define the following

$$A_1 = \frac{a_1}{b_1 s^2 + c_1 s + d_1} = G_{i\_\varphi\#1}(s),$$
$$A_2 = \frac{a_2}{b_2 s^2 + c_2 s + d_2} = G_{i\_\varphi\#2}(s) \tag{12.10}$$

From the control loop in Fig. 12.6, the phase shift angle perturbations can be written as

$$\begin{aligned}&\hat{\varphi}_1(s) = E_1\hat{v}_o + F_1\hat{i}_o + H_1\hat{i}_{in1}, \hat{\varphi}_2(s) = E_2\hat{v}_o + F_2\hat{i}_o + H_2\hat{i}_{in2}\\ &E_1 = E_2 = -K_{fo}G_vG_i, F_1 = F_2 = K_oG_i,\\ &H_1 = H_2 = -K_{fi}(K_{p\_i} + \tfrac{K_{i\_i}}{s}) = \tfrac{ls+m}{s}\end{aligned} \tag{12.11}$$

where

$$l = -K_{fi}K_{p\_i}, \quad m = -K_{fi}K_{i\_i} \tag{12.12}$$

From (12.5) and (12.9–12.12), the transfer function from the total LVS current to the difference between the two LVS currents is written by

$$\frac{\hat{i}_{in1} - \hat{i}_{in2}}{\hat{i}_{in}} = \frac{(A_1E_1 - A_2E_2)V_{in}/I_o + (A_1F_1 - A_2F_2)V_{in}/V_o}{(1 - A_1H_1)(1 - A_2H_2)} \tag{12.13}$$

In combination with expressions of A_1, E_1, F_1, H_1, A_2, E_2, F_2, and H_2, the characteristic equation can be derived as

$$\begin{aligned}G(s) = {} & b_1b_2s^6 + (b_1c_2 + b_2c_1)s^5 + (c_1c_2 + b_1d_2 - b_1a_2l + b_2d_1 - b_2a_1l)s^4\\ & + (-a_1b_2m - a_2b_1m + c_2d_1 - a_1c_2l + c_1d_2 - a_2c_1l)s^3\\ & + (-a_1c_2m - a_2c_1m + d_1d_2 - a_1d_2l - a_2d_1l + a_1a_2l^2)s^2\\ & + (-a_1d_2m - a_2d_1m + 2a_1a_2lm)s + a_1a_2m^2\end{aligned} \tag{12.14}$$

As can be seen in (12.14), the feed-forward coefficient K_o does not exist in the characteristic equation. This implies that the system reliability is not affected by different feed-forward coefficient K_o. As can be seen in (12.14), $b_1b_2 > 0$. For the system to be stable, the first of the Routh array must have the same sign. So the proper selection of different compensator parameters can be got by (12.14).

12.2.4 System Stability Analysis

The system is stable when the load variation or source voltage variation is large. The root locus of a single current-fed DC–DC converter is given to show the stability. From the small-signal model block diagram illustrated in Fig. 12.6, the transfer function for one module can be derived as

$$\frac{\hat{v}_{o}(s)}{\hat{v}_{ref}(s)} = \frac{G_{vi}(s)G_{v}(s)G_{i}(s)G_{i\_\varphi}(s)}{K_{fo}G_{vi}(s)G_{v}(s)G_{i}(s)G_{i\_\varphi}(s) + K_{fi}G_{i}(s)G_{i\_\varphi}(s) + 1} \tag{12.15}$$

Then, the system characteristic equation can be obtained as

$$q_4 s^4 + q_3 s^3 + q_2 s^2 + q_1 s + q_0 = 0 \tag{12.16}$$

where

$$\begin{aligned}
q_4 &= \omega L L_r C_d \\
q_3 &= 2L[(1-D)\Phi - \frac{\Phi^2}{4\pi}] \\
q_2 &= 4K_{fo}\frac{V_{in}Z_{out}}{V_o}K_{p_v}K_{p_i}(1-D)(1-D-\frac{\Phi}{2\pi})NV_o \\
&\quad + 4K_{fi}K_{p_i}(1-D)(1-D-\frac{\Phi}{2\pi})NV_o + (1-D)^2\omega L_r \\
q_1 &= 4K_{fo}\frac{V_{in}Z_{out}}{V_o}(1-D)(1-D-\frac{\Phi}{2\pi})NV_o(K_{p_v}K_{i_i} + K_{i_v}K_{p_i}) \\
&\quad + 4K_{fi}K_{i_i}(1-D)(1-D-\frac{\Phi}{2\pi})NV_o \\
q_0 &= 4K_{fo}\frac{V_{in}Z_{out}}{V_o}(1-D)(1-D-\frac{\Phi}{2\pi})NV_o K_{i_v}K_{i_i}
\end{aligned}$$

Figure 12.7 illustrates the root locus curves when the output power P_o changes at v_{in}= 18 V based on (12.16) and the system specifications given in Table 12.1. The four roots of the system characteristic equation are illustrated in Fig. 12.7a. The two roots that are closed to the imaginary axis are overlapped, and they are shown in detail in Fig. 12.7b, c, and d. As can be seen, when the output power changes from 100 W to 2 kW, all the system poles are in the left-half plane. That means the system is stable at different loads.

Figure 12.8 illustrates the root locus curves when the output power P_o changes at v_{in}= 28 V. As can be seen, all the system poles are in the left-half plane, the system is stable at different loads at v_{in}= 28 V.

In conclusion, if the controller parameters are designed properly, the system is stable when v_{in} is in the range of 18–28 V and P_o is lower than the rated 4 kW output.

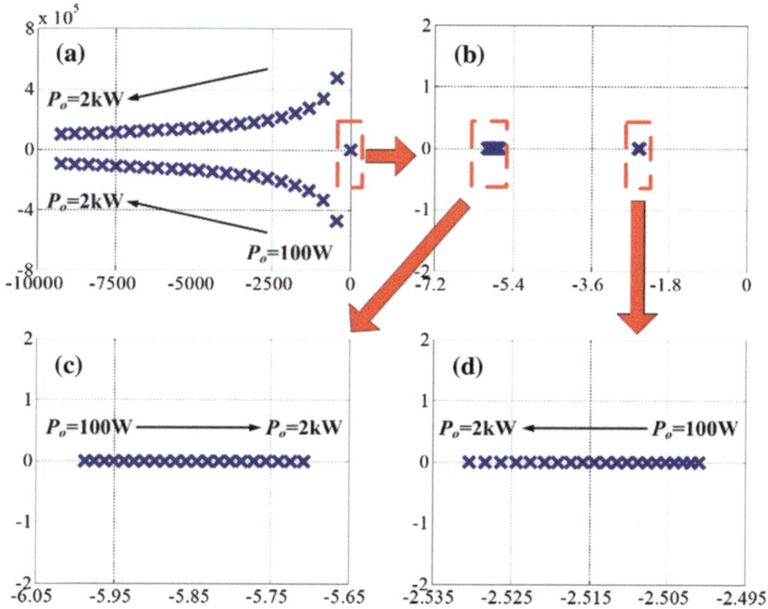

Fig. 12.7 Root locus when P_o changes at 18 V input

Table 12.1 Calculation specifications

Rated power P_o	4 kW	$L_1 = L_2 = L$	11 μH
HVS (v_o)	300 V	C_d, C_{o1}	30/22 μF
LVS (v_{in})	18–28 V	N_1:N_2	2:10
Switching f	50 kHz	L_r	1.0 μH

12.3 Feed-Forward Effect on the Dynamic Performance

12.3.1 Design of the Feed-Forward Coefficient K_o

According to (12.7–12.8), if $1 - M(s) = 0$, the output impedance is zero, so the feed-forward coefficient K_o should be

$$K_o = \frac{K_{fo}G_v(s)Z_{\text{out}}(s)}{T(s)} = \frac{(V_o/V_{\text{in}})[1 + K_{fi}G_i(s)G_{i\_\varphi}(s)]}{G_i(s)G_{i\_\varphi}(s)} \tag{12.17}$$

As can be seen in (12.17), the expression of feed-forward coefficient K_o is rather complicated and hard to be implemented if the output impedance is controlled to be zero ideally.

Actually, to simplify the design, the value of $1 - M(s)$ can be chosen to be a very small value. Figure 12.9 illustrates the magnitude curves for the closed-loop output impedance without and with the feed-forward using different coefficients. As

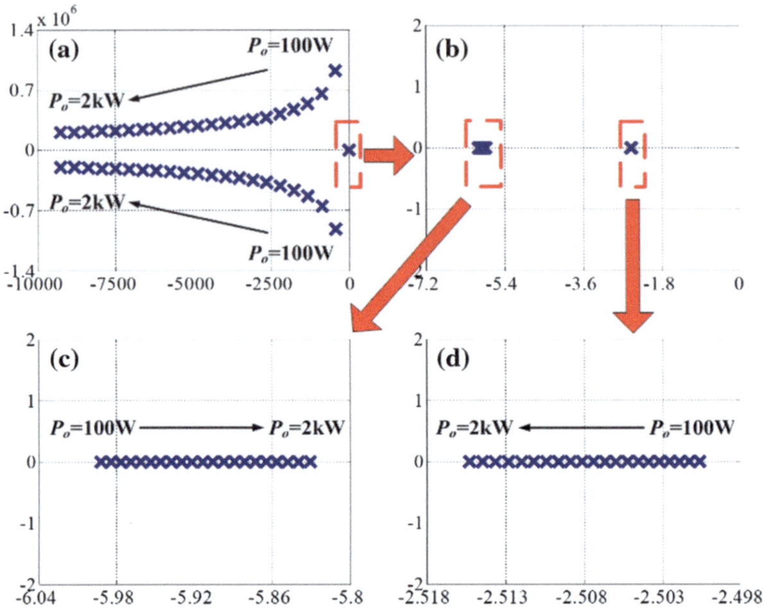

Fig. 12.8 Root locus when P_o changes at 28 V input

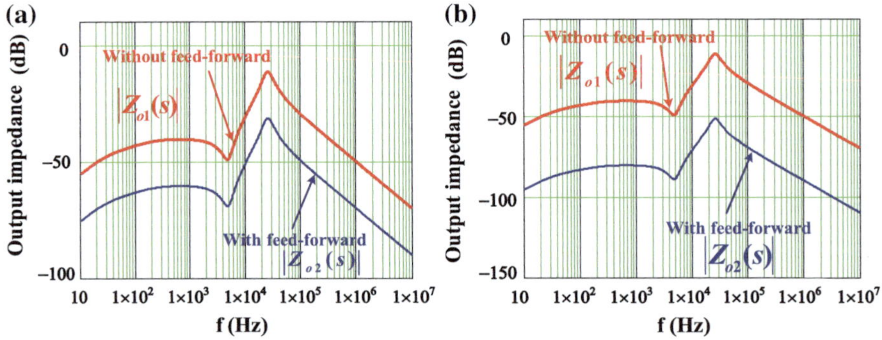

Fig. 12.9 Magnitude curves: **a** $1 - M(s) = 0.1$; **b** $1 - M(s) = 0.01$

can be seen in Fig. 12.9a and b, when the feed-forward is effective, the output impedance can be reduced dramatically. As illustrated, the output impedance drops as the value of $1 - M(s)$ decreases.

According to (12.8), the magnitude curve of 1 and $G_i(s)G_{i\_\varphi}(s)$ are shown in the Fig. 12.10. As seen, when the frequency is in the low-frequency range, the value $|G_i(s)G_{i\_\varphi}(s)| \gg |1|$. Because the load is the motor, the perturbation of the load current variations $\hat{i}_o$ mainly comes from the low-frequency range. So the feed-forward coefficient K_o can be optimized and selected approximately as

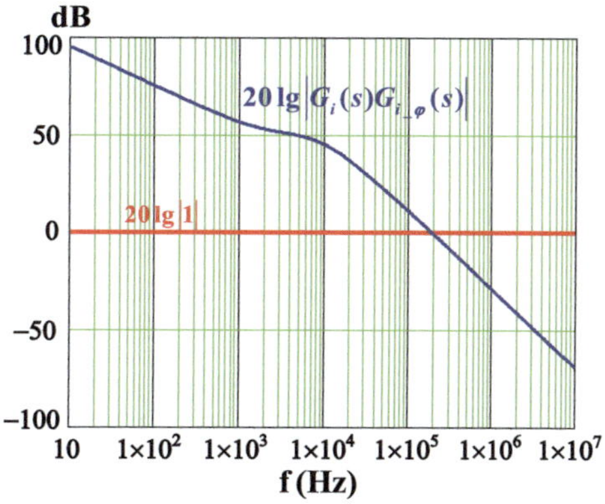

Fig. 12.10 Magnitude–frequency curve of 1 and $G_i(s)$ $G_{i\_\varphi}(s)$

$$K_o \approx (V_o/V_{in})K_{fi} = K_{op} \tag{12.18}$$

Substituting (12.18) into (12.8) yields

$$M(s) = \frac{K_o T(s)}{K_{fo} G_v(s) Z_{out}(s)} \approx \frac{K_{fi} G_i(s) G_{i\_\varphi}(s)}{1 + K_{fi} G_i(s) G_{i\_\varphi}(s)} = M_{op}(s) \tag{12.19}$$

12.3.2 Feed-Forward Effect on the Dynamic Performance

As can be seen in Fig. 12.11a, if K_o has been selected as the optimized value K_{op}, it is easy to be implemented in practical application. When the feed-forward control is effective, the output impedance can be reduced dramatically. As can be seen in Fig. 12.11b, if the feed-forward coefficient K_o is greater than or less than the optimal K_{op}, the output impedance is not the minimum in the low-frequency region. This shows that K_{op} is the optimal feed-forward coefficient. It is well known that the output voltage is less sensitive to the load current variation as output impedance becomes smaller. So the feed-forward loop can improve the dynamic performance.

12.3.3 Simulation Verification

Figure 12.12 shows the simulation results of one module in facing load transients and braking using the PSIM with and without the feed-forward control. As illustrated in Fig. 12.12a, better dynamic response in facing step load change from 200 W to 3 kW can be obtained with the feed-forward control. As can be seen in Fig. 12.12b, with the feed-forward control, the DC bus voltage spike can be

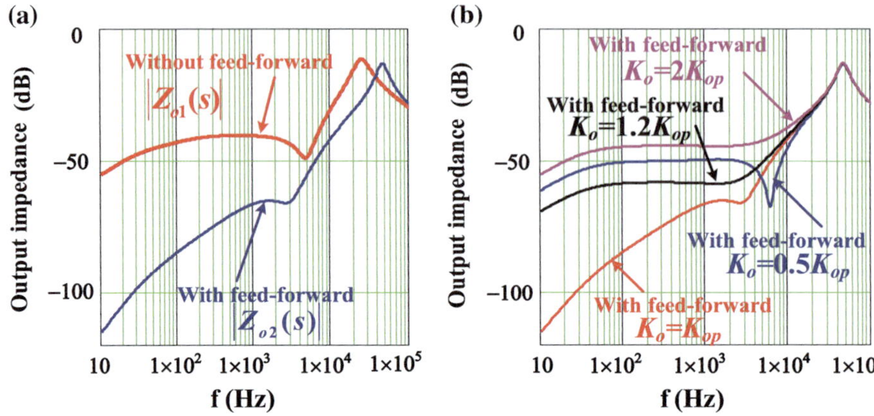

Fig. 12.11 Magnitude curves of output impedance Z_{o2}: **a** $M(s)$ = $M_{op}(s)$, **b** with different K_o

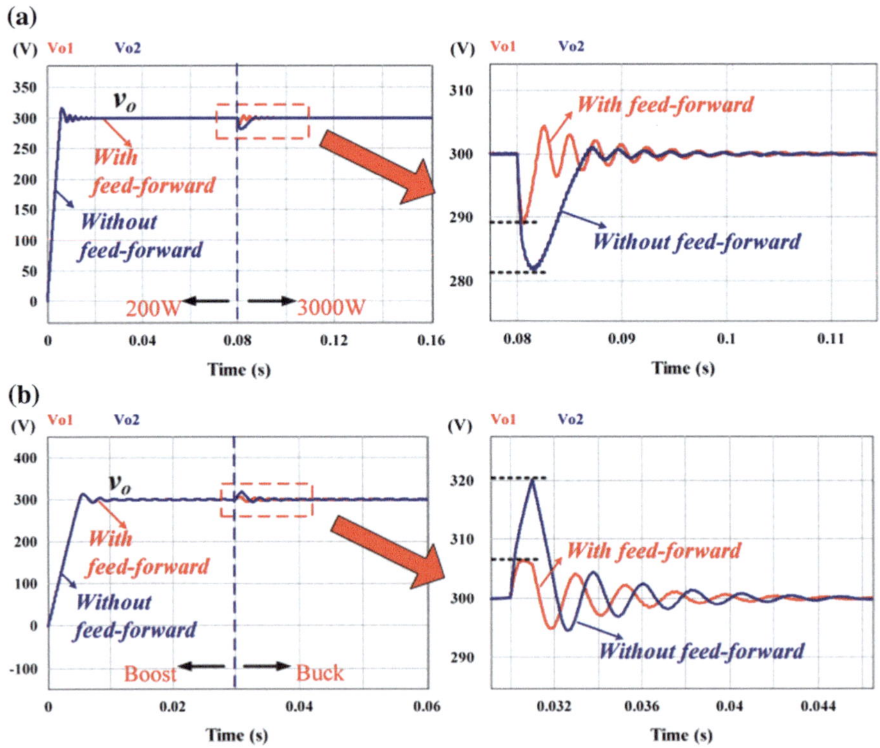

Fig. 12.12 Simulation results with feed-forward and without feed-forward: **a** load change from 200 W to 3 kW, **b** braking

suppressed compared to which without the feed-forward control. Thus, the feed-forward control can improve the system dynamic performance.

12.4 Leakage Inductance Effect on the Steady State and Dynamic Performance

12.4.1 Leakage Inductance Value Optimal Design and Its Effect on the Steady-State Performance

If the leakage inductance value is designed to be very small, leakage inductance current may be very high especially during the start-up procedure. Figure 12.13 shows the leakage inductance current curve during the start-up procedure since the output voltage is rather small initially. It should be noted that the worst case is that the duty cycle of the primary transformer voltage v_{ab} is 50% while the output voltage is zero.

As can be seen in Fig. 12.13, the voltage across the leakage inductance is that $L_r \cdot \mathrm{d}i/\mathrm{d}t = V_{cd} = N_1V_o/N_2 = 60$ V. The peak current can be expressed by

$$I_{\text{peak}} = \frac{V_{cd}}{L_r} \times \frac{T_s}{2} \times \frac{1}{2} = \frac{V_{cd}T_s}{4L_r} \tag{12.20}$$

Since the LVS switch Q_1 is two IXFN360N15T2 in parallel connection, the rated current of the MOSFET is 310 A. Thus, the maximum peak current is designed to be the half of the rating current

$$I_{\text{peak}} = \frac{V_{cd}T_s}{4L_r} < 310\ \text{A} \Rightarrow L_r > 0.96\ \mu\text{H} \tag{12.21}$$

Therefore, to suppress the rush current especially at the start-up process, the leakage inductance value L_r is designed no less than 1.0 μH. But different leakage inductance values may cause different current stress. As can be seen in Fig. 12.3, the leakage inductance current i_{Lr} RMS and peak values of one module can be written by

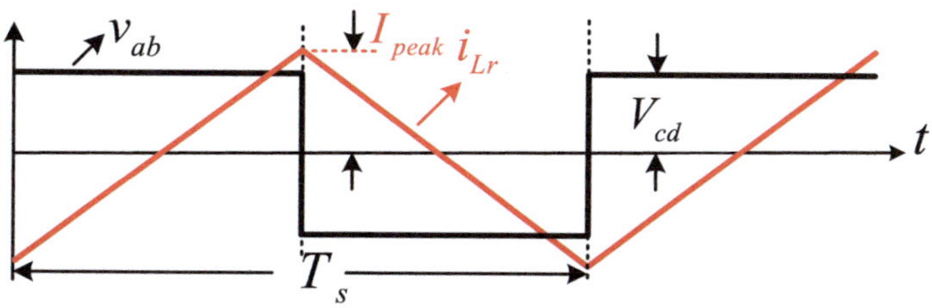

Fig. 12.13 Leakage inductance current during the start-up

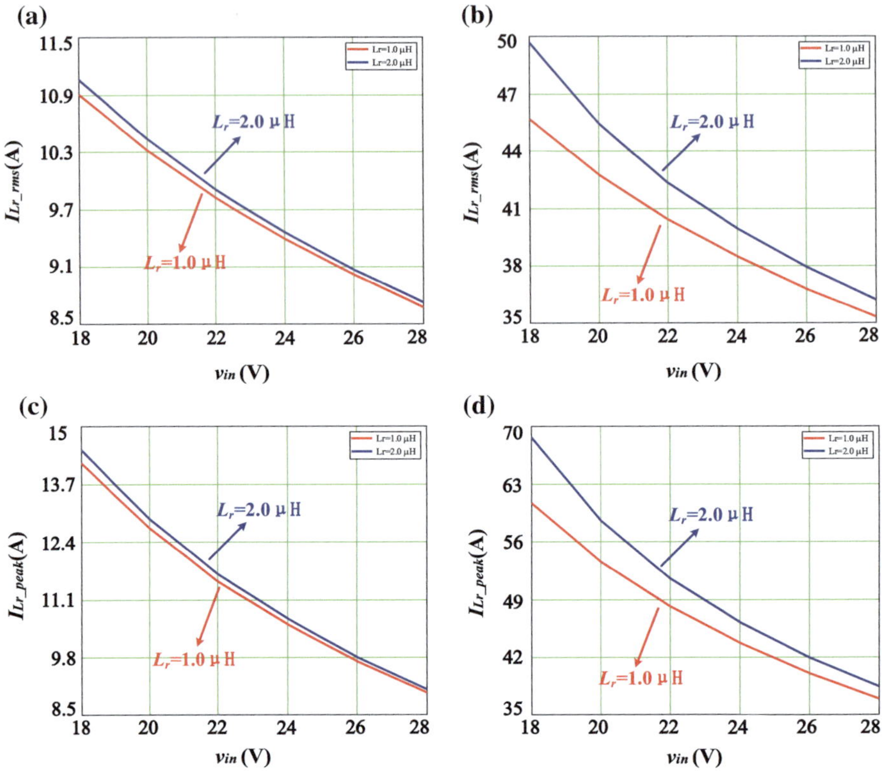

Fig. 12.14 Comparison of current i_{Lr} in boost mode: **a** RMS value at 500 W output, **b** RMS value at 2 kW output, **c** peak value at 500 W output, **d** peak value at 2 kW output

$$\begin{cases} I_{Lr\_\text{RMS}} = \sqrt{\frac{1}{2\pi}\int_0^{2\pi} i_{Lr}(\theta)^2 \text{d}\theta} = \frac{V_{cd}\varphi}{\omega L_r}\sqrt{\frac{6\pi - 6D\pi - \varphi}{3\pi}} \\ I_{Lr\_\text{peak}} = \frac{V_{cd}\varphi}{\omega L_r} \end{cases} \tag{12.22}$$

Based on (12.22), the RMS and peak values of current i_{Lr} versus the LVS voltage v_{in} in boost mode is illustrated in Fig. 12.14. As seen, when L_r = 1.0 μH, the RMS and peak values of the leakage inductance current can be reduced significantly under the same conditions including load, LVS voltage and HVS voltage.

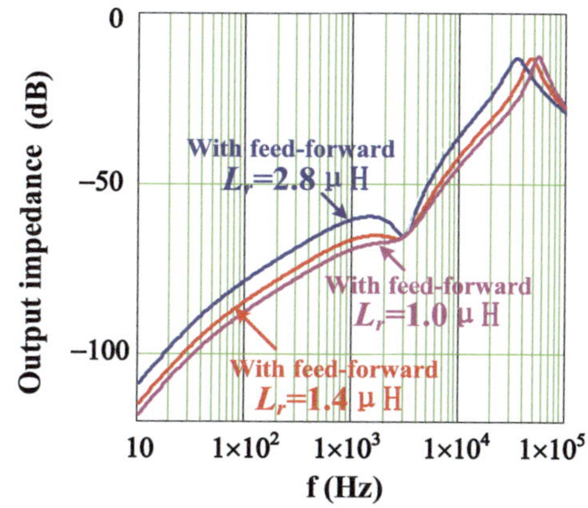

Fig. 12.15 Magnitude curves of output impedance Z_{o2} with different L_r

12.4.2 Feed-Forward Effect on the Dynamic Performance

According to (12.8) and (12.19), Fig. 12.15 gives the comparison of magnitude curves of the closed-loop output impedance at different leakage inductance. As can be seen in Fig. 12.15, as the equivalent leakage inductance value decreases, the output impedance value drops as well. It is well known that the output voltage is less sensitive to the load current variation as output impedance becomes smaller. Thus, the smaller the leakage inductance is, the better the dynamic performance is.

Figure 12.16 shows the simulation results of one module during the start-up process, load transients and braking stage using the PSIM with different leakage inductance values. In Fig. 12.16a, better dynamic response in facing step load change from 200 W to 3 kW can be obtained when the leakage inductance value is smaller. As can be seen in Fig. 12.16b, the DC bus voltage spike is lower during the braking stage with smaller leakage inductance value. Thus, the converter is preferred to operate with smaller leakage inductance to achieve better dynamic performance.

Figure 12.17 illustrates the circuit AC sweep results by PSIM simulation. As seen, the crossover frequency varies from 400 to 790 Hz as the leakage inductance value decreases from 2.0 to 1.0 μH. Thus, smaller leakage inductance leads to better dynamic performance.

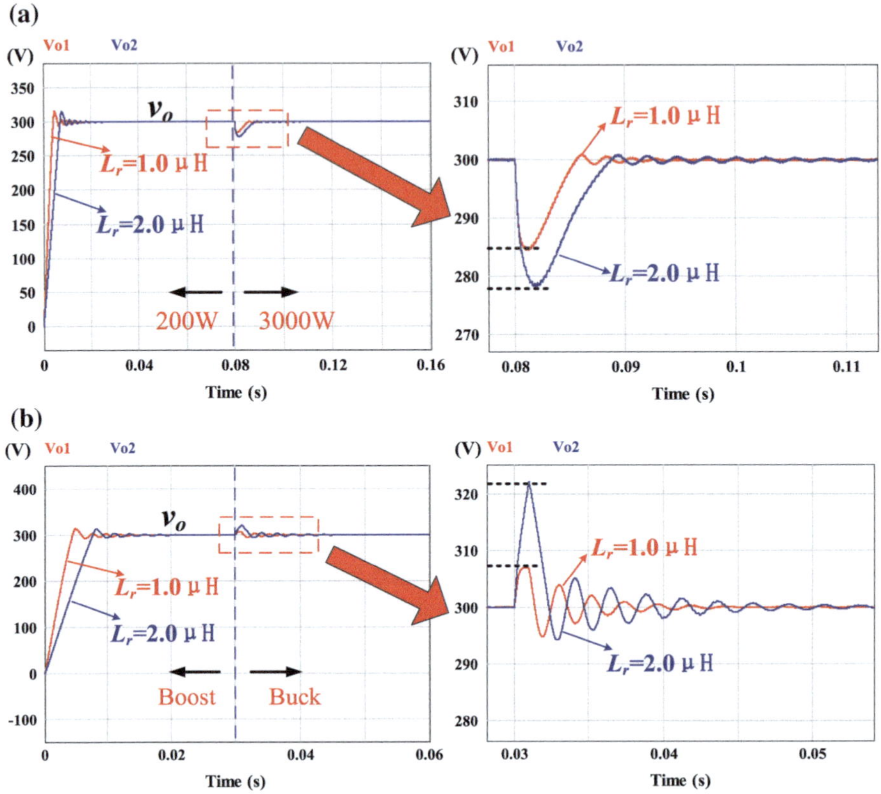

Fig. 12.16 Simulation results with L_r = 1.0 μH and L_r = 2.0 μH: **a** load change from 200 W to 3 kW, **b** braking

12.5 Experimental Verifications

12.5.1 *Prototype*

A 4 kW experimental prototype has been built in order to verify the effectiveness of the converter with the proposed control. Figure 12.18 gives the laboratory prototype picture.

12.5.2 *Steady-State Operation*

Figure 12.19 gives the experimental results of module #1 at rated load (2 kW) in boost mode with the minimum and maximum input voltages. As can be seen, with the proposed DPDPS control, the leakage inductance peak current is reduced, and

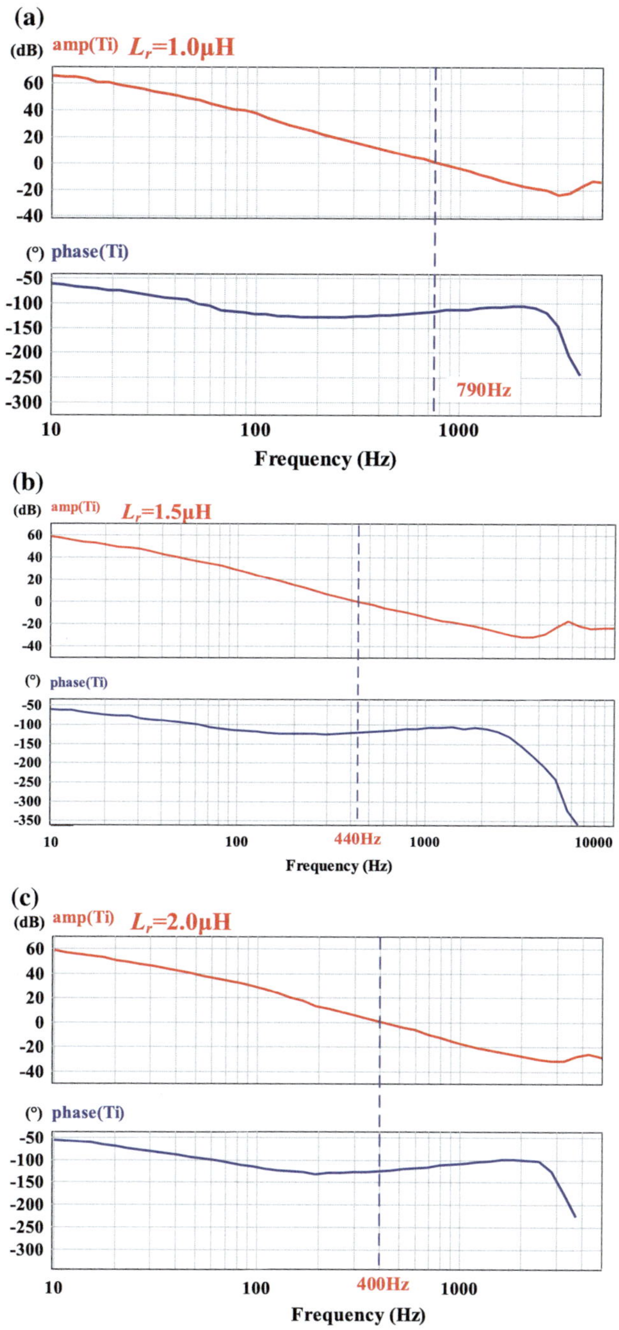

Fig. 12.17 Compensated voltage regulator loop gains using circuit AC sweep: **a** L_r = 1.0 μH, **b** L_r = 1.5 μH, **c** L_r = 2.0 μH

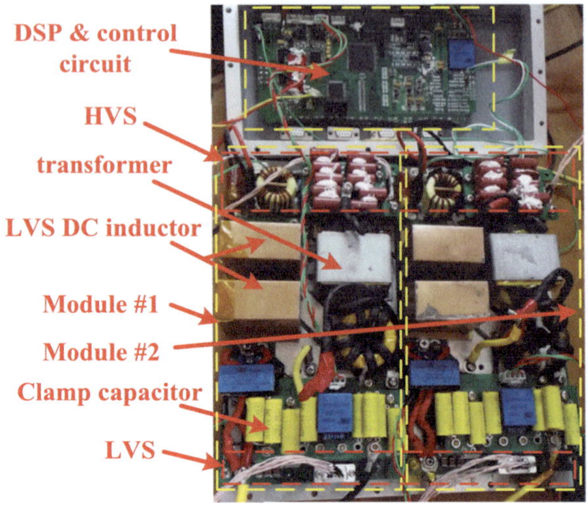

Fig. 12.18 Prototype picture

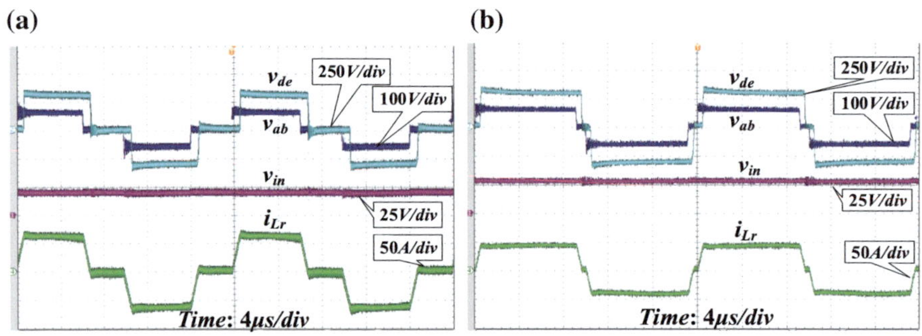

Fig. 12.19 Experimental results of module #1 with DPDPS control in boost mode at 2 kW output: **a** with v_{in}= 18 V; **b** with v_{in}= 28 V

the circulation loss can be minimized because the leakage inductance current remains as zero during the circulation stage.

Figure 12.20 gives the experimental results of module #1 at rated load (2 kW) in buck mode with the minimum and maximum output voltages. As can be seen, with the proposed DPDPS control, the leakage inductance peak current is reduced, and the circulation loss can be minimized.

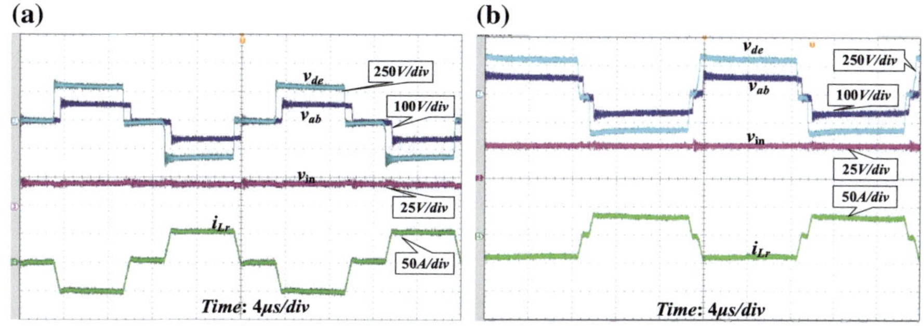

Fig. 12.20 Experimental results of module #1 with DPDPS control in buck mode at 2 kW output: **a** with v_{in} = 18 V; **b** with v_{in} = 28 V

12.5.3 Soft Switching Waveforms

With the proposed control strategy, all the switches on the LVS can achieve ZVS turn-on. For the HVS, half of the switches can be turned on at ZVS, while the other switches can be turned on at ZCS. The experimental results in Fig. 12.21 have verified the theoretical analysis.

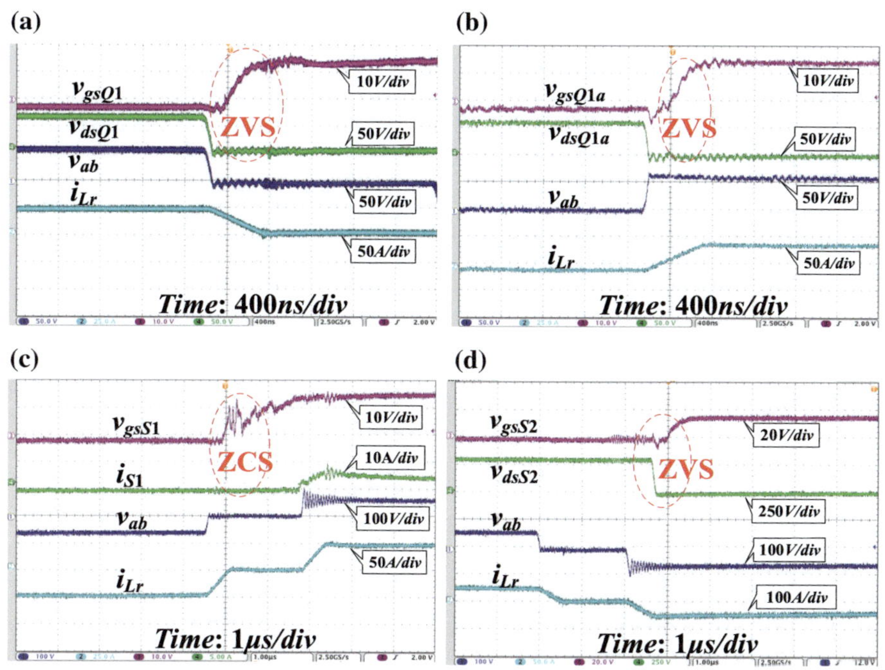

Fig. 12.21 Soft-switching waveforms of module #1 with DPDPS control in the boost mode at 24 V input and 2 kW output condition: **a** ZVS of Q_1; **b** ZVS Q_{1a}; **c** ZVS of S_1; **d** ZCS of S_2

Figure 12.21 gives the soft switching states for the switches at full load (2 kW) with 24 V input in the boost mode. As shown in Fig. 12.21a and b, both Q_1 and Q_{1a} can be turned on with ZVS. As can be seen in Fig. 12.21c and d, for the HVS switches, the leading leg switch S_1 can achieve ZCS turn-on while the lagging switch S_2 can achieve ZVS turn-on. The soft switching waveforms are similar in the buck mode.

12.5.4 Dynamic Performance with Inverter Driven AC Motor

Figure 12.22 shows the experimental results of motor drive, without and with feed-forward, respectively. v_o is the HVS voltage of the DC–DC converter, and i_o is the HVS total current. i_{in} is the LVS total current. As seen, during the braking period, the braking kinetic energy is converted into the electrical energy, making the HVS voltage increase. This makes the power flow from the HVS to LVS and charges the battery. As can be seen in Fig. 12.22a and d, due to the feed-forward control, the voltage drop is obviously reduced from 125 to 25 V and the dynamic

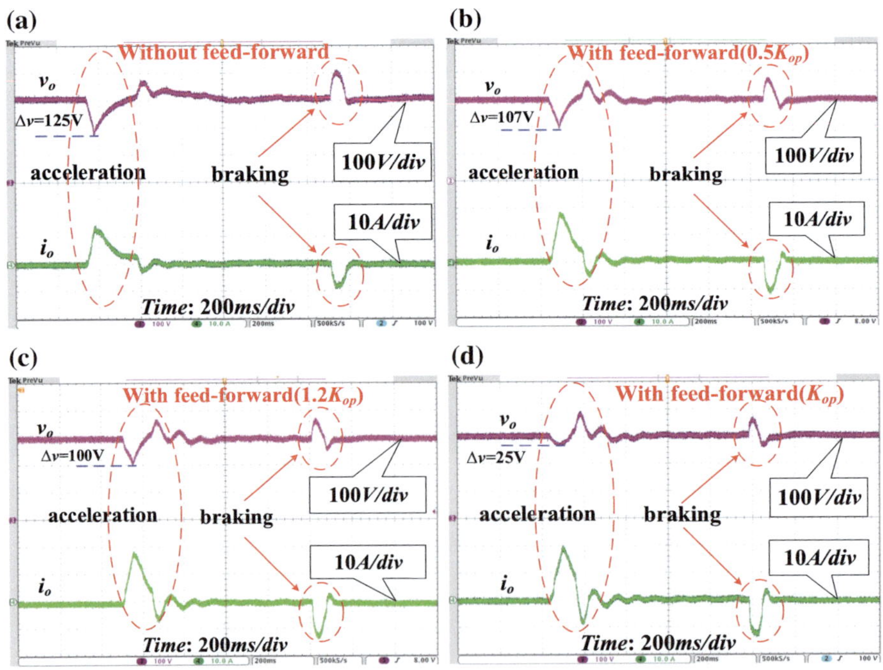

Fig. 12.22 Experimental results with motor drive: **a** without feed-forward, **b** with feed-forward coefficient $0.5K_{op}$, **c** with feed-forward coefficient $1.2K_{op}$, **d** with feed-forward coefficient K_{op}

performance is better and greatly improved. As can be seen in Fig. 12.22b and c, when the feed-forward coefficient K_o ($0.5K_{op}$ or $1.2K_{op}$) is greater than or less than the optimal value K_{op}, the voltage drop is obviously reduced from 125 to 107 V and 100 V, respectively. The dynamic performance is not the best.

12.5.5 Experimental Results of Current Sharing

As can be seen in Fig. 12.23a, the output currents of the two modules are equal to each other during the acceleration and braking periods. It implies excellent power sharing can be achieved even in transients. As can be seen in Fig. 12.23b, with the interleaving PWM modulation, the LVS current ripple can be minimized, increasing the battery lifetime.

12.5.6 Efficiency

The efficiency of the system at different loads and different source voltages is shown in Fig. 12.24. At rated load at 28 V input, the efficiency is over 95%. Since the conduction loss is larger at low battery voltage (18 V) than high battery voltage (28 V), the efficiency is higher at high source voltage (28 V) at heavy load.

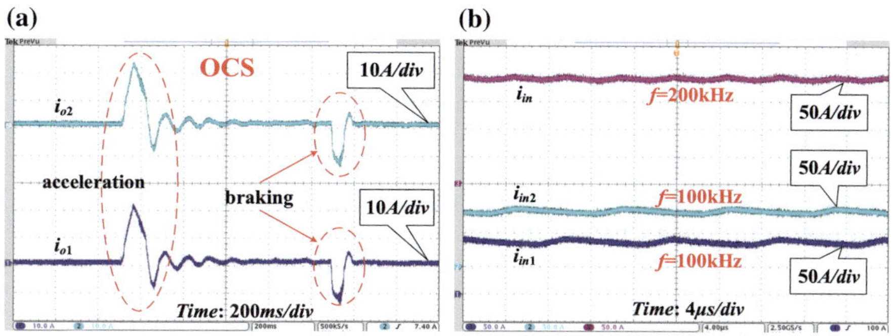

Fig. 12.23 Experimental results of current sharing: **a** OCS between two modules; **b** the ripple of input current

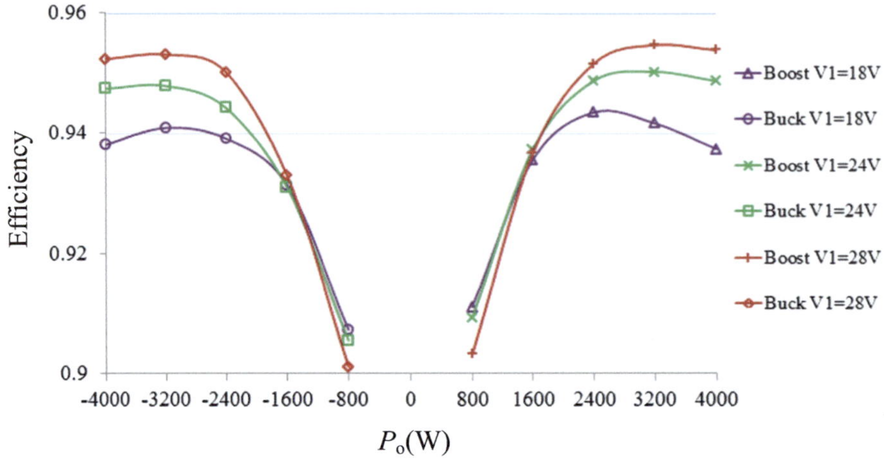

Fig. 12.24 Conversion efficiency

12.6 Conclusion

The chapter proposes parallel-connected current-fed modular bidirectional DC–DC converters for electrical drive powered by low-voltage battery. Based on the small-signal modeling employing the DPDPS control, the feed-forward control with the optimal coefficient can improve the system dynamic performance significantly. If the leakage inductance value can ensure the safe operation of the system, the smaller value can improve the system dynamic performance and reduce the RMS and peak current. With the proposed current-sharing strategy, excellent power sharing can be obtained not only under steady state but also in transients including the motor acceleration and braking regeneration. The ripple for the charge/discharge current of the battery can be minimized significantly with the interleaving technology not only within each module but also among all the constituent modules. The validity of the control strategy and converter has been verified by experimental results of a 4 kW prototype.

References

1. Chen W, Ruan X (2008) Modularization structure for series-parallel connected converters. In: IEEE Applied Power Electronics Conference and Exposition. IEEE, pp 1531–1535
2. Pan X, Rathore AK (2013) Novel interleaved bidirectional snubberless soft-switching current-fed full-bridge voltage doubler for fuel-cell vehicles. IEEE Trans Power Electron 28 (12):5535–5546
3. Shi J, Zhou L, He X (2012) Common-duty-ratio control of input-parallel output-parallel (IPOP) connected DC–DC converter modules with automatic sharing of currents. IEEE Trans Power Electron 27(7):1–10

4. Shi J, Liu T, Cheng J, He X (2015) Automatic current sharing of an input-parallel output-parallel (IPOP)-connected DC–DC converter system with chain-connected rectifiers. IEEE Trans Power Electron 30(6):2997–3016
5. Cheng J, Shi J, He X (2012) A novel input-parallel output-parallel connected DC–DC converter modules with automatic sharing of currents. In: IEEE Power Electronics and Motion Control Conference, pp 1871–1876
6. Sha D, You F, Wang X (2016) A high efficiency current fed semi dual active bridge DC–DC converter for low input voltage applications. IEEE Trans Ind Electron 63(4):2155–2164
7. Wen H, Xiao W, Su B (2014) Nonactive power loss minimization in a bidirectional isolated DC–DC converter for distributed power systems. IEEE Trans Ind Electron 61(12):6822–6831
8. Li H, Peng F (2004) Modeling of a new ZVS bi-directional DC–DC converter. IEEE Trans Aerosp Electron Syst 17(3):272–283
9. Zhang X, Guo F, Yao C, Li H, Xu P, Wang J (2014) Small-signal modeling and controller design of an isolated quasi-switched-capacitor DC/DC converter. In: IEEE Applied Power Electronics Conference and Exposition (APEC), pp 1032–1038
10. Cho Y, Lai J (2013) High-efficiency multiphase DC–DC converter for fuel-cell-powered truck auxiliary power unit. IEEE Trans Veh Technol 62(6):2421–2429
11. Pan X, Rathore AK (2015) Small-signal analysis of naturally commutated current-fed dual active bridge converter and control implementation using cypress PSoC. IEEE Trans Veh Technol 64(11):4996–5005
12. Zhu G, Ruan X, Zhang L, Wang X (2015) On the reduction of second harmonic current and improvement of dynamic response for two-stage single-phase inverter. IEEE Trans Power Electron 30(2):1028–1041
13. Zhang L, Ruan X, Ren X (2015) Second-harmonic current reduction and dynamic performance improvement in the two-stage inverters: an output impedance perspective. IEEE Trans Ind Electron 62(1):394–404
14. Yao C, Ruan X, Wang X (2015) Automatic mode-shifting control strategy with input voltage feed-forward for full-bridge-boost DC–DC converter suitable for wide input voltage range. IEEE Trans Power Electron 30(3):1668–1682
15. Yang Y, Ruan X, Zhang L, He J, Ye Z (2014) Feed-forward scheme for an electrolytic capacitor-less ac/dc led driver to reduce output current ripple. IEEE Trans Power Electron 29 (10):394–404
16. Yao C, Ruan X, Cao W, Chen P (2014) A two-mode control scheme with input voltage feed-forward for the two-switch buck-boost DC–DC converter. IEEE Trans Power Electron 29(4):2037–2048

MIX
Papier aus verantwortungsvollen Quellen
Paper from responsible sources
FSC® C105338

If you have any concerns about our products,
you can contact us on
ProductSafety@springernature.com

In case Publisher is established outside the EU,
the EU authorized representative is:
Springer Nature Customer Service Center GmbH
Europaplatz 3, 69115 Heidelberg, Germany

Printed by Libri Plureos GmbH
in Hamburg, Germany